COMPUTATIONAL PLASTICITY

Computational Methods in Applied Sciences

Volume 7

Series Editor

E. Oñate
International Center for Numerical Methods in Engineering (CIMNE)
Technical University of Catalunya (UPC)
Edificio C-1, Campus Norte UPC
Gran Capitán, s/n
08034 Barcelona, Spain
onate@cimne.upc.edu
www.cimne.com

Computational Plasticity

Edited by

Eugenio Oñate

Universitat Politècnica de Catalunya,
Barcelona, Spain

and

Roger Owen

University of Wales,
Swansea, United Kingdom

Springer

A C.I.P. Catalogue record for this book is available from the Library of Congress.

ISBN 978-90-481-7672-4(HB)
ISBN 978-1-4020-6577-4 (e-book)

Published by Springer,
P.O. Box 17, 3300 AA Dordrecht, The Netherlands.

www.springer.com

Printed on acid-free paper

Table of Contents

Preface

This book contains state of the art contributions in the field of computational plasticity. This topic encompasses a wide spectrum of areas in non-linear computational solid mechanics and modelling of materials and their industrial applications.

Despite the apparent activity in the field, the ever increasing rate of development of new engineering materials required to meet advanced technological needs poses fresh challenges in the field of constitutive modelling. The complex behaviour of such materials demands a closer interaction between numerical analysts and material scientists in order to produce thermodynamically consistent models which provide a response in keeping with fundamental micromechanical principles and experimental observations. This necessity for collaboration is further highlighted by the continuing remarkable developments in computer hardware which makes the numerical simulation of complex deformation responses increasingly possible.

The book contains 14 invited contributions written by distinguished authors who participated in the VIII International Conference on Computational Plasticity held at CIMNE/UPC (www.cimne.com), Barcelona, Spain, from 5th to 8th September 2005. The meeting was one of the Thematic Conferences of the European Community on Computational Methods in Applied Sciences (ECCOMAS, www.eccomas.org).

The different chapters of this book present recent progress and future research directions in the field of computational plasticity. A common line of many contributions is that a stronger interaction between the phenomenological and micromechanical modelling of plasticity behaviour is apparent and the use of inverse identification techniques is also more prominent. The development of adaptive strategies for plasticity problems continues to be a challenging goal, while it is interesting to note the permanence of element modelling as a research issue. Industrial forming processes, geomechanics, steel and concrete structures form the core of the applications of the different numerical methods presented in the book.

The book includes contributions sent directly by the authors and the editors cannot accept responsibility for any inaccuracies and opinions contained in the text.

The editors would like to take this opportunity for thanking all authors for submitting their contributions. We also express our gratitude to Maria Jesús Samper from CIMNE for her excellent work in the edition of this volume. Many thanks finally to ECCOMAS and Springer for accepting the publication of this book.

Eugenio Oñate
Universitat Politècnica de Catalunya
Barcelona, Spain

Roger Owen
University of Wales
Swansea, United Kingdom

A Multi-Scale Continuum Theory for Heterogeneous Materials

Franck Vernerey, Cahal McVeigh, Wing Kam Liu and Brian Moran

Department of mechanical engineering, Northwestern University,
2145 Sheridan Road, Evanston, IL 60208-3111, USA
w-liu@northwestern.edu

Summary. For the design of materials, it is important to faithfully model the physics due to interactions at the microstructural scales [18, 17, 19]. While brute-force modeling of all the details of the microstructure is too costly, current homogenized continuum models suffer from their inability to sufficiently capture the correct physics - especially where localization and failure are concerned.
To overcome this limitation, a multi-scale continuum theory is proposed so that kinematic variables representing the deformation at various scales are incorporated. The method of virtual power is then used to derive a system of coupled governing equations, each equation representing a particular scale and its interactions with the macro-scale. A constitutive relation is then introduced to preserve the underlying physics associated with each scale. The inelastic behavior is represented by multiple yield functions, each representing a particular scale of microstructure, but collectively coupled through the same set of internal variables. The proposed theory is applied to model porous metals and high strength steel. For the high strength steel the microstructure of interest consists of two populations of inclusions at distinct scales, in an alloy matrix.

1 Multi-Physics Multi-Scale Material Model

1.1 Kinematics and Virtual Power

In the spririt of the work performed by Cosserat [2], Mindlin [11] and Germain [10], the building blocks of the multi-scale continuum theory are a set of kinematic variables which represent the motion within the material's microstructure. These variables can capture the heterogeneous deformation due to the micro-mechanisms occuring at each scale. For an N-scale material, in addition to the macro-velocity field $\mathbf{v}$ we introduce N-1 independent micro-velocity gradients $\{\mathbf{L}^I\}_{1\leq I\leq N-1}$. In the context of a first gradient theory, the internal virtual power density is written as a linear combination of the various

Eugenio Oñate and Roger Owen (eds.), Computational Plasticity, 1–11.

kinetic variables and their gradients. Integrating over the body Ω, the virtual internal power is written as:

$$\delta P_{int} = \int_{\Omega} \left(\boldsymbol{\sigma} : \delta\mathbf{D} + \sum_{I=1}^{N-1} \left\{ \overline{\boldsymbol{\beta}}^{I} : (\delta\mathbf{L}^{I} - \delta\mathbf{L}) + \overline{\overline{\boldsymbol{\beta}}}^{I} \vdots \delta\mathbf{L}^{I}\overleftarrow{\nabla} \right\} \right) d\Omega. \quad (1)$$

where $\boldsymbol{\sigma}$ is the Cauchy stress, $\overline{\boldsymbol{\beta}}^{I}$ is the Ith micro-stress and $\overline{\overline{\boldsymbol{\beta}}}^{I}$ is the Ith micro-stress couple. The Ith micro-stresses are interpreted as a stress redistribution arising from the presence of heterogeneities within the Ith microstructure. The kinetic virtual power, external virtual power, governing equations and boundary conditions can be found here [4].

1.2 Constitutive Relation

For convenience, we introduce the generalized stress $\boldsymbol{\Sigma}$ and strain $\underline{\mathcal{D}}$ measures in a vector form containing the components of the macroscopic quantities as well as the various microscopic quantities defined in the previous section. Hence, we write:

$$\boldsymbol{\Sigma} = \left[\boldsymbol{\sigma}\; \overline{\boldsymbol{\beta}}^{1}\; \overline{\overline{\boldsymbol{\beta}}}^{1} \cdots \overline{\boldsymbol{\beta}}^{N-1}\; \overline{\overline{\boldsymbol{\beta}}}^{N-1} \right].$$
$$\underline{\mathcal{D}} = \left[\mathbf{D}\; \left[\mathbf{L}^{1} - \mathbf{L}\right] \mathbf{L}^{1}\overleftarrow{\nabla} \cdots \left[\mathbf{L}^{N-1} - \mathbf{L}\right] \mathbf{L}^{N-1}\overleftarrow{\nabla} \right].$$

The generalized rate of deformation $\underline{\mathcal{D}}$ is decomposed into an elastic part $\underline{\mathcal{D}}^{e}$ and a plastic part $\underline{\mathcal{D}}^{p}$:

$$\underline{\mathcal{D}} = \underline{\mathcal{D}}^{e} + \underline{\mathcal{D}}^{p}. \quad (2)$$

In the elastic regime, it is possible to introduce a generalized elastic matrix $\mathbf{C}$ such that the generalized stress increment $\boldsymbol{\Sigma}^{\triangledown}$ is related to the elastic part of the rate of deformation $\underline{\mathcal{D}}^{e}$ as follows:

$$\boldsymbol{\Sigma}^{\triangledown} = \mathbf{C} \cdot \underline{\mathcal{D}}^{e} \quad (3)$$

where $\boldsymbol{\Sigma}^{\triangledown}$ is an objective stress rate. The plastic response is written in a multi-physics plastic potential framework. We introduce a yield function ϕ^{I} with the Ith micro-scale of interest. The plastic flow at each scale is determined from the corresponding yield function in the context of associative plasticity.

2 Application to Porous metals

Substantial effort has been dedicated to the study and development of micro-mechanical models applicable to the plastic deformation and failure of porous metals. Two distinct mechanisms contribute to failure. The first stage involves the independent growth of voids subject to a remote hydrostatic stress. The

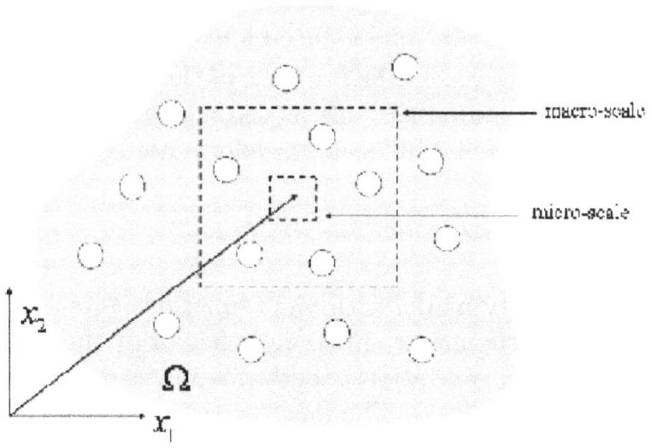

Fig. 1. Multi-scale decomposition for a two-scale porous material

second occurs when the voids a large enough to trigger a necking instability of the inter-void ligament. The second stage of failure is therefore characterized by an acceleration in the loss of load carrying capacity of the material as void coalescence occurs.
The use of traditional continuum methods to treat failure of porous metals is limited due to their inability to handle localization problems. To overcome this problem, the outlined multi-scale continuum theory is used to describe the response of porous metals after the onset of localization. The additional degrees of freedom introduced through the microfields can represent the inhomogeneous strains arising due to void interactions and coalescence.

2.1 Multi-Scale Model

The multiscale description of porous metal is based on the definition of two domains, representing the micro- and macro-scales, respectively (Fig. 1)

The kinematics are expressed in terms of the velocity $\mathbf{v}$ associated with the macro-scale and the rate of micro-deformation $\mathbf{D}^1$. The generalized stress and rate of deformation for the two-scale porous metal can therefore be written as:

$$\boldsymbol{\Sigma} = \left[\boldsymbol{\sigma} \;\; \overline{\boldsymbol{\beta}}^1 \;\; \overline{\overline{\boldsymbol{\beta}}}^1\right], \qquad \underline{\mathcal{D}} = \left[\mathbf{D} \quad (\mathbf{D}^1 - \mathbf{D}) \quad \mathbf{D}^1\overleftarrow{\nabla}\right], \tag{4}$$

where $\mathbf{D}$ is the symmetric part of the macro-velocity gradient, $\boldsymbol{\sigma}$ is the macro-stress, and $\overline{\boldsymbol{\beta}}^1$ and $\overline{\overline{\boldsymbol{\beta}}}^1$ are the micro-stress and micro-stress couple, respectively. From (1), the internal power is written as:

$$\delta P_{int} = \int_{\Omega} \left(\boldsymbol{\sigma} : \delta\mathbf{D} + \left\{\overline{\boldsymbol{\beta}}^1 : (\delta\mathbf{D}^1 - \delta\mathbf{D}) + \overline{\overline{\boldsymbol{\beta}}}^1 \vdots \, \delta\mathbf{D}^1\overleftarrow{\nabla}\right\}\right) d\Omega. \tag{5}$$

A physical interpretation is provided as follows: the macro-stress σ represents the macroscopic response that undergoes a loss of load carrying capacity due to void growth and coalescence. The presence of the micro-stress $\overline{\beta}^1$ represents the stress redistributions around voids as they grow. It provides a resistance of the matrix material to highly localized deformations.

2.2 Constitutive Relation

A two-potential plasticity model with yield functions $\Phi(\boldsymbol{\sigma})$ and $\Phi^1(\overline{\beta}^1, \overline{\overline{\beta}}^1)$ is employed to represent the material plastic behavior at the macro and micro-scale, respectively. Homogenization is performed over two averaging domains (macro and meso) as described. The macro-stress $\boldsymbol{\sigma}$ is determined by an averaging operation over the RVE $\bar{\Omega}$ (macro domain). The macroscopic plastic deformation is based on an extension of the von-Mises plasticity model that incorporates the effects of the porosity f. A standard yield function accounting for softening mechanisms due to void growth is given as follows:

$$\Phi = \Phi\left(\boldsymbol{\sigma}, f, \varepsilon\right) = 0, \tag{6}$$

where the material constants which appear in the function Φ are determined through cell modeling computer simulations. Microscopic plastic deformation is defined as the plastic deformation of the matrix material after the onset of void coalescence. The plastic model is taken in the form of J_2 flow plasticity modified to account for the effects of higher order stress. The microscopic yield function Φ^1 is given by:

$$\Phi^1 = \Phi^1(\overline{\beta}^1, \overline{\overline{\beta}}^1, \mathcal{E}^1) \tag{7}$$

where $\mathcal{E}^1$ is the microscopic effective plastic strain.

2.3 Numerical Results

We multi-scale model is now implemented and compared to a direct numerical sumilation (DNS) of a porous metal bar loaded in tension. In the DNS, the explicit microstructure is represented by a periodic distribution of circular voids of diameter $1\mu m$ and volume fraction 3%. An imperfection is introduced by making the void in the center of the specimen slightly larger than others. As localization is expected to occur around the imperfection, voids are only modeled in the central region. For the two-scale continuum simulation, the finite element discretisation is fine enough to capture size effects at the scale of the voids. An imperfection, in the form of a higher initial average porosity, is introduced in the center of the specimen to initiate localization. The comparison of the macroscopic plastic strain distribution for the DNS and the two-scale continuum is depicted in Fig. 2. The results are displayed at four different times t_1, t_2, t_3 and t_4. At time $t = t_1$, the material response reaches

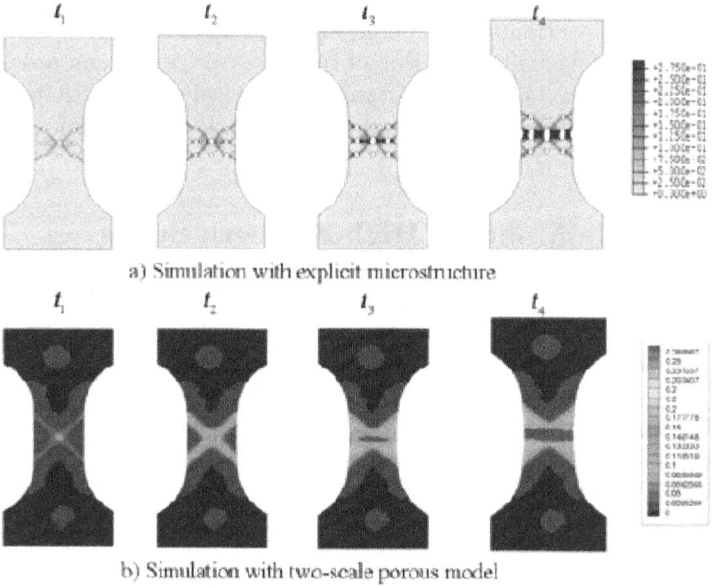

Fig. 2. Snapshots of the distribution of equivalent plastic strain in a specimen in tension for (a) the explicit microstructure (b) the equivalent two-scale continuum

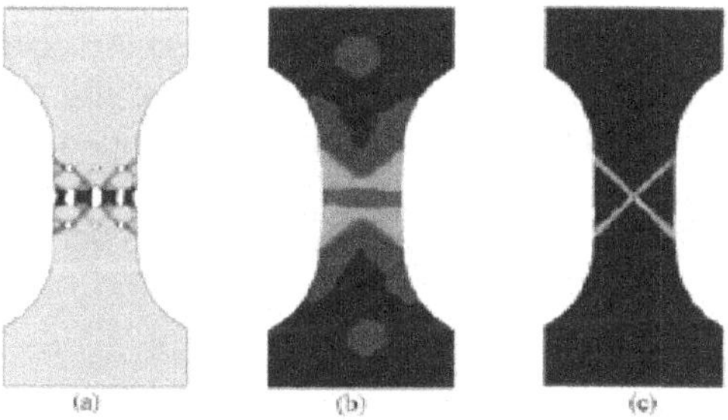

Fig. 3. Comparison of the distribution of plastic strain for (a) the explicit microstructure (b) the two-scale material (c) conventional continuum model (one-scale porous model)

the macroscopic instability point (softening of the macroscopic constitutive relation due to void growth), and deformation localizes in a shear band at 45 degrees to the tensile direction. The micro-stresses $\overline{\beta}^1$ and $\overline{\overline{\beta}}^1$ act to stabilize the void growth at time $t = t_2$. The onset of void coalescence occurs at time $t = t_3$ in a plane perpendicular to the loading direction. This mechanism is modeled by a yielding of the micro-stresses and therefore a decrease in resistance to micro-deformation. Consequently, a plane of highly localized plastic strain arises in the horizontal direction as shown at time t_4. The behavior of the two-scale continuum compares well with the behavior of the explicit microstructure. The main attraction of the two-scale continuum resides in the gain of computational time. For instance, the typical element size for the two-scale model is ten times as large as that used for the DNS (1 μm and 0.1 μm respectively).

3 Three-scale Model of High-Strength Steels

In the previous example, two scales of analysis where considered. Physically speaking, these were the average macroscale behavior of a population of voids and the microscale behavior defined by the coalescence between neighboring voids. We now consider a material in which three distinct scales can be considered. In high strength steels, previously examined within the hierarchical methodology [14, 12, 13, 15, 16, 3], the embedded particles are divided into two groups: the primary particles with the diameter around microns and the secondary particles in the range of 10 to 100 nanometers. By adding alloys during metallurgical processes, the desirable dispersed secondary particles, such as M2C carbides, are formed through precipitation or dissolved from primary particles, which blocks the dislocation path; thus, the strength of steel rise. Whereas the primary particles, such as nitrides, usually have lower toughness as compared with the iron matrix. The formation of primary particles is inevitable during manufacture process. A key-issue in steel design is to establish the relationships between the micro/nano-structures of steel and its macro-scale mechanical properties. To this end, we use the proposed multi-scale continuum theory. Three mechanisms are considered here: (i)the failure at the level of primary particles, (ii) the failure at the level of secondary particles and (iii) the failure by necking of the ligaments between micro-voids [1].

3.1 Multi-Scale Model

The three levels of microstructure are shown in Fig. 4. The macro-scale contains primary particles (large square particles) and secondary particles (small circular particles). The micro-scale contains only secondary particles and the sub-micro scale represents the matrix material. Following the framework described before, we introduce the macro-velocity $\mathbf{v}$, the micro-velocity gradient $\mathbf{D}^1$ and the sub-micro velocity gradient $\mathbf{D}^2$. The internal power for a three scale material is written as:

$$\delta P_{int} = \int_{\Omega} \boldsymbol{\sigma} : \delta\mathbf{D} + \left\{ \overline{\boldsymbol{\beta}}^1 : (\delta\mathbf{D}^1 - \delta\mathbf{D}) + \overline{\overline{\boldsymbol{\beta}}}^1 \vdots\, \delta\mathbf{D}^1 \overleftarrow{\nabla} \right\} + \left\{ \overline{\boldsymbol{\beta}}^2 : (\delta\mathbf{D}^2 - \delta\mathbf{D}) + \overline{\overline{\boldsymbol{\beta}}}^2 \vdots\, \delta\mathbf{D}^2 \overleftarrow{\nabla} \right\} d\Omega. \tag{8}$$

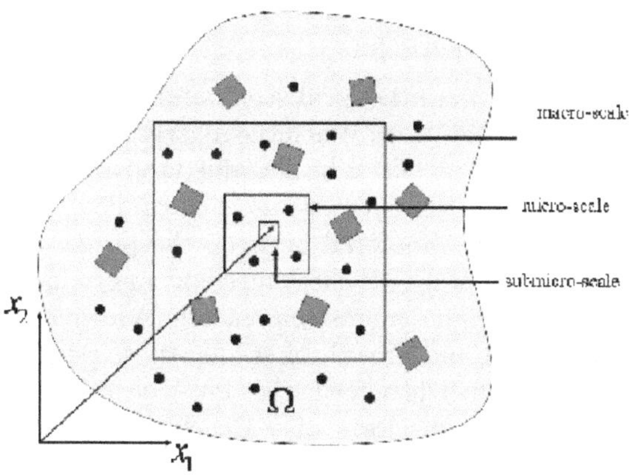

Fig. 4. Proposed multi-scale framework for the modelling of high strength steel

A physical interpretation of the various terms can be given as follows. The quantity $\mathbf{D}^1 - \mathbf{D}$ is the inhomogeneous deformation resulting from the localization that occurs from the nucleation of large voids (from primary particles) and the inhomogeneous deformation $\mathbf{D}^2 - \mathbf{D}$ is the result of the localization that occurs from nucleation of micro-voids (from secondary particles). The micro-stress at a scale is interpreted as the stress exerted by the microstructure to redistribute plastic deformation around the voids at that scale.

3.2 Constitutive Relation

The failure of High Strength Steels involves mechanisms which involve the microstructure at each scale. Three yield functions are introduced: the macro-potential $\Phi(\boldsymbol{\sigma})$ and the first and second micro-potential $\Phi^1(\overline{\boldsymbol{\beta}}^1, \overline{\overline{\boldsymbol{\beta}}}^1)$ and $\Phi^2(\overline{\boldsymbol{\beta}}^2, \overline{\overline{\boldsymbol{\beta}}}^2)$, respectively. Each of the above functions is associated with a failure mechanism summarized as follows: (i) void nucleation and growth at primary particles followed by strain localization of width approximately equal to the diameter of a 'primary' void. (ii) void nucleation and growth at primary particles followed by strain localization of width approximately equal to the diameter of a 'secondary' void. (iii) secondary-void coalescence by necking of the inter-void ligament and localization of the deformation in the neck.
The macroscopic yield function Φ describes the evolution of the macro-stress and plastic strains. It represents the plastic deformation and response of the RVE and can be derived within the hierarchical methodology. The form of the constitutive relation at the macro-scale is such that:

$$\Phi = \Phi\left(\boldsymbol{\sigma}, f, \varepsilon\right) \tag{9}$$

where f and ε are the porosity and effective plastic strain, respectively. This function accounts for the loss of load carrying capacity of the material due to the cumulative effects of void nucleation, growth and coalescence from both the primary and secondary particles.

The micro-stresses ($\overline{\beta}^1, \overline{\overline{\beta}}^1$) are the local stress and moment generated in the matrix by larger voids as they grow. When the matrix material fails, the micro-couple and micro-stress can no longer be sustained to the same extent i.e. they yield. The yielding point of the micro-stress is controlled by the degradation and strain localization in the material between the primary particles, which is driven by the formation of a void sheet mechanism originating at the scale of the secondary particles. The microscopic plastic deformation is therefore a function of the nucleation and growth of secondary particle nucleated voids (void sheet). As such, volumetric deformation must be accounted for. A simple model can be developed in the form of a Drucker-Prager plasticity model, modified to account for the effects of higher order stress. It is written in the form:

$$\Phi^1 = \Phi^1(\overline{\beta}^1, \overline{\overline{\beta}}^1, \mathcal{E}^1) \tag{10}$$

where $\mathcal{E}^1$ is the microscopic effective plastic strain.

The submicro-stresses provide a measure of the local stress distribution around the secondary particle nucleated voids during void growth. This mechanism is similar to that described in the previous section for porous materials. The yielding point of the submicro-stresses corresponds to the inability of the matrix material to transmit stress moments after ligament instability. Submicroscopic plastic deformation is subsequently defined as the plastic deformation of the matrix material after the onset of void coalescence (necking of the ligament). The yield function is written in the form of a plasticity model modified to account for the effects of higher order stress:

$$\Phi^2 = \Phi^2(\overline{\beta}^2, \overline{\overline{\beta}}^2, \mathcal{E}^2) \tag{11}$$

where $\mathcal{E}^2$ is the sub-microscopic effective plastic strain.

3.3 Numerical Results

A two-dimensional plane strain analysis of a shear test was employed to evaluate the performance of the three-scale material model in shear. The behavior of high strength steel in pure shear has been investigated experimentally by Cowie et al. [5]. Two computations are performed here: (a) direct numerical simulation (DNS) of the shear test with explicit modeling of particles and (b) a computation of the same model using the three-scale steel model. A comparison is then drawn between the two simulations.

The stages of shear failure are explained as follows.

(a) Debonding of primary particles, and strain localization between the primary particle nucleated voids. In the multi-scale model this corresponds to the yielding of the macro-stress and the formation of a shear band of

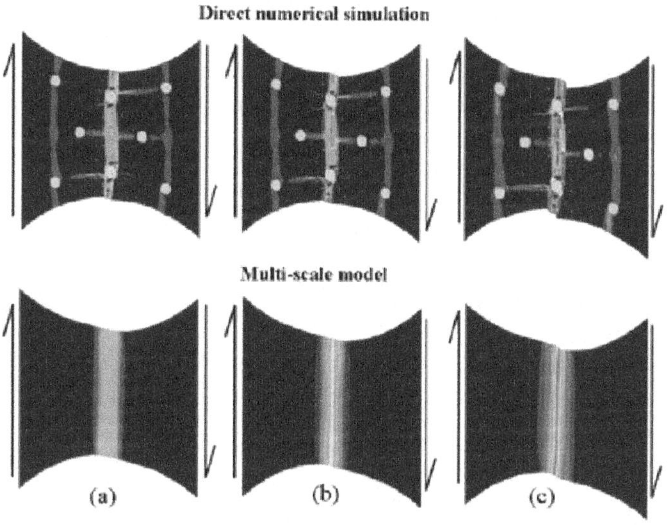

Fig. 5. Contours of effective plastic strain at different stages of deformation

width related to the primary particle size. During this stage the micro-stress remains in the elastic regime and is responsible for sustaining the observed size effect.

(b) Shear driven nucleation and growth at the scale of the secondary particles leads to a terminal shear instability. In the multi-scale simulation this mechanism is modeled through the yielding of the micro-stress. The size effect resulting from the micro-stress can no longer be sustained. The formation of a shear band of width related to the secondary particle size is then observed. This size effect arises from the existence of a sub-microstress that remains in the elastic regime.

(c) As deformation increases, coalsecence of secondary particle nucleated voids will occur, leading to a rapid decrease in the load carrying capacity of the material. In the multiscale model this is captured through a yielding of the sub-microstress. No size effect is then captured. This is consistent with the final failure of a material.

4 Conclusion

The outlined multiscale model offers several advantages over traditional continuum theory. It can be used to model a material which contains an arbitrary number of scales of microstructure. The internal interactions in the microstructure and the resulting size effects can be accounted for at each scale within the mathematical framework. The mechanisms resulting in heterogeneous deformation at each scale are modeled individually through the introduction of mutliple plastic potentials, while being coupled through the internal state variables. This theory has been successfully applied to describe the behavior of a porous metal and a high strength steel. This multi-scale framework has been successfully used in [4] to capture the fracture toughness of a high strength steel in terms of the underlying microstructure. Improvements in efficency may be possible by employing the Bridging Scale Method [7, 8, 9, 6].

Acknowledgements

The authors gratefully acknowledge the support of the ONR D3D Digital Structure Consortium (award N00014-05-C-0241) and the National Science Foundation.

References

1. McVeigh C, Vernerey F, Liu WK, Moran B and Olson GB (2006) An interactive microvoid shear localization mechanism in high strength steels. Submitted to Journal of Mechanics and Physics of Solids
2. Cosserat E and Cosserat F (1909) Theorie des corps deformables. Paris: A Hermann et Fils
3. Horstemeyer MF, Lathrop J, Gokhale AM and Dighe M (2000) Modeling stress state dependent damage evolution in a cast al-si-mg aluminium ally. Theoretical and applied fracture mechanics 33:31–47
4. Vernerey FJ (2006) Multi-Scale Continuum theory for material with microstructure. PhD thesis, Northwestern University
5. Cowie JG, Azrin M and Olson GB (1989) Microvoid formation during shear deformation of ultrahigh strength steels. Metallurgical transactions 20A:143–153
6. Wagner G, Karpov EG and Liu Wing Kam (2004) Molecular dynamics boundary conditions for regular crystal lattices. Computer Method in Applied Mechanics and Engineering 193:1579–1601
7. Kadowaki H and Liu WK (2004) Bridging multi-scale method for localization problems. Computer Methods in Applied Mechanics and Engineering 193:3267–3302
8. Kadowaki H and Liu WK (2005) A multiscale approach for the micropolar continuum model. Computer modeling in Engineering and Science 7:269–282
9. Wagner GJ and Liu WK (2003) Coupling of atomistic and continuum simulations using a bridging scale decomposition. Journal of Computational Physics 190:249–274
10. Germain P (1973) The method of virtual power in continuum mechanics. part 2 microstructure. SIAM Journal of Applied Mathematics 25:556–575
11. Mindlin RD (1964) Micro-structure in linear elasticity. Arch. Ration. Mech. Anal 16:15–78

12. Hao S, Moran B, Liu WK and Olson GB (2003) A hierarchical multi-physics constitutive model for steels design. Journal of Computer-Aided Materials Design 10:99–142
13. Hao S, Liu WK and Chang CT (2000) Computer implementation of damage models by finite element and meshfree methods. Computer methods in Applied Mechanics and Engineering 187:401–440
14. Hao S, Liu WK, Moran B, Vernerey F and Olson GB (2004) Multiple-scale constitutive model and computational framework for the design of ultra-high strength, high toughness steels. Comput. Methods Appl. Mech. Engrg 193(17-20):1865–1908
15. Tvergaard V (1982) Ductile fracture by cavity nucleation between larger voids. J. Mech. Pys. Solids 30:265–286
16. Tvergaard V (1988) 3d-analysis of localization failure in a ductile material containing two size-scales of spherical particles. Engineering Fracture Mechanics 31:421–436
17. Liu WK, Karpov EG and Park HS (2005) Nano Mechanics and Materials: Theory, Multiscale Methods and Applications, Wiley
18. Liu WK, Karpov EG, Zhang S and Park HS (2004) An introduction to computational nanomechanics and materials. Computer Methods in Applied Mechanics and Engineering 193:1529–1578
19. Liu WK, Hao S, Vernerey FJ, Kadowaki H, Park H and Qian D (2003) Multiscale analysis and design in heterogeneous system. VII International Conference on Computational Plasticity (COMPLAS VII) 7

Towards a Model for Large Strain Anisotropic Elasto-Plasticity

F.J. Montáns[1] and K.J. Bathe[2]

[1] Universidad de Castilla-La Mancha. E.T.S. de Ingenieros Industriales C/ Camilo José Cela s/n. 13071-Ciudad Real. Spain. fco.montans@uclm.es
[2] Massachusetts Institute of Technology. Dept. of Mechanical Engineering 77 Massachusetts Avenue. Cambridge, U.S.A. kjb@mit.edu

Summary. The modeling of large strain anisotropic elasto-plasticity requires that the elastic response can be anisotropic, the yielding is governed by anisotropic yield functions, the hardening is anisotropic and the principal anisotropic elastic and yield directions can align themselves to more favorable stress directions during the response. For general finite element analysis, the model also needs to be macroscopically-based and computationally effective. We have worked towards such a model based on using the decomposition of the deformation gradient into elastic and plastic parts, logarithmic strains, exponential mapping and the plastic spin as an internal variable. The objective of this presentation is to give basic theoretical considerations and a computational framework for this anisotropic elasto-plasticity model. We also present some numerical results.

1 Introduction

Computational large strain plasticity is now well established for the case of isotropic elasto-plasticity, see for example References [1–4]. In these formulations, the use of hyperelasticity [5] was important both conceptually —to not dissipate energy during elastic cycles, see Reference [6]— and algorithmically —to avoid the integration of objective stress rates. Also, the use of the Lee multiplicative decomposition of the deformation gradient into an elastic and a plastic part [7, 8] —previously used by Bilby et al. [9]— and the formulation of the incremental equations in terms of logarithmic strains and Kirchhoff stresses resulted into a remarkable simplicity for the computational return mapping algorithm [10, 11]. Indeed, the backward integration of the rate equation, via the exponential product formula, rendered the incremental algorithm in the logarithmic strain space identical to the small strain return mapping algorithm. This simplicity allowed for an extension of both the theory and the algorithm to anisotropic mixed hardening [11]. Both "unrotated" [10, 11] and spatial formulations and algorithms followed, sometimes formulated in the principal stress/strain space [12, 13]. Of course, all these formulations are

Eugenio Oñate and Roger Owen (eds.), Computational Plasticity, 13–36.

completely equivalent, see Reference [14]. However, if kinematic hardening is considered, special care must be used because some plastic spin effects may be inadvertently introduced, resulting in completely different response predictions, see Reference [15]. Alternatively, a special but not standard type of kinematic hardening may be considered [16].

Computational plasticity formulations based on logarithmic strains and work-conjugate stresses have now been used extensively and have also been extended successfully to applications in soil mechanics, see Reference [17]. Common to the different formulations are the following ingredients: isotropic hyperelasticity, exponential mapping and the additive return in the logarithmic strain space.

In this framework, the algorithm of Eterović and Bathe [11] seems to be the first implicit algorithm to allow for some anisotropy. This anisotropy is given by the kinematic hardening rule. The algorithm is consistently linearized in Reference [15], where it is noted that anisotropic yield functions may be directly employed in the formulation.

However, all the large strain computational plasticity algorithms based on logarithmic strains are using one hypothesis: the elastic strains and the work-conjugate stress measures commute. This property only holds in isotropic elasticity and hence the previous formulations are not applicable in anisotropic elasticity. While some publications appeared recently on computational anisotropic plasticity, see for example [18–21], the published algorithms are not developed following the successful framework of using the deformation gradient decomposition into an elastic and a plastic part, logarithmic strains and exponential mapping.

In many cases the effect of the elastic anisotropy is considerably smaller than the effect of the plastic anisotropy and, hence, elastic isotropy is assumed with an anisotropic yield function, see for example Reference [22]. The algorithm presented in [15] is then directly applicable. An important observation discussed in this reference is that, then, the plastic spin does not enter in the dissipation equation and as a consequence, any plastic rotation is possible. Hence, a constitutive hypothesis may be chosen. However, for general anisotropic elasto-plasticity the plastic spin requires special attention. In order to understand the nature and practical effects of the plastic spin, researchers have proposed several ad-hoc constitutive equations, see for example [23–31].

A possible effect of the plastic spin is to cause rotations of the elastic and plastic anisotropy directions during loading. These rotations take place at different strain levels depending on the type of material considered [22], as observed and measured experimentally, see References [32–36]. Some materials preserve an almost isotropic elastic behaviour as aluminum [37, 38], and also only show a small rotation of the preferred plasticity orthotropy directions [22, 34, 35]. In these References, approximately 5° of rotation is measured for 20% of strains. Other metals such as steel contain both a substantial elastic and plastic anisotropy and show a significant rotation of the orthotropy directions, see References [32, 33, 36]. In these works a realignment of the anisotropy directions in steel sheets to new directions during the loading is experimentally observed for strain levels of about 10%. Considering a thermodynamical framework of continua, it seems natural to relate the development of these rotations to the plastic flow and to a decrease of the stored energy.

In the following sections we present a model and an algorithm to describe the plastic behaviour of metals with both elastic and plastic anisotropies, and consider the rotation of the elastic and plastic anisotropy directions. This theory is an extension of the formulation presented in References [11] and [15]. The theory employs the multiplicative decomposition of the deformation gradient, logarithmic strain measures, the exponential mapping formula and the plastic spin to govern the rotations of the elastic and plastic anisotropy directions.

2 Kinematics and Incremental Integrations

In the derivation of the continuum formulation and the algorithm for the stress calculation we follow the notation used in References [1], [2] and [39]. We include in this section both the continuum and algorithmic formulations since they are closely related.

2.1 Kinematics of Deformation: Multiplicative Decomposition and Strain Rate Tensors

The Lee decomposition yields the following multiplicative decomposition for the deformation gradient

$$ {}_0^t\boldsymbol{X} = {}_{p(t)}^{\ t}\boldsymbol{X}\ {}_{\ 0}^{p(t)}\boldsymbol{X} := {}_0^t\boldsymbol{X}^e\ {}_0^t\boldsymbol{X}^p \tag{1} $$

where, conceptually, $p\,(t)$ is a configuration with the elastic strains relaxed (stress-free configuration). Hence, ${}_{p(t)}^{\ t}\boldsymbol{X} \equiv {}_0^t\boldsymbol{X}^e$ and ${}_{\ 0}^{p(t)}\boldsymbol{X} \equiv {}_0^t\boldsymbol{X}^p$. Of course the decomposition of Equation (1) is unique if ${}_0^t\boldsymbol{X}^p$ is known, for example, via the integration of the plastic path.

The spatial velocity gradient ${}^t\boldsymbol{l} = \partial\ {}^t\boldsymbol{v}/\partial\ {}^t\boldsymbol{x}$, where ${}^t\boldsymbol{v}\,(\ {}^t\boldsymbol{x})$ is the time derivative of the displacements, is

$$ {}^t\boldsymbol{l} = {}_0^t\dot{\boldsymbol{X}}\ {}_0^t\boldsymbol{X}^{-1} \tag{2} $$

and its symmetric part, the spatial deformation rate tensor, is

$$ {}^t\boldsymbol{d} = \tfrac{1}{2}\left({}_0^t\dot{\boldsymbol{X}}\ {}_0^t\boldsymbol{X}^{-1} + {}_0^t\boldsymbol{X}^{-T}\ {}_0^t\dot{\boldsymbol{X}}^T \right) \tag{3} $$

In view of Eq.(1), Eq.(2) can be decomposed as

$$ \boldsymbol{l} = \boldsymbol{l}^e + \boldsymbol{l}^p = \dot{\boldsymbol{X}}^e \left(\boldsymbol{X}^e\right)^{-1} + \boldsymbol{X}^e \left[\dot{\boldsymbol{X}}^p \left(\boldsymbol{X}^p\right)^{-1}\right] \left(\boldsymbol{X}^e\right)^{-1} \tag{4} $$

In the equations to follow we frequently omit the time left-indices when a confusion is hardly possible. The tensor

$$\boldsymbol{L}^p := \dot{\boldsymbol{X}}^p \left(\boldsymbol{X}^p\right)^{-1} \tag{5}$$

is the modified plastic velocity gradient. The tensor $\boldsymbol{L}^p$ operates in the intermediate stress-free configuration. The symmetric part of $\boldsymbol{L}^p$ is the modified plastic deformation rate tensor

$$\boldsymbol{D}^p := \tfrac{1}{2}\left[\dot{\boldsymbol{X}}^p \left(\boldsymbol{X}^p\right)^{-1} + \left(\boldsymbol{X}^p\right)^{-T} \dot{\boldsymbol{X}}^{pT}\right] \tag{6}$$

while its skew part is the modified plastic spin.

$$\boldsymbol{W}^p := \tfrac{1}{2}\left[\dot{\boldsymbol{X}}^p \left(\boldsymbol{X}^p\right)^{-1} - \left(\boldsymbol{X}^p\right)^{-T} \dot{\boldsymbol{X}}^{pT}\right] \tag{7}$$

On the other hand, if we define $\boldsymbol{L}$, the modified velocity gradient as the pull-back of $\boldsymbol{l}$ to the intermediate configuration, we arrive at

$$\boldsymbol{L} = \boldsymbol{L}^e + \boldsymbol{C}^e \boldsymbol{L}^p \tag{8}$$

where $\boldsymbol{C}^e \equiv \boldsymbol{X}^{e\,T} \boldsymbol{X}^e$ is the right Cauchy-Green deformation tensor in the stress-free configuration. Note that some of our definitions differ from those sometimes found in the literature. For example $\boldsymbol{D}^p$ is sometimes defined as $\boldsymbol{D}^p = sym\left(\boldsymbol{C}^e \boldsymbol{L}^p\right)$, see for example [40–42].

2.2 Integration of the Plastic Deformation Gradient

This section follows the same procedure presented in Reference [15]. From Equation (5), the evolution of the plastic deformation gradient tensor is given by the differential equation

$${}^{t}_{0}\dot{\boldsymbol{X}}^p = {}^{t}\boldsymbol{L}^p \, {}^{t}_{0}\boldsymbol{X}^p \tag{9}$$

whose backward-Euler exponential solution is given by

$${}^{t+\Delta t}_{\ \ 0}\boldsymbol{X}^p = \exp\left(\Delta t \quad {}^{t+\Delta t}\boldsymbol{L}^p\right) \, {}^{t}_{0}\boldsymbol{X}^p \tag{10}$$

where the exponential function of a matrix $\exp\left(\Delta t \ {}^{t+\Delta t}\boldsymbol{L}^p\right)$ for small steps such that $\left\|\Delta t \ {}^{t+\Delta t}\boldsymbol{L}^p\right\| \ll 1$ can be approximated by

$$\exp\left(\Delta t \ {}^{t+\Delta t}\boldsymbol{L}^p\right) \simeq \boldsymbol{I} + \Delta t \ {}^{t+\Delta t}\boldsymbol{L}^p \tag{11}$$

Using

$$\Delta t \ {}^{t+\Delta t}\boldsymbol{L}^p = \Delta t \ {}^{t+\Delta t}\boldsymbol{D}^p + \Delta t \ {}^{t+\Delta t}\boldsymbol{W}^p \tag{12}$$

we have that for small steps the following property holds

$$\exp\left(\Delta t \ {}^{t+\Delta t}\boldsymbol{L}^p\right) \simeq \exp\left(\Delta t \ {}^{t+\Delta t}\boldsymbol{D}^p\right) \exp\left(\Delta t \ {}^{t+\Delta t}\boldsymbol{W}^p\right) \tag{13}$$

yielding the following update formulas:

$$ {}^{t+\Delta t}_{\ 0}\boldsymbol{X}^{p\ -1} = {}^{t}_{0}\boldsymbol{X}^{p\ -1} \exp\left(-\Delta t\ {}^{t+\Delta t}\boldsymbol{W}^{p}\right) \exp\left(-\Delta t\ {}^{t+\Delta t}\boldsymbol{D}^{p}\right) \tag{14} $$

$$ {}^{t+\Delta t}_{\ 0}\boldsymbol{X}^{e} = \boldsymbol{X}^{e}_{*} \exp\left(-\Delta t\ {}^{t+\Delta t}\boldsymbol{W}^{p}\right) \exp\left(-\Delta t\ {}^{t+\Delta t}\boldsymbol{D}^{p}\right) \tag{15} $$

where the tensor $\boldsymbol{X}^{e}_{*} := {}^{t+\Delta t}_{\ t}\boldsymbol{X}\ {}^{t}_{0}\boldsymbol{X}^{e}$ is the trial elastic deformation gradient (i.e. with the plastic state frozen). The polar decomposition theorem applied to the trial elastic deformation gradient yields

$$ \boldsymbol{X}^{e}_{*} = \boldsymbol{R}^{e}_{*}\ \underleftarrow{\boldsymbol{U}}^{e}_{*} \tag{16} $$

where $\underleftarrow{\boldsymbol{U}}^{e}_{*}$ is the trial (elastic) right stretch tensor and $\boldsymbol{R}^{e}_{*}$ is the trial rotation tensor.

We now define the following incremental ("plastic") rotation

$$ {}^{t+\Delta t}_{\ t}\boldsymbol{R}^{w} = \exp\left(\Delta t\ {}^{t+\Delta t}\boldsymbol{W}^{p}\right) \tag{17} $$

Using (15) and defining $\underleftarrow{\boldsymbol{C}}^{e}_{*} := \boldsymbol{X}^{e\ T}_{*}\ \boldsymbol{X}^{e}_{*}$ (the trial right Cauchy-Green deformation tensor) we obtain

$$ \underleftarrow{\boldsymbol{C}}^{e}_{*} = {}^{t+\Delta t}_{\ t}\boldsymbol{R}^{wT} \exp\left(\Delta t\ {}^{t+\Delta t}\boldsymbol{D}^{p}\right)\ {}^{t+\Delta t}_{\ 0}\boldsymbol{C}^{e} \exp\left(\Delta t\ {}^{t+\Delta t}\boldsymbol{D}^{p}\right)\ {}^{t+\Delta t}_{\ t}\boldsymbol{R}^{w} \tag{18} $$

We now define the logarithmic strains by

$$ {}^{t+\Delta t}_{\ 0}\boldsymbol{E}^{e} := \tfrac{1}{2}\log\ {}^{t+\Delta t}_{\ 0}\boldsymbol{C}^{e} \quad \text{and} \quad \underleftarrow{\boldsymbol{E}}^{e}_{*} := \tfrac{1}{2}\log\ \underleftarrow{\boldsymbol{C}}^{e}_{*} \tag{19} $$

The following tensors are the measures updated to the stress-free configuration

$$ \boldsymbol{C}^{e}_{*} := {}^{t+\Delta t}_{\ t}\boldsymbol{R}^{w}\ \underleftarrow{\boldsymbol{C}}^{e}_{*}\ {}^{t+\Delta t}_{\ t}\boldsymbol{R}^{wT} \tag{20} $$

$$ \boldsymbol{E}^{e}_{*} := {}^{t+\Delta t}_{\ t}\boldsymbol{R}^{w}\ \underleftarrow{\boldsymbol{E}}^{e}_{*}\ {}^{t+\Delta t}_{\ t}\boldsymbol{R}^{wT} \tag{21} $$

For future expressions we note that $\underleftarrow{(\cdot)}$ are the quantities $(\cdot)$ rotated to the rotationally-frozen configuration (defined below), i.e.

$$ \underleftarrow{(\cdot)} = {}^{t+\Delta t}_{\ t}\boldsymbol{R}^{wT}\ (\cdot)\ {}^{t+\Delta t}_{\ t}\boldsymbol{R}^{w} \tag{22} $$

Using Equations (11) and (19), Equation (18) may be written as

$$ \boldsymbol{E}^{e}_{*} \simeq {}^{t+\Delta t}_{\ 0}\boldsymbol{E}^{e} + \Delta t\ {}^{t+\Delta t}\boldsymbol{D}^{p} \tag{23} $$

with the additional restriction that $\| \boldsymbol{E}^{e}_{*} \| \ll 1$, i.e. the elastic strains and incremental steps are only moderately large. This restriction is typically fulfilled in metal plasticity. Alternatively, expression (23) may be written as

$$ {}^{t+\Delta t}_{\ 0}\underleftarrow{\boldsymbol{E}}^{e} \simeq \underbrace{\underleftarrow{\boldsymbol{E}}^{e}_{*}}_{\text{known}} - \Delta t\ {}^{t+\Delta t}\underleftarrow{\boldsymbol{D}}^{p} \tag{24} $$

In this expression the plastic rotations given by ${}^{t+\Delta t}_{t}\boldsymbol{R}^w$ are not included. We have emphasized that, since $\boldsymbol{R}^e_*$ is the rotation known from the polar decomposition, $\underleftarrow{\boldsymbol{U}}^e_*$ and $\underleftarrow{\boldsymbol{E}}^e_*$ are the stretch and logarithmic strain tensors also known from the polar decomposition. Hence, we can also write

$$\underleftarrow{\mathcal{L}}\ \boldsymbol{E}^e \simeq \underleftarrow{\mathcal{L}}\ \boldsymbol{E}^e_* - \ \boldsymbol{D}^p \tag{25}$$

which is the large strain anisotropic counterpart of the small strain isotropic additive decomposition $\dot{\boldsymbol{\varepsilon}}^e = \dot{\boldsymbol{\varepsilon}} - \dot{\boldsymbol{\varepsilon}}^p$. In this sense $\Delta t\ {}^{t+\Delta t}\underleftarrow{\boldsymbol{D}}^p$ is seen as the plastic strains increment in a configuration rotated ${}^{t+\Delta t}_{t}\boldsymbol{R}^{wT}$ from the stress free configuration. We will call this configuration the "rotationally-frozen configuration" or the "*trial* stress-free configuration". By the symbol $\underleftarrow{\mathcal{L}}$ we define the derivative

$$\underleftarrow{\mathcal{L}}\,(\cdot) = {}^{t+\Delta t}_{t}\boldsymbol{R}^w\ \frac{d}{dt}\underbrace{\left[\ {}^{t+\Delta t}_{t}\boldsymbol{R}^{wT}\ (\cdot)\ {}^{t+\Delta t}_{t}\boldsymbol{R}^w\right]}_{\underleftarrow{(\cdot)}}\ {}^{t+\Delta t}_{t}\boldsymbol{R}^{wT} = \frac{d}{dt}(\cdot) + (\cdot)\,\boldsymbol{W}^p - \boldsymbol{W}^p\,(\cdot) \tag{26}$$

2.3 Rotationally-Frozen Configuration

The following discussion is given to provide some insight into the use of configurations and rotations.

The rotation component of the trial elastic gradient tensor is obtained by use of the right polar decomposition theorem and formulas (15) and (23) as:

$$\begin{aligned}
\boldsymbol{R}^e_* &= {}^{t+\Delta t}_{0}\boldsymbol{R}^e \exp\left({}^{t+\Delta t}_{0}\boldsymbol{E}^e\right) \exp\left(\Delta t\ {}^{t+\Delta t}\boldsymbol{D}^p\right) \exp\left(\Delta t\ {}^{t+\Delta t}\boldsymbol{W}^p\right) \exp\left(-\boldsymbol{E}^e_*\right) \\
&= {}^{t+\Delta t}_{0}\boldsymbol{R}^e\ {}^{t+\Delta t}_{t}\boldsymbol{R}^w \exp\left({}^{t+\Delta t}_{0}\underleftarrow{\boldsymbol{E}}^e\right)\ {}^{t+\Delta t}_{t}\boldsymbol{R}^{wT} \\
&\qquad {}^{t+\Delta t}_{t}\boldsymbol{R}^w \exp\left(\Delta t\ {}^{t+\Delta t}\underleftarrow{\boldsymbol{D}}^p\right)\ {}^{t+\Delta t}_{t}\boldsymbol{R}^{wT}\ {}^{t+\Delta t}_{t}\boldsymbol{R}^w \exp\left(-\underleftarrow{\boldsymbol{E}}^e_*\right) \\
&\simeq {}^{t+\Delta t}_{0}\boldsymbol{R}^e\ {}^{t+\Delta t}_{t}\boldsymbol{R}^w
\end{aligned} \tag{27}$$

which yields

$${}^{t+\Delta t}_{0}\boldsymbol{R}^e \simeq \boldsymbol{R}^e_*\ {}^{t+\Delta t}_{t}\boldsymbol{R}^{wT} \tag{28}$$

Assume that at a given point, we have a deformation gradient ${}^t_0\boldsymbol{X} = {}^t_0\boldsymbol{X}^e\ {}^t_0\boldsymbol{X}^p$. For the next step we have an incremental deformation gradient ${}^{t+\Delta t}_{t}\boldsymbol{X}$ and a trial state given by

$$\text{trial state:}\left\{\begin{array}{l}
{}^{t+\Delta t}_{0}\boldsymbol{X} = \boldsymbol{X}^e_*\ {}^t_0\boldsymbol{X}^p \\
\boldsymbol{X}^e_* = {}^{t+\Delta t}_{t}\boldsymbol{X}\ {}^t_0\boldsymbol{X}^e = \boldsymbol{R}^e_*\ \underleftarrow{\boldsymbol{U}}^e_* = {}^{t+\Delta t}_{0}\boldsymbol{R}^e\ {}^{t+\Delta t}_{t}\boldsymbol{R}^w\ \underleftarrow{\boldsymbol{U}}^e_*
\end{array}\right. \tag{29}$$

If ${}^{t+\Delta t}_{t}\boldsymbol{R}^w \neq \boldsymbol{I}$ then the trial stress-free configuration given by a rotation $\boldsymbol{R}^{eT}_*$ from the spatial configuration (*the one actually known from* ${}^{t+\Delta t}_{0}\boldsymbol{X}$) is not the actual stress free configuration at the end of the step. That stress-free configuration is given by ${}^{t+\Delta t}_{\ 0}\boldsymbol{R}^{eT}$, only known after integration. Fortunately, Equations (24) and (25) show that the additive algorithm also holds in the configuration in which the plastic rotation ${}^{t+\Delta t}_{\ t}\boldsymbol{R}^{w}$ has not been applied yet.

From a continuum perspective, using the following definitions

$$\boldsymbol{w}^e_* = \dot{\boldsymbol{R}}^e_* \ \boldsymbol{R}^{eT}_*, \quad \boldsymbol{w}^e = \dot{\boldsymbol{R}}^e \ \boldsymbol{R}^{eT}, \quad \boldsymbol{W}^p = \dot{\boldsymbol{R}}^w \ \boldsymbol{R}^{wT} \ \text{and} \ \boldsymbol{w}^p = \ \boldsymbol{R}^e \ \boldsymbol{W}^p \ \boldsymbol{R}^{eT} \tag{30}$$

performing the time derivative of Equation (27) and multiplying by $\boldsymbol{R}^{eT}_* = \boldsymbol{R}^{wT} \ \boldsymbol{R}^{eT}$ to transfer the Equation to the spatial configuration, we obtain

$$\dot{\boldsymbol{R}}^e_* \ \boldsymbol{R}^{eT}_* = \ \dot{\boldsymbol{R}}^e \ \boldsymbol{R}^{eT} + \ \boldsymbol{R}^e \ \dot{\boldsymbol{R}}^w \ \boldsymbol{R}^{wT} \ \boldsymbol{R}^{eT} \tag{31}$$

i.e.

$$\boldsymbol{w}^e_* = \boldsymbol{w}^e + \boldsymbol{w}^p \tag{32}$$

3 Free Energy Function and Dissipation Inequality

The global free energy ψ is assumed to admit the additive decomposition into a stored energy function $\mathcal{W}$ and a hardening potential $\mathcal{H}$

$$\psi\left(\boldsymbol{X}^e, \boldsymbol{X}^i, \boldsymbol{X}^i\text{-history}\right) = \ \mathcal{W}\left(\boldsymbol{X}^e, \boldsymbol{X}^i\text{-history}\right) + \mathcal{H}\left(\ \boldsymbol{X}^i, \boldsymbol{X}^i\text{-history}\right) \tag{33}$$

where $\boldsymbol{X}^i$ are internal variables which can be defined by a gradient of some local internal displacements. Obviously, by objectivity, the stored energy function should depend only on the stretch part of $\boldsymbol{X}^e$, since it must remain unaffected by rigid body motions.

We assume that the solid deforms in such a way that the total potential energy is minimum and that the dissipation energy is maximum including the effect of plastic spin. We assume that the rotation of the anisotropy directions is also related to a change of the potential energy. Hence, special considerations arise for the stored energy function and the hardening potential. We incorporate these considerations in the energy function and hardening potential proposed below.

3.1 Stored Energy Function: Orthotropic Hyperelasticity Based on Logarithmic Strain Measures

In the case of isotropic *elasticity,* the right Cauchy deformation tensor $\boldsymbol{C}^e$ and the second Piola-Kirchhoff stress tensor $\boldsymbol{S}$ commute. Then, the Mandel stress tensor $\boldsymbol{\Xi} := \boldsymbol{C}^e\boldsymbol{S}$ and the symmetric part of the Mandel stress tensor, $\boldsymbol{\Xi}_s$, results in

$$\boldsymbol{\Xi}_s = \boldsymbol{\Xi} = \tfrac{1}{2}\left(\boldsymbol{C}^e\boldsymbol{S} + \boldsymbol{S}\boldsymbol{C}^e\right) = \boldsymbol{U}^e\boldsymbol{S}\boldsymbol{U}^e \tag{34}$$

which in terms of the spatial Kirchhoff stress tensor $\boldsymbol{\tau}$ may be written as

$$\boldsymbol{\Xi}_s = \boldsymbol{U}^e\boldsymbol{X}^{e-1}\boldsymbol{\tau}\boldsymbol{X}^{e-T}\boldsymbol{U}^e = \boldsymbol{R}^{eT}\boldsymbol{\tau} \ \boldsymbol{R}^e = \bar{\boldsymbol{\tau}} \tag{35}$$

where $\bar{\boldsymbol{\tau}}$ is the rotated Kirchhoff stress tensor [2]. Assuming isotropic elasticity, the governing strain energy function can be expressed as

$$\begin{aligned}\mathcal{W} &= U(J) + \mu\, \boldsymbol{E}^{ed} : \boldsymbol{E}^{ed} \\ &= U(J) + \mu\, \boldsymbol{E}^{e} : \mathbb{P} : \boldsymbol{E}^{e}\end{aligned} \tag{36}$$

where $\boldsymbol{E}^e := \ln \boldsymbol{U}^e$ are the logarithmic strains, $\boldsymbol{E}^{ed} := \mathbb{P} : \boldsymbol{E}^e$ are the deviatoric logarithmic strains, $\mathbb{P} := \mathbb{I} - \frac{1}{3}\boldsymbol{I} \otimes \boldsymbol{I}$ is the deviatoric projector tensor and $\mathbb{I}$ and $\boldsymbol{I}$ are, respectively, the fourth and second order identity tensors. The two terms in Equation (36) represent the volumetric and deviatoric strain energies respectively. The rotated Kirchhoff stress tensor is obtained as (see for example Reference [15])

$$\boldsymbol{\Xi} \equiv \boldsymbol{\Xi}_s \equiv \bar{\boldsymbol{\tau}} = \frac{\partial \mathcal{W}}{\partial \boldsymbol{E}^e} = JU'(J) + 2\mu\, \mathbb{P} : \boldsymbol{E}^e \tag{37}$$

whereas the skew-symmetric part of the Mandel stress tensor, $\boldsymbol{\Xi}_w$, vanishes.

As also mentioned in the introduction, in anisotropic plasticity, the *elastic* properties are frequently considered as isotropic given that experimentally much smaller deviations from isotropy are observed for *elastic* properties than for *plastic* properties. In this work we wish to relax the assumption of isotropy for the elastic properties and allow moderate anisotropic elasticity, and hence we use the following function:

$${}^t\mathcal{W} = U\left({}^tJ\right) + \mu\, {}^t\boldsymbol{E}^e : {}^t\mathbb{A}^d : {}^t\boldsymbol{E}^e \tag{38}$$

where ${}^t\mathbb{A}^d$ is the orthotropy structural tensor which has the same characteristic space as $\mathbb{P}$ (commutes with $\mathbb{P}$) and is, in general, close to $\mathbb{P}$. Note that in this case the volumetric component is assumed as isotropic reducing the number of independent constants to seven. Of course, a general anisotropic tensor ${}^t\mathbb{A}$ (with nine constants) may also be employed with the logarithmic strain. In this case the stored energy function is of the type

$${}^t\mathcal{W} = \tfrac{1}{2}\, {}^t\boldsymbol{E}^e : {}^t\mathbb{A} : {}^t\boldsymbol{E}^e \tag{39}$$

where ${}^t\mathbb{A}^{-1}$ in the principal material orthotropy axes $\{{}^t\boldsymbol{P}_i,\ i = 1,2,3\}$, may be written as —see for example Reference [1]—

$${}^t\mathbb{A}^{-1} = \begin{bmatrix} 1/E_a & -\nu_{ba}/E_b & -\nu_{ca}/E_c & 0 & 0 & 0 \\ -\nu_{ab}/E_a & 1/E_b & -\nu_{cb}/E_c & 0 & 0 & 0 \\ -\nu_{ac}/E_a & -\nu_{bc}/E_b & 1/E_c & 0 & 0 & 0 \\ 0 & 0 & 0 & 1/G_{ab} & 0 & 0 \\ 0 & 0 & 0 & 0 & 1/G_{bc} & 0 \\ 0 & 0 & 0 & 0 & 0 & 1/G_{ca} \end{bmatrix}_{\{{}^t\boldsymbol{P}_i\}} \tag{40}$$

where E_a, E_b, E_c are the Young's moduli in the principal directions, ν_{ba}, ν_{ca}, ν_{cb}, ν_{ab}, ν_{ac}, ν_{bc} are the Poisson ratios and G_{ab}, G_{bc}, G_{ca} are the shear moduli. These constants need to satisfy certain conditions, see e.g. [1].

In Equations (38) and (39), the strain tensor, and thus also the elastic anisotropy tensor, are defined in the unrotated configuration (i.e. with the elastic rotations given by $\boldsymbol{R}^e$ removed). Between time t and time $t + \Delta t$, this configuration and all objects defined in this configuration rotate by the amount ${}^{t+\Delta t}_{\ t}\boldsymbol{R}^w$ due to the plastic spin—see Equation (28). And during the plastic flow, also the lattice rotation takes place, resulting in an additional rotation ${}^{t+\Delta t}_{\ t}\boldsymbol{R}^A$ for the anisotropy directions. For example, the resulting energy of Equation (39) at $t + \Delta t$ is

$$
\begin{aligned}
{}^{t+\Delta t}\mathcal{W} &= \tfrac{1}{2}\ {}^{t+\Delta t}\underleftarrow{\boldsymbol{E}}^e : {}^{t}\widetilde{\mathbb{A}} : {}^{t+\Delta t}\underleftarrow{\boldsymbol{E}}^e \\
&= \tfrac{1}{2}\ {}^{t+\Delta t}\underleftarrow{\boldsymbol{E}}^e : {}^{t+\Delta t}\underleftarrow{\mathbb{A}} : {}^{t+\Delta t}\underleftarrow{\boldsymbol{E}}^e \\
&= \tfrac{1}{2}\ {}^{t+\Delta t}\boldsymbol{E}^e : {}^{t+\Delta t}\mathbb{A} : {}^{t+\Delta t}\boldsymbol{E}^e
\end{aligned}
\tag{41}
$$

where by ${}^{t}\widetilde{\mathbb{A}} =: {}^{t+\Delta t}\underleftarrow{\mathbb{A}}$ we imply that the tensor has rotated by ${}^{t+\Delta t}_{\ t}\boldsymbol{R}^A$ with respect to ${}^{t}\mathbb{A}$. The rotation tensor ${}^{t+\Delta t}_{\ t}\boldsymbol{R}^A$ represents the rotation of the elastic anisotropy tensors. Therefore, the energy expression in the stress-free configuration is obtained by the rotation ${}^{t+\Delta t}_{\ t}\boldsymbol{R}^w$ and the rotation ${}^{t+\Delta t}_{\ t}\boldsymbol{R}^A$.

The derivative of the stored energy function in absence of plastic rotations is

$$
\dot{\mathcal{W}}\Big|_{apr} = \underleftarrow{\boldsymbol{E}}^e : \underleftarrow{\mathbb{A}} : \dot{\underleftarrow{\boldsymbol{E}}}^e \tag{42}
$$

$$
= \boldsymbol{E}^e : \mathbb{A} : \underleftarrow{\mathcal{L}}\boldsymbol{E}^e \tag{43}
$$

The tensor $\underleftarrow{\boldsymbol{T}} = \partial\mathcal{W} \,/\, \partial\underleftarrow{\boldsymbol{E}}^e\Big|_{apr} = \underleftarrow{\mathbb{A}} : \underleftarrow{\boldsymbol{E}}^e$ is defined as the symmetric logarithmic stress tensor (or generalized Kirchhoff stress tensor) in the rotation-free configuration. If the lattice structure is not fixed, the variation of the strain energy function in the rotation-free configuration is

$$
\dot{\mathcal{W}} = \underleftarrow{\boldsymbol{T}} : \dot{\underleftarrow{\boldsymbol{E}}}^e + \tfrac{1}{2}\underleftarrow{\boldsymbol{E}}^e : \dot{\underleftarrow{\mathbb{A}}} : \underleftarrow{\boldsymbol{E}}^e \tag{44}
$$

The second term in Equation (44) is the work variation due to a rotation of the lattice, which after plastic flow will be in a more favorable orientation. Hence this term must be negative. Defining $\underleftarrow{\boldsymbol{W}}^A := \dot{\boldsymbol{R}}^A\ \boldsymbol{R}^{A\,T}$ as the structural anisotropy spin tensor and using ${}^{t+\Delta t}\underleftarrow{\boldsymbol{T}} = {}^{t+\Delta t}\underleftarrow{\mathbb{A}} : {}^{t+\Delta t}\underleftarrow{\boldsymbol{E}}^e$, the second term of Equation (44) can be written in symbolic notation as

$$
\boxed{\tfrac{1}{2}\,\underleftarrow{\boldsymbol{E}}^e : \dot{\underleftarrow{\mathbb{A}}} : \underleftarrow{\boldsymbol{E}}^e} = \tfrac{1}{2}\left(\underleftarrow{\boldsymbol{E}}^e\underleftarrow{\boldsymbol{W}}^A - \underleftarrow{\boldsymbol{W}}^A\underleftarrow{\boldsymbol{E}}^e\right) : \underleftarrow{\boldsymbol{T}} + \tfrac{1}{2}\,\underleftarrow{\boldsymbol{T}} : \left(\underleftarrow{\boldsymbol{E}}^e\underleftarrow{\boldsymbol{W}}^A - \underleftarrow{\boldsymbol{W}}^A\underleftarrow{\boldsymbol{E}}^e\right)
$$

$$
= \underleftarrow{\boldsymbol{T}} : \left(\underleftarrow{\boldsymbol{E}}^e\underleftarrow{\boldsymbol{W}}^A - \underleftarrow{\boldsymbol{W}}^A\underleftarrow{\boldsymbol{E}}^e\right)
$$

$$
\boxed{= \underleftarrow{\boldsymbol{T}}_w : \underleftarrow{\boldsymbol{W}}^A} \tag{45}
$$

where $\underleftarrow{\boldsymbol{T}}_w$ is a skew tensor defined as

$$\underleftarrow{\boldsymbol{T}}_w := \underleftarrow{\boldsymbol{E}}^e \, \underleftarrow{\boldsymbol{T}} - \underleftarrow{\boldsymbol{T}} \, \underleftarrow{\boldsymbol{E}}^e \tag{46}$$

Of course we can define equivalent measures in the unrotated configuration:

$$\boldsymbol{T} := {}^{t+\Delta t}_{\ \ \ t}\boldsymbol{R}^w \, \underleftarrow{\boldsymbol{T}} \, {}^{t+\Delta t}_{\ \ \ t}\boldsymbol{R}^{wT} = \mathbb{A} : \boldsymbol{E}^e \quad \text{and} \quad \boldsymbol{T}_w := \boldsymbol{E}^e \, \boldsymbol{T} - \boldsymbol{T} \, \boldsymbol{E}^e \tag{47}$$

or in the spatial configuration

$${}^{t+\Delta t}\boldsymbol{t} := {}^{t+\Delta t}_{\ \ \ 0}\boldsymbol{R}^e \, {}^{t+\Delta t}\boldsymbol{T} \, {}^{t+\Delta t}_{\ \ \ 0}\boldsymbol{R}^{eT} = \boldsymbol{R}^e_* \, \underleftarrow{\boldsymbol{T}} \, \boldsymbol{R}^{eT}_* \quad \text{and} \quad \boldsymbol{t}_w := \boldsymbol{e}^e \boldsymbol{t} - \boldsymbol{t}\boldsymbol{e}^e \tag{48}$$

where $\boldsymbol{e}^e := \ln \boldsymbol{V}^e$ and $\boldsymbol{V}^e$ is the stretch tensor obtained from the left polar decomposition theorem.

Finally we obtain

$$\boxed{\dot{\mathcal{W}} = \underleftarrow{\boldsymbol{T}} : \dot{\underleftarrow{\boldsymbol{E}}}^e + \underleftarrow{\boldsymbol{T}}_w : \underleftarrow{\boldsymbol{W}}^A} \tag{49}$$

$$\boxed{\dot{\mathcal{W}} = \boldsymbol{T} : \underleftarrow{\mathcal{L}} \boldsymbol{E}^e + \boldsymbol{T}_w : \boldsymbol{W}^A} \tag{50}$$

Here $\boldsymbol{T}_w$ is determined once $\boldsymbol{T}$ and $\boldsymbol{E}^e$ are known, i.e., *it is not an independent or internal variable*. The second term in the variation of the stored energy function is the variation due to the rotation of the anisotropy axes, and depends on the non-coaxiality of stresses and elastic strains.

Figure 1 refers to the different rotations and configurations used. The spatial configuration for strains is known from the material configuration by ${}^{t+\Delta t}_{\ \ \ 0}\boldsymbol{X}$. From this spatial configuration we can directly obtain the trial unrotated configuration (rotationally frozen) by the rotation $\boldsymbol{R}^{eT}_*$, obtained from the polar decomposition Eq.(16) of the trial elastic gradient (29_2). Then, as shown below, we obtain the plastic correction ${}^{t+\Delta t}_{\ \ \ t}\boldsymbol{R}^w$ for the trial unrotated configuration, which has the final value given by Equation (28). Due to the rotation ${}^{t+\Delta t}_{\ \ \ t}\boldsymbol{R}^w$ the preferred elastic and plastic anisotropy directions rotate by the additional amount ${}^{t+\Delta t}_{\ \ \ t}\boldsymbol{R}^A$ in the form given below. Since we are assuming moderate incremental steps, both ${}^{t+\Delta t}_{\ \ \ t}\boldsymbol{R}^w$ and ${}^{t+\Delta t}_{\ \ \ t}\boldsymbol{R}^A$ may be considered to commute.

3.2 Hardening Potential

Consider a hardening potential of a similar form to that of the stored energy function

$$\mathcal{H} = \mathcal{H}^{kin} + \mathcal{H}^{iso} + \mathcal{H}^w = \tfrac{1}{2}\bar{h} \ \left[(1-M)\, \tfrac{3}{2} \, \boldsymbol{E}^i : \mathbb{H} : \boldsymbol{E}^i + M \ \zeta^2\right] + \tfrac{1}{2}K_w \xi^2 \tag{51}$$

$$= \tfrac{1}{2}\bar{h} \ \left[(1-M)\, \tfrac{3}{2} \, \underleftarrow{\boldsymbol{E}}^i : \underleftarrow{\mathbb{H}} : \underleftarrow{\boldsymbol{E}}^i + M \ \zeta^2\right] + \tfrac{1}{2}K_w \xi^2 \tag{52}$$

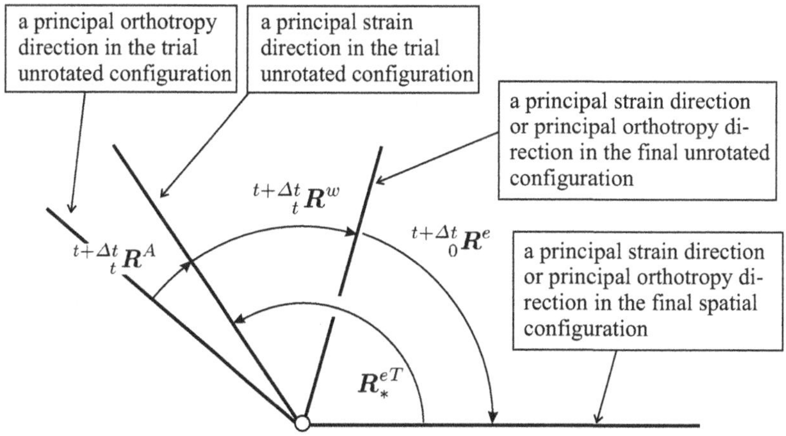

Fig. 1. Configurations involved in the stress-integration algorithm

where M is the mixed hardening parameter, $\frac{2}{3}H := \bar{h}(1-M)$ is the effective kinematic hardening modulus and $K := \bar{h}M$ is the isotropic hardening modulus. The parameter $\bar{h}$ plays the role of effective hardening modulus. The parameter K_w is a hardening for couple-stresses. Eq.(51) corresponds to a SPM description of hardening, see Reference [43]. However, for constant $\bar{h}$ it coincides with the SPS method. In (51) we have included the possibility of anisotropic kinematic hardening through the use of an anisotropy tensor $\mathbb{H}$, similar to $\mathbb{A}^d$. The tensor $\underleftarrow{\mathbb{H}}$, rotates at the speed given by the internal spin tensor $\boldsymbol{W}^H$ for similar reasons as those given for the stored energy function.

We define the *internal overstresses* as

$$\kappa := \frac{\partial \psi}{\partial \zeta} \quad \text{and} \quad \kappa_w := \frac{\partial \psi}{\partial \xi} \tag{53}$$

Hence,

$$\kappa = \frac{\partial \psi}{\partial \zeta} = \frac{\partial \mathcal{H}}{\partial \zeta} = K\zeta \quad \text{and} \quad \kappa_w = \frac{\partial \psi}{\partial \xi} = \frac{\partial \mathcal{H}}{\partial \xi} = K_w \xi \tag{54}$$

or

$$\dot{\kappa} = \frac{\partial^2 \psi}{\partial \zeta^2}\dot{\zeta} = \frac{\partial^2 \mathcal{H}}{\partial \zeta^2}\dot{\zeta} = K\dot{\zeta} \quad \text{and} \quad \dot{\kappa}_w = \frac{\partial^2 \psi}{\partial \xi^2}\dot{\xi} = \frac{\partial^2 \mathcal{H}}{\partial \xi^2}\dot{\xi} = K_w \dot{\xi} \tag{55}$$

We *define* the *internal backstress* as

$$\underleftarrow{\boldsymbol{B}}_s = \left.\frac{\partial \psi}{\partial \underleftarrow{\boldsymbol{E}}^i}\right|_{apr} = \left.\frac{\partial \mathcal{H}}{\partial \underleftarrow{\boldsymbol{E}}^i}\right|_{apr} = H \underleftarrow{\mathbb{H}} : \underleftarrow{\boldsymbol{E}}^i \tag{56}$$

Then, the derivative of the hardening potential is

$$\dot{\mathcal{H}} = \underleftarrow{\boldsymbol{B}}_s : \dot{\underleftarrow{\boldsymbol{E}}}^i + \tfrac{1}{2} H \, \underleftarrow{\boldsymbol{E}}^i : \dot{\underleftarrow{\mathbb{H}}} : \underleftarrow{\boldsymbol{E}}^i + \kappa\dot{\zeta} + \kappa_w \dot{\xi} \tag{57}$$

and following the same steps as for the stored energy function

$$\tfrac{1}{2}H \; \underleftarrow{\boldsymbol{E}}^i : \dot{\underleftarrow{\mathbb{H}}} : \underleftarrow{\boldsymbol{E}}^i = \underleftarrow{\boldsymbol{B}}_w : \underleftarrow{\boldsymbol{W}}^H \tag{58}$$

where $\underleftarrow{\boldsymbol{B}}_w$ is a skew tensor defined as

$$\underleftarrow{\boldsymbol{B}}_w := \underleftarrow{\boldsymbol{E}}^i \; \underleftarrow{\boldsymbol{B}}_s - \underleftarrow{\boldsymbol{B}}_s \; \underleftarrow{\boldsymbol{E}}^i \tag{59}$$

Finally we have

$$\begin{aligned} \dot{\mathcal{H}} &= \underleftarrow{\boldsymbol{B}}_s : \dot{\underleftarrow{\boldsymbol{E}}}^i + \underleftarrow{\boldsymbol{B}}_w : \underleftarrow{\boldsymbol{W}}^H + \kappa\dot{\zeta} + \kappa_w\dot{\xi} \\ &= \boldsymbol{B}_s : \underleftarrow{\mathcal{L}} \boldsymbol{E}^i + \boldsymbol{B}_w : \boldsymbol{W}^H + \kappa\dot{\zeta} + \kappa_w\dot{\xi} \end{aligned} \tag{60}$$

However, we note that equations (51), (54) and (56) may not be formally adequate because they are defined in terms of *total* internal strains and, as the plastic strains, they are path dependent. Hence directly assuming (60), (55) and the rate form of (56) is more appropriate, and (51) should be taken just for motivation purposes. Furthermore, Equation (33) should formally be assumed in rate form, and in the derivations to follow only the rate form will be used.

4 Mapping Tensors from Quadratic to Logarithmic Strain Space

In large strain plasticity, logarithmic strain measures frequently yield simple and natural descriptions. Of course, these strains may be used in any configuration simply using the proper stretch tensor to obtain them. The following relationship holds:

$$\boldsymbol{E}^e = \boldsymbol{R}^{e\,T} \boldsymbol{e}^e \boldsymbol{R}^e \quad \text{with} \quad \boldsymbol{E}^e = \ln \boldsymbol{U}^e, \;\; \boldsymbol{e}^e = \ln \boldsymbol{V}^e \tag{61}$$

Hence, it is noted that for logarithmic strain tensors, the push-forward and pull-back operations are performed with the rotation part of the deformation gradient alone. One may say that the stress-free configuration and the "unrotated" configuration are coincident in the logarithmic strain space. Obviously, since the logarithmic strain tensors and the Almansi and Green strains are all unique for a given deformation gradient, there exist a one-to-one mapping between them. For example

$$\boldsymbol{E}^e = \mathbb{M}_{\boldsymbol{A}}^{\boldsymbol{E}} : \boldsymbol{A}^e \tag{62}$$

where if the spectral forms of the strain tensors are

$$\boldsymbol{E}^e = \sum_{i=1}^{3} \ln \lambda_i^e \; \boldsymbol{N}_i \otimes \boldsymbol{N}_i, \quad \boldsymbol{A}^e = \sum_{i=1}^{3} \tfrac{1}{2} \left(\lambda_i^{e\;2} - 1 \right) \; \boldsymbol{N}_i \otimes \boldsymbol{N}_i \tag{63}$$

$\mathbb{M}_{\boldsymbol{A}}^{\boldsymbol{E}}$ can be written as

$$\mathbb{M}_{\boldsymbol{A}}^{\boldsymbol{E}} = \sum_{i=1}^{3} \frac{2\ln\lambda_i^e}{\lambda_i^{e\ 2}-1} \boldsymbol{N}_i \otimes \boldsymbol{N}_i \otimes \boldsymbol{N}_i \otimes \boldsymbol{N}_i \tag{64}$$

as it is straightforward to verify. Conversely

$$\mathbb{M}_{\boldsymbol{E}}^{\boldsymbol{A}} = \sum_{i=1}^{3} \frac{\lambda_i^{e\ 2}-1}{2\ln\lambda_i^e} \boldsymbol{N}_i \otimes \boldsymbol{N}_i \otimes \boldsymbol{N}_i \otimes \boldsymbol{N}_i \tag{65}$$

is such that $\boldsymbol{A}^e = \mathbb{M}_{\boldsymbol{E}}^{\boldsymbol{A}} : \boldsymbol{E}^e$. In a similar way, there is a one-to-one mapping between the deformation rate tensor and the time-derivative of the logarithmic strains. These mapping tensors may be found to be (see Reference [51])

$$\mathbb{M}_{\boldsymbol{D}}^{\dot{\boldsymbol{E}}} = \frac{\partial \boldsymbol{E}^e}{\partial \boldsymbol{A}^e} = \sum_{i=1}^{3} \frac{1}{\lambda_i^{e\ 2}} \boldsymbol{M}_i \otimes \boldsymbol{M}_i + \sum_{i=1}^{3}\sum_{j\neq i} 2\frac{\ln\lambda_j^e - \ln\lambda_i^e}{\lambda_j^{e\ 2} - \lambda_i^{e\ 2}}\ \boldsymbol{M}_i \overset{s}{\odot} \boldsymbol{M}_j \tag{66}$$

and

$$\mathbb{M}_{\dot{\boldsymbol{E}}}^{\boldsymbol{D}} = \frac{\partial \boldsymbol{A}^e}{\partial \boldsymbol{E}^e} = \sum_{i=1}^{3} \lambda_i^{e\ 2} \boldsymbol{M}_i \otimes \boldsymbol{M}_i + \sum_{i=1}^{3}\sum_{j\neq i} \frac{1}{2}\frac{\lambda_j^{e\ 2} - \lambda_i^{e\ 2}}{\ln\lambda_j^e - \ln\lambda_i^e}\ \boldsymbol{M}_i \overset{s}{\odot} \boldsymbol{M}_j \tag{67}$$

where

$$\boldsymbol{M}_i := \boldsymbol{N}_i \otimes \boldsymbol{N}_i \tag{68}$$

$$\boldsymbol{M}_i \overset{s}{\odot} \boldsymbol{M}_j := \tfrac{1}{4}\left(\boldsymbol{N}_i \otimes \boldsymbol{N}_j + \boldsymbol{N}_j \otimes \boldsymbol{N}_i\right) \otimes \left(\boldsymbol{N}_i \otimes \boldsymbol{N}_j + \boldsymbol{N}_j \otimes \boldsymbol{N}_i\right) \equiv \boldsymbol{M}_j \overset{s}{\odot} \boldsymbol{M}_i \tag{69}$$

These tensors have major and minor symmetries and represent the one-to-one mappings relating deformation rates as

$$\dot{\boldsymbol{E}}^e = \mathbb{M}_{\boldsymbol{D}}^{\dot{\boldsymbol{E}}} : \boldsymbol{D}^e \quad \text{and} \quad \boldsymbol{D}^e = \mathbb{M}_{\dot{\boldsymbol{E}}}^{\boldsymbol{D}} : \dot{\boldsymbol{E}}^e \tag{70}$$

respectively. Furthermore, in the rotation-frozen configuration

$$\underleftarrow{\dot{\boldsymbol{E}}}^e = \underleftarrow{\mathbb{M}}_{\boldsymbol{D}}^{\dot{\boldsymbol{E}}} : \underleftarrow{\boldsymbol{D}}^e \ \text{and}\ \underleftarrow{\boldsymbol{D}}^e = \underleftarrow{\mathbb{M}}_{\dot{\boldsymbol{E}}}^{\boldsymbol{D}} : \underleftarrow{\dot{\boldsymbol{E}}}^e \tag{71}$$

Also, in the stress-free configuration

$$\underleftarrow{\mathcal{L}}\boldsymbol{E}^e = \mathbb{M}_{\boldsymbol{D}}^{\dot{\boldsymbol{E}}} : \underleftarrow{\mathcal{L}}\boldsymbol{A}^e \ \text{and}\ \underleftarrow{\mathcal{L}}\boldsymbol{A}^e = \mathbb{M}_{\dot{\boldsymbol{E}}}^{\boldsymbol{D}} : \underleftarrow{\mathcal{L}}\boldsymbol{E}^e \tag{72}$$

For future use, we define two fourth order mapping tensors

$$\underleftarrow{\mathbb{W}}^M := \frac{1}{2}\left(\underleftarrow{\boldsymbol{C}}^e \overset{3}{\cdot} \underleftarrow{\mathbb{M}}_{\boldsymbol{D}}^{\dot{\boldsymbol{E}}} - \underleftarrow{\boldsymbol{C}}^e \overset{4}{\cdot} \underleftarrow{\mathbb{M}}_{\boldsymbol{D}}^{\dot{\boldsymbol{E}}}\right) \tag{73}$$

and

$$\underleftarrow{\mathbb{S}}^{M} := \frac{1}{2}\left(\underleftarrow{\boldsymbol{C}}^{e} \overset{3}{\cdot} \underleftarrow{\mathbb{M}}_{\boldsymbol{D}}^{\dot{\boldsymbol{E}}} + \underleftarrow{\boldsymbol{C}}^{e} \overset{4}{\cdot} \underleftarrow{\mathbb{M}}_{\boldsymbol{D}}^{\dot{\boldsymbol{E}}}\right) \tag{74}$$

where by $\left(\overset{n}{\cdot}\right)$ we imply the contraction of the $n - index$ of the fourth order tensor with the second index of the second order tensor. Then, it can be shown that if we define

$$\underleftarrow{\boldsymbol{K}} := \underleftarrow{\boldsymbol{S}} : \underleftarrow{\mathbb{M}}_{\dot{\boldsymbol{E}}}^{\boldsymbol{D}} \quad \text{so that } \underleftarrow{\boldsymbol{S}} =: \underleftarrow{\boldsymbol{K}} : \underleftarrow{\mathbb{M}}_{\boldsymbol{D}}^{\dot{\boldsymbol{E}}} \tag{75}$$

we obtain

$$\underleftarrow{\boldsymbol{\Xi}} := \underleftarrow{\boldsymbol{C}}^{e}\underleftarrow{\boldsymbol{S}} = \underleftarrow{\boldsymbol{C}}^{e}\left(\underleftarrow{\boldsymbol{K}} : \underleftarrow{\mathbb{M}}_{\boldsymbol{D}}^{\dot{\boldsymbol{E}}}\right) = \underleftarrow{\boldsymbol{K}} : \left(\underleftarrow{\mathbb{S}}^{M} + \underleftarrow{\mathbb{W}}^{M}\right) \tag{76}$$

and

$$\underleftarrow{\boldsymbol{K}}_{w} := \underleftarrow{\boldsymbol{K}} : \underleftarrow{\mathbb{W}}^{M} = \underleftarrow{\boldsymbol{E}}^{e}\underleftarrow{\boldsymbol{K}} - \underleftarrow{\boldsymbol{K}}\underleftarrow{\boldsymbol{E}}^{e} \equiv \underleftarrow{\boldsymbol{\Xi}}_{w} \tag{77}$$

$$\underleftarrow{\boldsymbol{\Xi}}_{s} = \underleftarrow{\boldsymbol{K}} : \underleftarrow{\mathbb{S}}^{M} \tag{78}$$

The tensor $\boldsymbol{K}$ is actually the *generalized Kirchhoff stress tensor* $\boldsymbol{T}$, see also below, and hence the conversion to the symmetric part of the Mandel stress tensor $\boldsymbol{\Xi}_s$ is given by Equation (78).

5 Dissipation Inequality

The stress power in the reference volume may be expressed in the intermediate configuration as

$$\mathcal{P} \equiv \boldsymbol{S} : \boldsymbol{L} = \boldsymbol{S} : (\boldsymbol{L}^{e} + \boldsymbol{C}^{e}\boldsymbol{L}^{p}) \tag{79}$$

$$= \boldsymbol{S} : (\boldsymbol{D}^{e} + \boldsymbol{W}^{e}) + \boldsymbol{S} : \boldsymbol{C}^{e}(\boldsymbol{D}^{p} + \boldsymbol{W}^{p}) \tag{80}$$

where $\boldsymbol{S}$ is the pull-back of the Kirchhoff stress $\boldsymbol{\tau}$ to the stress-free configuration. Since $\boldsymbol{S}$ is symmetric the product $\boldsymbol{S} : \boldsymbol{W}^e = 0$, i.e. the modified elastic spin (which also contains the rigid-body spin) produces no work. Thus, in a rotationally-frozen configuration we are left with

$$\mathcal{P} \equiv \underleftarrow{\boldsymbol{S}} : \underleftarrow{\boldsymbol{L}} = \underleftarrow{\boldsymbol{K}} : \dot{\underleftarrow{\boldsymbol{E}}}^{e} + \underleftarrow{\boldsymbol{S}} : \underleftarrow{\boldsymbol{C}}^{e}\left(\underleftarrow{\boldsymbol{D}}^{p} + \underleftarrow{\boldsymbol{W}}^{p}\right) \tag{81}$$

where we used (71) and (75). Alternatively, in the stress-free configuration

$$\boldsymbol{S} : \boldsymbol{L} = \boldsymbol{K} : \underleftarrow{\mathcal{L}}\boldsymbol{E}^{e} + \boldsymbol{S} : \boldsymbol{C}^{e}(\boldsymbol{D}^{p} + \boldsymbol{W}^{p}) \tag{82}$$

$$= \boldsymbol{K} : \underleftarrow{\mathcal{L}}\boldsymbol{E}^{e} + \boldsymbol{C}^{e}\boldsymbol{S} : (\boldsymbol{D}^{p} + \boldsymbol{W}^{p}) \tag{83}$$

Using $\boldsymbol{\Xi} = \boldsymbol{C}^e\boldsymbol{S}$, the stress power can be written as

$$\boldsymbol{S} : \boldsymbol{L} = \boldsymbol{K} : \underleftarrow{\mathcal{L}}\boldsymbol{E}^{e} + (\boldsymbol{\Xi}_{s} + \boldsymbol{\Xi}_{w}) : (\boldsymbol{D}^{p} + \boldsymbol{W}^{p})$$

$$= \boldsymbol{K} : \underleftarrow{\mathcal{L}}\boldsymbol{E}^{e} + \boldsymbol{\Xi}_{s} : \boldsymbol{D}^{p} + \boldsymbol{\Xi}_{w} : \boldsymbol{W}^{p} \tag{84}$$

Thus, the symmetric Mandel stress tensor produces power on the modified plastic strain rate, whereas the skew-symmetric Mandel tensor produces power on the modified plastic spin. This last work is due to the kinematic coupling produced by the Lee decomposition and the possible rotation of elastic anisotropy axes. In the case of isotropy or deformation through the orthotropy axes, the term vanishes. Neglecting the effect of temperature, the dissipation inequality from the second law of the thermodynamics is

$$\dot{\mathcal{D}} = \mathcal{P} - \dot{\psi} \geq 0 \tag{85}$$

where $\dot{\psi}$ is the free energy function rate, assumed to be $\dot{\psi} = \dot{\mathcal{W}} + \dot{\mathcal{H}}$. Thus using (49) and (60)

$$\dot{\psi} = \boldsymbol{T} : \underset{\leftarrow}{\mathcal{L}} \boldsymbol{E}^e + \boldsymbol{T}_w : \boldsymbol{W}^A + \boldsymbol{B}_s : \underset{\leftarrow}{\mathcal{L}} \boldsymbol{E}^i + \boldsymbol{B}_w : \boldsymbol{W}^H + \kappa\dot{\zeta} + \kappa_w\dot{\xi} \tag{86}$$

and

$$\begin{aligned} \dot{\mathcal{D}} = & (\boldsymbol{K} - \boldsymbol{T}) : \underset{\leftarrow}{\mathcal{L}} \boldsymbol{E}^e + \boldsymbol{\Xi}_s : \boldsymbol{D}^p + \boldsymbol{\Xi}_w : \boldsymbol{W}^p \\ & - \boldsymbol{T}_w : \boldsymbol{W}^A - \boldsymbol{B}_s : \underset{\leftarrow}{\mathcal{L}} \boldsymbol{E}^i - \boldsymbol{B}_w : \boldsymbol{W}^H - \kappa\dot{\zeta} - \kappa_w\dot{\xi} \geq 0 \end{aligned} \tag{87}$$

Since the equality must hold for pure elastic deformations,

$$\boldsymbol{T} = \boldsymbol{K} \tag{88}$$

and, in consequence,

$$\boldsymbol{T}_w = \boldsymbol{K}_w \equiv \boldsymbol{\Xi}_w \tag{89}$$

The reduced (plastic) dissipation inequality is now

$$\dot{\mathcal{D}}^p = \boldsymbol{\Xi}_s : \boldsymbol{D}^p + \boldsymbol{\Xi}_w : \boldsymbol{W}^d - \boldsymbol{B}_s : \underset{\leftarrow}{\mathcal{L}} \boldsymbol{E}^i - \boldsymbol{B}_w : \boldsymbol{W}^H - \kappa\dot{\zeta} - \kappa_w\dot{\xi} \geq 0 \tag{90}$$

where we defined the *dissipative* spin tensor in the unrotated configuration as

$$\boldsymbol{W}^d := \boldsymbol{W}^p - \boldsymbol{W}^A \tag{91}$$

We note that if $\boldsymbol{W}^p = \boldsymbol{W}^A$ then the skew part of the Mandel stress tensor does not contribute to the dissipation function. On the other hand, since $\boldsymbol{W}^A$ is assumed to be a function of $\boldsymbol{W}^p$, if $\boldsymbol{W}^p = \boldsymbol{0}$ then $\boldsymbol{W}^A = \boldsymbol{0}$ and no dissipation takes place either due to the skew part of the Mandel stress tensor. We will assume that the following relationship holds

$$\boldsymbol{W}^A = \rho \boldsymbol{W}^p \tag{92}$$

where ρ is a material scalar parameter. Then

$$\boldsymbol{W}^d = (1 - \rho)\, \boldsymbol{W}^p \tag{93}$$

We assume now –without loss of generality– that the elastic region is enclosed by two yield functions $f_s(\boldsymbol{\Xi}_s, \boldsymbol{B}_s, \kappa)$ and $f_w(\boldsymbol{\Xi}_w, \boldsymbol{B}_w, \kappa_w)$, the Lagrangian for the constrained problem is $L = \dot{\mathcal{D}}^p - \dot{t} f_s - \dot{\gamma} f_w$, where $\dot{t}$ and $\dot{\gamma}$

are the consistency parameter increments. Note also that $\underleftarrow{\mathcal{L}}\boldsymbol{E}^i \equiv \boldsymbol{D}^i$. If we claim that the principle of maximum dissipation holds, the stress and other internal variables are such that $\nabla L = 0$, i.e. for the yield function expressions given

$$\nabla L = 0 \Rightarrow \begin{cases} \dfrac{\partial L}{\partial \boldsymbol{\Xi}_s} = 0 \Rightarrow \boldsymbol{D}^p = \dot{t}\dfrac{\partial f_s}{\partial \boldsymbol{\Xi}_s} \text{ and } \dfrac{\partial L}{\partial \boldsymbol{\Xi}_w} = 0 \Rightarrow \boldsymbol{W}^d = \dot{\gamma}\dfrac{\partial f_w}{\partial \boldsymbol{\Xi}_w} \\ \dfrac{\partial L}{\partial \boldsymbol{B}_s} = 0 \Rightarrow \underleftarrow{\mathcal{L}}\boldsymbol{E}^i = -\dot{t}\dfrac{\partial f_s}{\partial \boldsymbol{B}_s} \text{ and } \dfrac{\partial L}{\partial \boldsymbol{B}_w} = 0 \Rightarrow \boldsymbol{W}^H = -\dot{\gamma}\dfrac{\partial f_w}{\partial \boldsymbol{B}_w} \\ \dfrac{\partial L}{\partial \kappa} = 0 \Rightarrow \dot{\zeta} = -\dot{t}\dfrac{\partial f_s}{\partial \kappa} \text{ and } \dot{\xi} = -\dot{\gamma}\dfrac{\partial f_w}{\partial \kappa_w} \end{cases} \tag{94}$$

These expressions are the associated flow and hardening rules for general elastoplasticity at finite strains. It is noted that if, as usual, the enclosure of the elastic region for the symmetric part is expressed in the form of $f_s(\boldsymbol{\Xi}_s - \boldsymbol{B}_s ...)$ then for associative plasticity the following relationship is automatically enforced

$$\underleftarrow{\mathcal{L}}\boldsymbol{E}^i \equiv \boldsymbol{D}^i = \boldsymbol{D}^p \tag{95}$$

Furthermore, $\boldsymbol{W}^i$ does not affect the dissipation function and can be freely prescribed. In view of Equation (95), and assuming that internal variables rotate as the plastic variables, we will set

$$\boldsymbol{W}^i = \boldsymbol{W}^p \tag{96}$$

and, as a consequence

$$\boldsymbol{X}^i = \boldsymbol{X}^p \tag{97}$$

The loading/unloading (complementary) Kuhn-Tucker conditions are, as usual

$$\dot{t} \geq 0, \quad f_s \leq 0 \quad \text{and} \quad \dot{t}f_s \equiv 0 \tag{98}$$

$$\dot{\gamma} \geq 0, \quad f_w \leq 0 \quad \text{and} \quad \dot{\gamma}f_w \equiv 0 \tag{99}$$

and the consistency conditions are

$$\dot{t}\dot{f}_s \equiv 0 \quad \text{and} \quad \dot{\gamma}\dot{f}_w \equiv 0 \tag{100}$$

The formulations presented herein and in Reference [44] show some similarities with some other works, see for example References [18, 20, 45–48], but there are also some significant differences; in particular we are using logarithmic strains in an incremental form.

6 Yield Functions

There is still much experimental work needed to establish the elastic domain and yield functions for the symmetric and skew parts of the Mandel stress tensor. From the current experimental evidence it is difficult to infer sound data about a macroscopic (continuum) elastic domain for the skew part of the Mandel stress tensor, and for the plastic spin evolution. Hence, at this point a "reasonable" proposition is necessary. An ad hoc extension of the small strains theory without plastic spin follows.

6.1 Yield Function for the Symmetric Part

For the symmetric part of the Mandel stress tensor the well-known Hill's quadratic yield criterion is assumed to hold, i.e. the yield function for $\boldsymbol{\Xi}_s$ is given by the expression (see for example Reference [1])

$$f_s := \frac{3}{2\kappa^2}\left(\boldsymbol{\Xi}_s - \boldsymbol{B}_s\right) : \mathbb{A}_s^p : \left(\boldsymbol{\Xi}_s - \boldsymbol{B}_s\right) - 1 = 0 \tag{101}$$

where $\mathbb{A}_s^p$ is the plastic anisotropy tensor which in this work we assume to have the same preferred anisotropy directions as the elastic anisotropy tensor. Given this function, the specific values of the internal variable increments are obtained from Equations (94) and (95) as

$$\boldsymbol{D}^i = \boldsymbol{D}^p = \dot{t}\frac{\partial f_s}{\partial \boldsymbol{\Xi}_s} = \frac{3}{\kappa^2}\mathbb{A}_s^p : \left(\boldsymbol{\Xi}_s - \boldsymbol{B}_s\right)\, \dot{t} \tag{102}$$

The internal isotropic variable rate is obtained as

$$\dot{\zeta} = -\dot{t}\frac{\partial f_s}{\partial \kappa} = \frac{2}{\kappa}\left(f_s + 1\right)\, \dot{t} \tag{103}$$

which, at the yield condition ($f_s = 0$) takes the value $\dot{\zeta} = 2\dot{t}\,/\kappa$. The physical meaning of $\dot{\zeta}$ is the effective plastic strain rate, see Reference [1].

6.2 Yield Function for the Skew Part

For the skew part, in this work we consider the simplest possible yield function, of the Mises type

$$f_w = \|\boldsymbol{\Xi}_w\| - \sqrt{2}\kappa_w \tag{104}$$

where κ_w is the allowed yield value, which may take the value of zero. From Equation (94), the specific flow variables take the form

$$\boldsymbol{W}^d = \dot{\gamma}\frac{\partial f_w}{\partial \boldsymbol{\Xi}_w} = \dot{\gamma}\hat{\boldsymbol{\Xi}}_w \tag{105}$$

$$\dot{\xi} = -\dot{\gamma}\frac{\partial f_w}{\partial \kappa_w} = \sqrt{2}\dot{\gamma} \tag{106}$$

where we defined the "direction" $\hat{\boldsymbol{\Xi}}_w := \boldsymbol{\Xi}_w / \|\boldsymbol{\Xi}_w\|$. The physical meaning of $\dot{\xi}$ is the effective *dissipative* rotation rate. Using Equations (92) and (93) we obtain

$$\boldsymbol{W}^p = \frac{1}{(1-\rho)}\dot{\gamma}\hat{\boldsymbol{\Xi}}_w \quad \text{and} \quad \boldsymbol{W}^A = \frac{\rho}{(1-\rho)}\dot{\gamma}\hat{\boldsymbol{\Xi}}_w \tag{107}$$

One important consequence of the function f_w defined in Equation (104) and the expression (107_2) is that if $1 > \rho > 0$ then $\boldsymbol{W}^A$ and $\boldsymbol{\Xi}_w \equiv \boldsymbol{T}_w$ have the same direction. But as noted just after Equation (44), the term $\boldsymbol{T}_w : \boldsymbol{W}^A$ should be negative. Hence possible values are $\rho > 1$ and $\rho < 0$. If $\rho > 1$ then the term $\boldsymbol{T}_w : \boldsymbol{W}^A$ is negative and $\boldsymbol{W}^p$ and $\boldsymbol{W}^A$ have the same direction. If $\rho < 0$ then the term $\boldsymbol{T}_w : \boldsymbol{W}^A$ is also negative and $\boldsymbol{W}^p$ and $\boldsymbol{W}^A$ have opposite direction. The actual rotation direction is not only determined by ρ, but also by the elastic anisotropy tensor because its shape may change the direction of $\boldsymbol{\Xi}_w$.

6.3 Coupling of Symmetric and Skew Parts

The yield function Equation (104) would mean an instantaneous rotation once $\|\boldsymbol{\Xi}_w\|$ is over the allowed value $\sqrt{2}\kappa_w$. However this is not consistent with experiments, in which progressive rotations are observed. Aside, in mechanics of single crystals this rotation is not independent of the ordinary (symmetric) plastic flow (Schmid's law). Hence, in this work we propose a viscoplasticity-like flow for the skew part in which the effective plastic strain plays the role of the time variable. This proposed expression is

$$\dot{\xi} = \left(\frac{< f_w >}{\eta} \right)^m \dot{\zeta} \tag{108}$$

where $< \cdot >$ is the Macauley bracket function, η is the "viscosity" material parameter with units of (couple-)stress and m is another material parameter. Hence, f_w may have values greater than zero which relax with plastic flow. In terms of consistency parameters, Equation (108) may be written as

$$\dot{\gamma} = \frac{\sqrt{2}}{\kappa} \left(\frac{< f_w >}{\eta} \right)^m \dot{t} \tag{109}$$

Hence, $\dot{\gamma}$ is zero if either $f_w \leq 0$ or $\dot{t} = 0$.

7 Numerical Example

In order to test the capabilities of the present theory in modelling the rotation of the anisotropy directions, we have carried out some numerical experiments. In these numerical tests, we aim for predictions of the experimental results reported in Reference [32]. In these experiments, a rotation of the material symmetry was observed when a steel sheet is strained in a direction that forms an angle θ with the rolling or prestrain direction. Details of the experiments are given in Reference [32]. Only small changes in the shape of the yield function were observed and hence the shape of the yield function can be assumed to remain constant, see also Reference [33].

However, unfortunately, in Reference [32] only the measured *plastic* anisotropy and its evolution are reported. Since our theory includes and indeed uses elastic anisotropy, we need to assume elastic anisotropy parameters. In a uniaxial test, a relevant degradation of Young's modulus and a variation of Poisson's ratio in the test direction has been reported [49]. Elastic anisotropy has also been measured in rolled steel, brass and aluminum, see for example [50]. We therefore assume the following elastic (only slightly anisotropic) material parameters: $E_a = 2.04 \times 10^{11}$ Pa, $E_b = 2.03 \times 10^{11}$ Pa, $E_c = 2.10 \times 10^{11}$ Pa, $\nu_{ab} = 0.3$, $\nu_{ac} = 0.3$, $\nu_{bc} = 0.3$, and $G_{ab} = 0.82 \times 10^{11}$ Pa. The yield stress κ_0 and Hill's yield function parameters have been reported in Reference [32], i.e. $f = 0.3613$, $h = 0.4957$, $g = 0.3535$ and we used $N = 1.175$ and $\kappa_0 = 23 \times 10^7$ Pa. The hardening has been considered as isotropic according to the formula $\kappa = \kappa_\infty - (\kappa_\infty - \kappa_0) \exp(-\delta\zeta) + \bar{h}\zeta$, for which the constants have been deduced from the experimental data, $\bar{h} = 3.5 \times 10^8$ Pa, $\kappa_\infty = 1.2\kappa_0$, $\delta = 30$. We also used the parameters $\kappa_w = 0$, $m = 2$, $\rho = -2$ and $\eta = 600$ Pa. All these parameters should really be chosen based on experimental results. However, we used the mentioned values and only adjusted η to match the experimental data. Details of the numerical implementation of the theory may be found in Reference [51].

The Young's modulus and the yield stress in the different directions, as well as their evolution are shown in Figure 2. In this figure we also compare the predicted yield stresses with the experimental data for the case of the applied load at $\theta = 30^o$ to the rolling direction. Figure 3 compares experimental data and computed results for $\theta = 30^o$, 45° and 60°. Of course, different elastic anisotropy constants (obtained experimentally) would change the predictions, but then also the material parameters η, ρ and m should be based on experimental results. An important feature of our formulation is that different rotation rates are obtained for different angles, and the predictions may not be symmetric for 30 and 60 degrees —in accordance with experimental results— even though the yield function is almost symmetric about the direction 44.7 degrees with the rolling direction. This is due to the selected shape for the anisotropy tensors.

8 Conclusions

We presented our research towards a model for anisotropic elasto-plasticity. The model shall represent possible anisotropic elasticity, anisotropic yield surfaces, hardening and the rotation of the elastic and plastic orthotropy directions during plastic flow. Both the continuum and time integration incremental formulations are simultaneously derived since incremental formulations give some insight into the continuum formulation. The model and the integration algorithm are derived using the multiplicative Lee decomposition of the total deformation gradient into an elastic and a plastic part. However, no total plas-

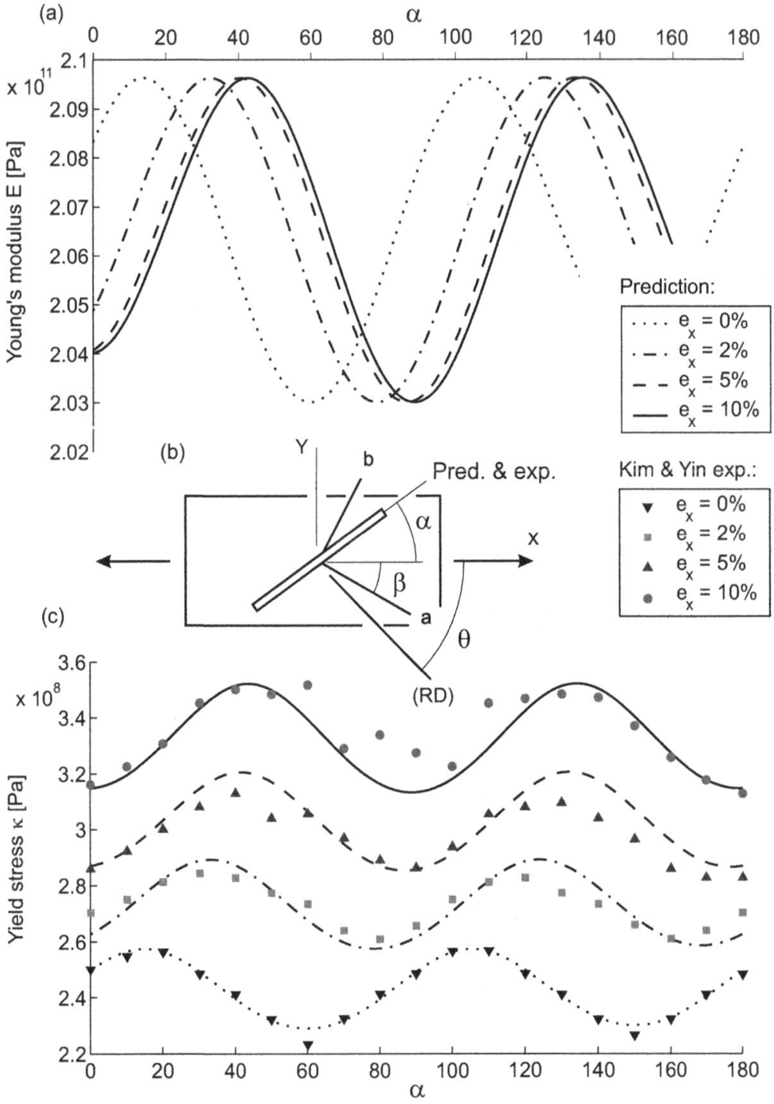

Fig. 2. (a) Prediction of the evolution of the Young's modulus profile at different spatial strains e_x for a uniaxial load at an angle of $\theta = 30^o$ with respect to the rolling direction (RD). (b) Angles involved in the example. Angle of the uniaxial load with the rolling direction (θ), angle of the principal direction a with the uniaxial load (β) —initially $\beta = \theta$—, angle of the Young's modulus and yield stress shown in the curves with the uniaxial load (α). (c) Comparison of the experimental data of [32] with the prediction of the evolution of the yield stress profile at different spatial strains e_x for a plane stress load at an angle of $\theta = 30^o$ with respect to the rolling direction

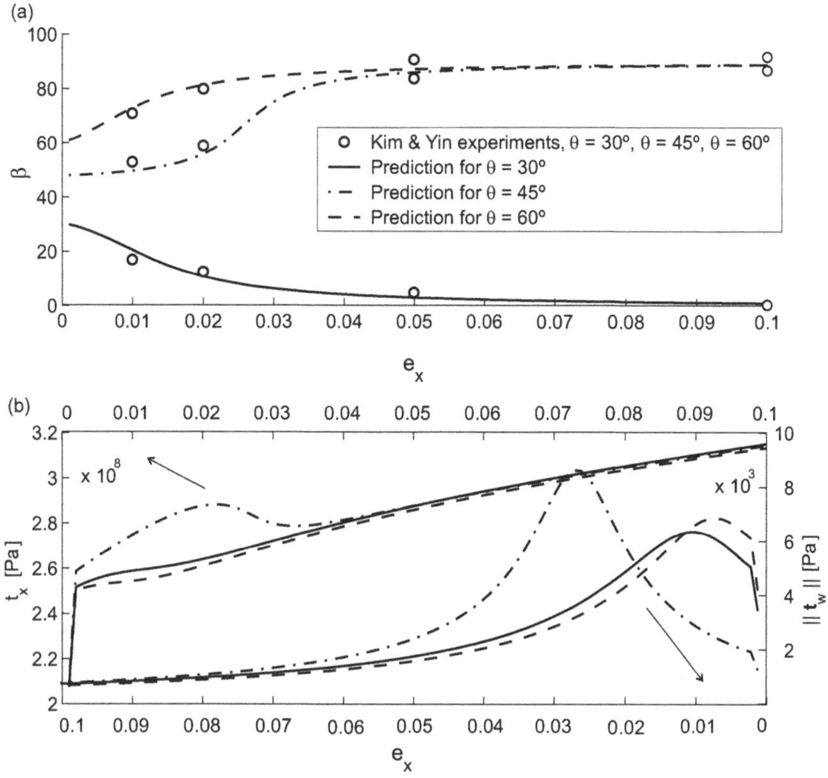

Fig. 3. (a) Prediction for the evolution of the principal orthotropy directions and comparison with the experimental values of Reference [32]. (b) Uniaxial stress and couple-stress evolutions

tic deformation measure is used. Plastic deformations and plastic rotations are considered only incrementally.

The stresses are directly obtained from the logarithmic elastic strains, with the model assuming linear elastic anisotropy and moderate elastic strains, but accounting for possible large plastic strains.

The model presented offers, in particular, possibilities to simulate the rotation of the material axes observed in experiments on anisotropic elasto-plastic materials, and as an example good correlation to the experimental data of Kim and Yin has been obtained.

However, clearly, the model must be studied much more. The sensitivity of the model predictions with respect to the model parameters needs to be identified, and comparisons of computed solutions with test data for many more and varied material tests need to be obtained. These studies will also identify the limitations of the model and whether the model could be simplified without significant loss of predictive capability.

References

1. Kojić M, Bathe KJ (2005) Inelastic analysis of solids and structures. Springer-Verlag, New York
2. Bathe KJ (1996) Finite element procedures. Prentice-Hall, New Jersey
3. Simó JC (1988) A framework for finite strain elastoplasticity based on maximum plastic dissipation and the multiplicative decomposition. Part I: Continuum formulation. Comp Meth Appl Mech Engrng 66: 199-219. Part II: Computational aspects. Comp Meth Appl Mech Engrng 68: 1-31
4. Simó JC, Hughes TJR (1998) Computational inelasticity. Springer-Verlag, New York
5. Bathe KJ, Ramm E, Wilson EL (1975) Finite element formulations for large deformation dynamic analysis. Int J Num Meth Engrg 9: 353-386
6. Kojić M, Bathe KJ (1987) Studies of finite element procedures—Stress solution of a closed elastic strain path with stretching and shearing using the updated Lagrangian Jaumann formulation. Comp Struct 26: 175-179
7. Lee EH (1969) Elastic-plastic deformations at finite strains. J Appl Mech ASME 36: 1-6
8. Lee EH, Liu DT (1967) Finite-strain elastic-plastic theory with application to plane wave analysis. J Appl Phys 38: 19-27
9. Bilby BA, Bullough R, Smith E (1955) Continuous distributions of dislocations: A new application of the methods of Non-Riemannian Geometry. Proc Roy Soc London Ser A 231: 263-273
10. Weber G, Anand L (1990) Finite deformation constitutive equations and a time integration procedure for isotropic hyperelastic-viscoplastic solids. Comp Meth Appl Mech Engrg 79: 173-202
11. Eterović AL, Bathe KJ (1990) A hyperelastic-based large strain elasto-plastic constitutive formulation with combined isotropic-kinematic hardening using the logarithmic stress and strain measures. Int J Num Meth Engrg 30: 1099-1114
12. Simó JC (1992) Algorithms for multiplicative plasticity that preserve the form of the return mappings of the infinitesimal theory. Comp Meth Appl Mech Engrg 99: 61-112
13. Perić D, de Souza EA (1999) A new model for Tresca plasticity at finite strains with an optimal parametrization in the principal space. Comp Meth Appl Mech Engrg 171: 463-489
14. Montáns FJ, Bathe KJ (2003) On the stress integration in large strain elasto-plasticity. In: Bathe KJ (ed) Computational fluid and solid mechanics 2003. Elsevier, Oxford
15. Montáns FJ, Bathe KJ (2005) Computational issues in large strain elasto-plasticity: An algorithm for mixed hardening and plastic spin. Int J Num Meth Engrg 63: 159-196
16. Cuitiño A, Ortiz M (1992) A material-independent method for extending stress update algorithms from small strain plasticity to finite plasticity with multiplicative kinematics. Engrg Comp 9: 437-451
17. Borja RI, Tamagnini C (1998) Cam–Clay plasticity, Part III: Extension of the infinitesimal model to include finite strains. Comp Meth Appl Mech Engrg 155, 77-95
18. Papadopoulus P, Lu J (1998) A general framework for the numerical solution of problems in finite elasto-plasticity. Comp Meth Appl Mech Engrg 159: 1-18

19. Car E, Oller S, Oñate E (2001) A large strain plasticity model for anisotropic materials—composite material application. Int J Plasticity 17: 1537-1463
20. Miehe C, Apel N, Lambrecht M (2002) Anisotropic additive plasticity in the logarithmic strain space: modular kinematic formulation and implementation based on incremental minimization principles for standard materials. Comp Meth Appl Mech Engrg 191: 5383-5426
21. Eidel B, Gruttmann F (2003) On the theory and numerics of orthotropic elastoplasticity at finite plastic strains. In: Bathe KJ (ed) Computational fluid and solid mechanics 2003. Elsevier, Oxford
22. Han CS, Lee MG, Chung K, Wagoner RH (2003) Integration algorithms for planar anisotropic shells with isotropic and kinematic hardening at finite strains. Comm Num Meth Engrg 19: 473-490
23. Loret B (1983) On the effects of plastic rotation in the finite deformation of anisotropic elastoplastic materials. Mech Mater 2: 287-304
24. Dafalias YF (1985) The plastic spin. J Appl Mech ASME 52: 865-871
25. Dafalias YF (1990) On the microscopic origin of the plastic spin. Acta Mech 82: 31-48
26. Anand L (1983) Elasto-viscoplasticity: Constitutive modelling and deformation processing. In Teodosiu C, Raphanel JL, Sidoroff F (eds) Large plastic deformations. AA Balkema, Rotterdam
27. Dafalias YF (1998) The plastic spin: necessity or redundancy? Int J Plasticity 14: 909-931
28. Kuroda M, Tvergaard V (2001) Plastic spin associated with a non-normality theory of plasticity. Eur J Mech A/Solids 20: 893-905
29. Khan AS, Huang S (1995) Continuum theory of plasticity. John Wiley & Sons, New York
30. Levitas V (1998) A new look at the problem of plastic spin based on stability analysis. J Mech Phys Solids 46: 557-590
31. Tong W, Tao H, Jiang X (2004) Modeling the rotation of orthotropic axes of sheet metals subjected to off-axis uniaxial tension. J Appl Mech ASME 71: 521-531
32. Kim KH, Yin JJ (1997) Evolution of anisotropy under plane stress. J Mech Phys Solids 45: 841-851
33. Kowalewski ZL, Sliwowski M (1997) Effect of cyclic loading on the yield surface evolution of 18G2A low-alloy steel. Int J Mech Sci 39: 51-68
34. Bunge HJ, Nielsen I (1997) Experimental determination of plastic spin in polycrystalline materials. Int J Plasticity 13: 435-446
35. Truong Qui HK, Lippmann H. (2001) Plastic spin and evolution of an anisotropic yield condition. Int J Mech Sci 43: 1969-1983
36. Boheler JP, Koss S (1991) Evolution of anisotropy in sheet steels subjected fo off-axes large deformation. In Brueler O, Mannl V, Najar J (eds) Advances in continuum mechanics. Springer, Berlin
37. R.H. Randal, C. Zener (1940) Internal friction of aluminum. Phys Rev 58: 472-473
38. Tam AC, Leung WP (1984) Measurement of small elastic anisotropy in solids using laser-induced ultrasonic pulses. Appl Phys L 45: 1040-1042
39. Dvorkin EN, Goldschmit MB (2006) Nonlinear continua. Springer-Verlag, New York

40. Morán B, Ortiz M, Shih CF (1990) Formulation of implicit finite element methods for multiplicative finite deformation plasticity. Int J Num Meth Engrg 29: 483-514
41. Anand L (1985) Constitutive equations for hot-working of metals. Int J Plasticity 1: 213-231
42. Simó JC (1986) On the computational significance of the intermediate configuration and hyperelastic relations in finite deformation elastoplasticity. Mech Mat 4: 439-451
43. Bathe KJ, Montáns FJ (2004) On modelling mixed hardening in computational plasticity. Comp Struct 82: 535-539
44. Montáns FJ, Bathe KJ (2005) Large strain anisotropic plasticity including effects of plastic spin. In: Bathe KJ (ed) Computational fluid and solid mechanics 2005. Elsevier, Oxford
45. Tsakmakis Ch (2004) Description of plastic anisotropy effects at large deformations—part I: restrictions imposed by the second law and the postulate of Il'iushin. Int J Plasticity 20: 167-198
46. Eidel B, Gruttmann F (2005) Anisotropic pile-up pattern at spherical indentation into a fcc single crystal—finite element analysis versus experiment. In: Bathe KJ (ed) Computational fluid and solid mechanics 2005. Elsevier, Oxford
47. Han C-S, Choi Y, Lee J-K, Wagoner RH (2002) A FE formulation for elasto-plastic materials with planar anisotropic yield functions and plastic spin. Int J Solids Struct 39: 5123-5141
48. Han C-S, Chung K, Wagoner RH, S-I Oh (2003) A multiplicative finite elasto-plastic formulation with anisotropic yield functions. Int J Plasticity 19: 197-211
49. Luo L, Ghosh AK (2003) Elastic and inelastic recovery after plastic deformation of DQSK steel sheet. J Engrg Mater Tech ASME 125: 237-246
50. Lauwagie T, Sol H, Roebben G, Heylen W, Shi Y (2002) Validation of the Resonalyser method: an inverse method for material identification. Proc Int Conf Noise Vibration Engrg, ISMA2002, Leuven, Belgium
51. Montáns FJ, Bathe KJ (in preparation) A framework for computational large strain plasticity—anisotropic elasticity, anisotropic yield functions and mixed hardening

Localized and Diffuse Bifurcations in Porous Rocks Undergoing Shear Localization and Cataclastic Flow

Ronaldo I. Borja

Department of Civil and Environmental Engineering, Stanford University, Stanford, CA 94305, USA
borja@stanford.edu

Summary. Under normal temperature porous rocks can fail either by shear strain localization or cataclastic flow. Shear localization results from the coalescence of microcracks leading to a tabular deformation band, whereas cataclastic flow is characterized by grain crushing and pore collapse resulting in a severely damaged but macroscopically homogeneous compacted continuum. In this paper we view the two types of instability as arising from two distinct bifurcation modes. The first mode, predicted from the singularity of the acoustic tensor, produces a strain rate jump tensor of rank one and defines a deformation band. The second mode, predicted from the singularity of the tangent constitutive operator, is diffuse and produces a full-rank strain rate jump tensor. After identifying the relevant bifurcation mode, we present a framework for capturing post-failure responses through constitutive branching. The post-collapse constitutive response features a cohesion softening-friction hardening applied either to an emerging fault for shear localization or to the bulk constitutive theory for diffuse pore collapse instability.

1 Introduction

Under normal laboratory testing conditions porous rocks tested under triaxial compression can fail either by shear strain localization or by cataclastic flow [1–4]. The relevant mode of failure depends on the confining pressure and temperature under which the rock sample is tested. Trends are depicted pictorially in Figs. 1 and 2 for the case of triaxial compression testing of Adamswiller sandstone (the details of the testing are given in [4]). At lower confining pressures (5 and 20 MPa) the rock sample first compacts and then dilates until it forms a well-defined shear band. This is characteristic of brittle failure. The specimen tested at 40 MPa exhibits multiple bands representing transition from localized to cataclastic failure. Multiple compaction bands are not atypical at this transition pressure, with the axial strain increasing in Fig. 1 due to porosity reduction as the thickness of the compaction bands increases (Wong, personal communication).

Eugenio Oñate and Roger Owen (eds.), Computational Plasticity, 37–53.

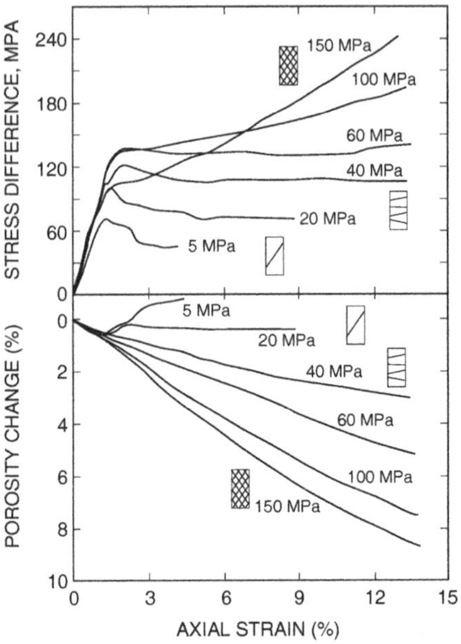

Fig. 1. Mechanical data for Adamswiller sandstone from six triaxial compression tests under different effective pressures as indicated. Samples tested at 5 and 20 MPa failed by shear localization; sample tested at 40 MPa exhibited multiple bands; samples tested at higher confining pressures failed by cataclastic flow. Figure reproduced from [4]

At higher confining pressures (60, 100, and 150 MPa) the sample compacts throughout as the mean normal stress increases up to a knee in the stress-strain curve (Fig. 2b), after which an accelerated volume compaction occurs with some hardening in the principal stress difference (Fig. 1) and mild increase in the mean normal stress (Fig. 2). The micromechanical processes may be considered brittle in all specimens due to the pervasive grain-scale microcracking. However, on a macroscopic scale the modes of failure may be described very differently in that shear localization manifests a brittle mode whereas cataclastic flow appears much more ductile. Though not reported for this particular sandstone, Hirth and Tullis [1] observed that cataclastic flow in a porous quartzite is only a transient behavior in that when the pores have collapsed any subsequent deformation could become more localized into a shear band. This typically occurs at much higher strains, of the order 25%, and subsequent volume deformation could eventually become dilative.

In this paper we employ classical bifurcation analysis to predict shear localization and cataclastic flow failure modes in rocks at macroscale. Shear localization is predicted from the singularity of the acoustic tensor as described by Rudnicki and Rice [5]. The upshot of the prediction is a planar band across

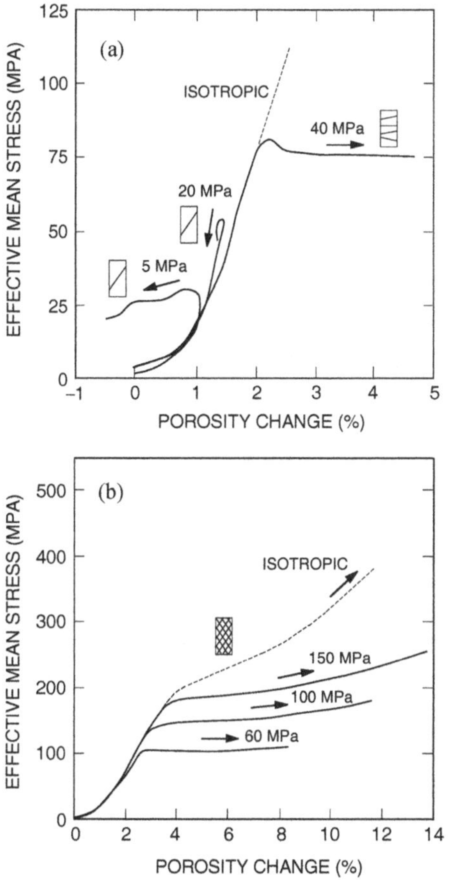

Fig. 2. Shear-induced dilation and compaction in Adamswiller sandstone: (a) samples tested at 5 and 20 MPa confining pressures localized in shear across a single band, sample tested at 40 MPa exhibited multiple bands; (b) samples tested at higher confining pressures failed by cataclastic flow. Figure reproduced from [4]

which the velocity gradient exhibits a uniform jump while the nominal traction rate vector remains continuous. In this case the jump in the velocity gradient is a rank-one tensor. Cataclastic flow does not involve a deformation band, so we predict the occurrence of this mode from the singularity of the tangent constitutive operator as described by Borja [6]. In this case the jump in the velocity gradient is a full-rank tensor obtained from the condition that the nominal stress rate remain continuous. We term a full-rank velocity gradient jump tensor as diffuse bifurcation, in contrast to 'global bifurcation' as used, for example, in [7]. A comparison of rank-one and full-rank velocity gradient tensors is shown in Fig. 3.

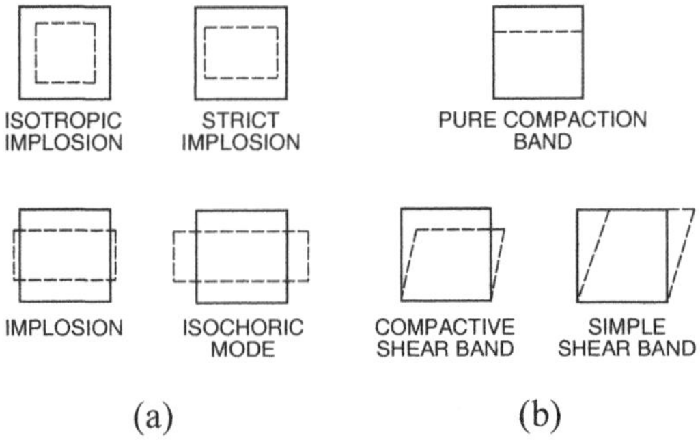

Fig. 3. Eigenmodes for bifurcations leading to pore collapse for: (a) singular tangent constitutive operator; (b) singular tangent acoustic tensor, see Reference [6]

Grain crushing and pore collapse in rocks lead to a drastically different material. For rocks experiencing shear localization, these processes are limited to a narrow zone called deformation band [8, 9]. In cataclastic flow, on the other hand, grain crushing and pore collapse are pervasive throughout the entire volume. Deformation at post-collapse is dominated by continued grain fracturing, as well as by rigid body rotation and relative translation of the fractured grains and fragments. Effectively, the 'new' material now behaves as a dense particulate medium possessing much different constitutive properties than those of the intact rock. In this paper we also present a framework for constitutive branching to allow the constitutive model to capture the new material properties in both shear localization and cataclastic flow failure regimes.

Notations and symbols used in this paper are as follows: bold-faced letters denote tensors and vectors; the symbol '$\cdot$' denotes an inner product of two vectors (e.g. $\boldsymbol{a} \cdot \boldsymbol{b} = a_i b_i$), or a single contraction of adjacent indices of two tensors (e.g. $\boldsymbol{c} \cdot \boldsymbol{d} = c_{ij} d_{jk}$); the symbol ':' denotes an inner product of two second-order tensors (e.g. $\boldsymbol{c} : \boldsymbol{d} = c_{ij} d_{ij}$), or a double contraction of adjacent indices of tensors of rank two and higher (e.g. $\boldsymbol{C} : \boldsymbol{\epsilon}^{\mathrm{e}} = C_{ijkl} \epsilon^{\mathrm{e}}_{kl}$); the symbol '$\otimes$' denotes a juxtaposition, e.g., $(\boldsymbol{a} \otimes \boldsymbol{b})_{ij} = a_i b_j$. Finally, for any symmetric second order tensors $\boldsymbol{\alpha}$ and $\boldsymbol{\beta}$, $(\boldsymbol{\alpha} \otimes \boldsymbol{\beta})_{ijkl} = \alpha_{ij}\beta_{kl}$, $(\boldsymbol{\alpha} \oplus \boldsymbol{\beta})_{ijkl} = \alpha_{jl}\beta_{ik}$, and $(\boldsymbol{\alpha} \ominus \boldsymbol{\beta})_{ijkl} = \alpha_{il}\beta_{jk}$.

2 Shear Localization

2.1 E-mode for Shear Localization

The condition for shear localization is given by

$$\boldsymbol{A}^{\mathrm{ep}} \cdot \boldsymbol{m} = \boldsymbol{0}\,, \qquad A^{\mathrm{ep}}_{ij} = N_A \mathcal{A}^{\mathrm{ep}}_{iAjB} N_B\,, \tag{1}$$

where $\boldsymbol{A}^{\mathrm{ep}}$ is the elastoplastic acoustic tensor, $\boldsymbol{\mathcal{A}}^{\mathrm{ep}} = \partial \boldsymbol{P} / \partial \boldsymbol{F}$ is the first tangential elastoplastic moduli tensor, $\boldsymbol{P}$ is the first Piola-Kirchhoff stress tensor, $\boldsymbol{F}$ is the deformation gradient, $\boldsymbol{m}$ is the unit vector in the direction of velocity jump across the band, and $\boldsymbol{N}$ (with component N_A) is the unit normal vector to the band in the undeformed configuration, see [10].

Determining the eigenmode (e-mode) provides an insight into the volume change response at bifurcation. Let $J = \det(\boldsymbol{F}) > 0$ denote the Jacobian of the motion, with material time derivative $\dot{J} = J \operatorname{div} \boldsymbol{v}$. The jump in $\dot{J}$ induced by a non-unique velocity field is given by

$$[\![\dot{J}]\!] = \dot{J} - \dot{J}^* = J \operatorname{div} [\![\boldsymbol{v}]\!] = J \operatorname{tr} [\![\boldsymbol{l}]\!] = J \operatorname{tr}([\![\dot{\boldsymbol{F}}]\!] \cdot \boldsymbol{F}^{-1}) . \tag{2}$$

Accordingly, we characterize the appropriate volume change mode from the following criteria:

$$\operatorname{sgn} [\![\dot{J}]\!] = \begin{cases} (-) & \Longrightarrow \quad \text{contraction/implosion;} \\ (+) & \Longrightarrow \quad \text{dilation/explosion.} \end{cases} \tag{3}$$

The transition condition $[\![\dot{J}]\!] = 0$ denotes an isochoric mode.

For condition (1) to provide a non-trivial solution we must have

$$\inf\{\mathcal{D}(\boldsymbol{N}) \in \Re | \mathcal{D}(\boldsymbol{N}) = \det(\boldsymbol{A}^{\text{ep}})\} = 0 , \tag{4}$$

where we view the field of determinants of $\boldsymbol{A}^{\text{ep}}$ as a set function of $\boldsymbol{N}$ and we want the infimum of this set to vanish at a critical orientation defined by the band normal $\boldsymbol{N}$. At this critical orientation the tensor $\boldsymbol{A}^{\text{ep}}$ is singular and the vector $\boldsymbol{m}$ defines its e-mode. We determine the correct algebraic sign of this e-mode from the conditions described in [9, 10].

For shear localization the form of the jump tensor $[\![\dot{\boldsymbol{F}}]\!]$ is given by

$$[\![\dot{\boldsymbol{F}}]\!] = \varphi \boldsymbol{m} \otimes \boldsymbol{N} , \tag{5}$$

where φ denotes the magnitude of the jump. Noting from Nanson's formula that

$$\boldsymbol{n} = (h/h_0) \boldsymbol{N} \cdot \boldsymbol{F}^{-1} , \tag{6}$$

where h and h_0 are the band thicknesses in the deformed and undeformed configurations, respectively, then we get

$$[\![\dot{J}]\!] = J \frac{h}{h_0} \operatorname{tr}(\boldsymbol{m} \otimes \boldsymbol{n}) = J \frac{h}{h_0} (\boldsymbol{m} \cdot \boldsymbol{n}) . \tag{7}$$

The shear band mode is then classified as either compactive or dilative depending on the sign of the vector dot product $\boldsymbol{m} \cdot \boldsymbol{n}$. Following [9], we classify a deformation band according to the following criteria:

$$\begin{cases} \boldsymbol{m} \cdot \boldsymbol{n} = 1 : & \text{pure dilation band;} \\ 0 < \boldsymbol{m} \cdot \boldsymbol{n} < 1 : & \text{dilatant shear band;} \\ \boldsymbol{m} \cdot \boldsymbol{n} = 0 : & \text{simple shear band;} \\ -1 < \boldsymbol{m} \cdot \boldsymbol{n} < 0 : & \text{compactive shear band;} \\ \boldsymbol{m} \cdot \boldsymbol{n} = -1 : & \text{pure compaction band.} \end{cases} \tag{8}$$

Note that the above definitions only pertain to the specific mode at bifurcation.

2.2 Constitutive Branching for Shear Localization

Slip weakening, a process describing the mechanical response of rocks from the onset of shear localization to the beginning of the residual state, provides a critical link between pre-localization and residual state behaviors. Developing a constitutive framework for this initial stage of slip instability requires knowledge of the residual state to which the slip weakening model will branch. For a shear band that is deforming by tangential frictional sliding we consider the classical Mohr-Coulomb friction law formulated in the Lagrangian description. Following [11], we take the nominal traction vector at material point X on the band as $\boldsymbol{t} = \boldsymbol{P} \cdot \boldsymbol{N}$, where $\boldsymbol{N}$ defines the unit band normal in the reference configuration (determined from the bifurcation analysis). We can resolve this traction into normal and tangential components at the current configuration as

$$\boldsymbol{t}(X,t) = t_N(X,t)\boldsymbol{n} + t_T(X,t)\boldsymbol{\xi}\,, \tag{9}$$

where $\boldsymbol{n} = \boldsymbol{N} \cdot \boldsymbol{F}^{-1}/\|\boldsymbol{N} \cdot \boldsymbol{F}^{-1}\|$, and t_N and t_T are normal and tangential components obtained from

$$t_N = \boldsymbol{n} \cdot \boldsymbol{P} \cdot \boldsymbol{N}\,, \qquad t_T = \|\boldsymbol{t}_T\|\,, \qquad \boldsymbol{\xi} = \boldsymbol{t}_T / t_T\,, \qquad \boldsymbol{t}_T = \boldsymbol{t} - t_N \boldsymbol{n}\,, \tag{10}$$

assuming planar sliding (some metric transformation may be required for sliding along curvilinear surfaces). Note that t_N (negative for compression) and t_T are resolved nominal stresses representing forces in the current configuration per unit undeformed area.

Next we write the Mohr-Coulomb friction law at residual state using the format of classical plasticity. Let μ denote the coefficient of friction; then we have

$$\Phi = t_T + \mu t_N \le 0\,, \qquad [\![\boldsymbol{v}]\!] = \dot{\zeta}\boldsymbol{\xi}\,, \qquad \dot{\zeta} \ge 0\,, \qquad \Phi\dot{\zeta} = 0\,. \tag{11}$$

In the above expressions Φ takes the role of the yield function at residual state, $[\![\boldsymbol{v}]\!]$ is the velocity jump across the band (evaluated from a non-associative flow rule), $\dot{\zeta}$ is the nonnegative slip rate, and the fourth expression is the Kuhn-Tucker condition. The band kinematics yields $[\![\dot{J}]\!] = J\,\mathrm{tr}[\![\boldsymbol{l}]\!] = 0$, since $[\![\boldsymbol{v}]\!] \cdot \boldsymbol{n} = 0$ for a band that moves by tangential sliding. Furthermore, the flow rule suggests that the velocity jump $[\![\boldsymbol{v}]\!]$ across the band is fully plastic, consistent with results obtained for strong discontinuity kinematics [12, 13].

We can also accommodate a variable coefficient of friction with this framework. For small slips and slow slip velocities the variation of the coefficient of friction μ may be described by the Dieterich-Ruina law [14–16]

$$\mu = \mu^* + A\ln(\dot{\zeta}/V^*) + B\ln(\theta/\theta^*) \tag{12}$$

$$\dot{\theta} = 1 - \theta\dot{\zeta}/D_c\,, \tag{13}$$

where $\dot{\zeta}$ is the magnitude of $[\![\boldsymbol{v}]\!]$, θ is a state variable, and A, B, μ^*, V^*, θ^*, and D_c are material parameters. The state variable θ has been linked to the changing set of frictional contacts and wear on the materials [16], and D_c is a characteristic slip required to replace a contact population representative of a previous sliding condition with a contact population created under a new sliding condition.

For zero and near-zero slip velocities, such as what occurs near the tip of a nucleating fault, the expression for μ as given by the above logarithmic function becomes singular. To circumvent this problem we view frictional sliding as a rate process and add backward jumps in the spirit of the Arrhenius law to obtain the regularized form [17]

$$\mu = A \sinh^{-1}\left[\frac{\dot{\zeta}}{2V^*}\exp\left(\frac{\mu^* + B\ln(\theta/\theta^*)}{A}\right)\right]. \tag{14}$$

This equation then predicts a zero coefficient of friction at zero slip velocity. The above variable friction model has been implemented in the context of strong discontinuity finite elements [18, 19]. Note that for large slips and high slip velocities, such as those encountered in earthquake fault nucleation

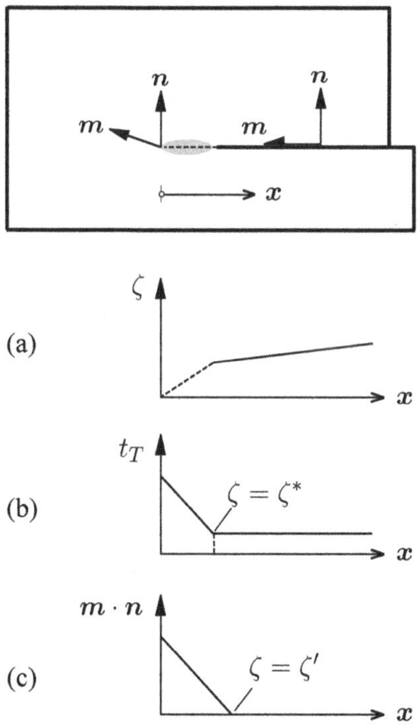

Fig. 4. Model for fault nucleation and propagation in rock: (a) fault nucleates (dashed line) over a characteristic distance; (b) slip weakening takes place over characteristic slip ζ^*; (c) initially dilatant shear band becomes isochoric over a characteristic slip ζ'

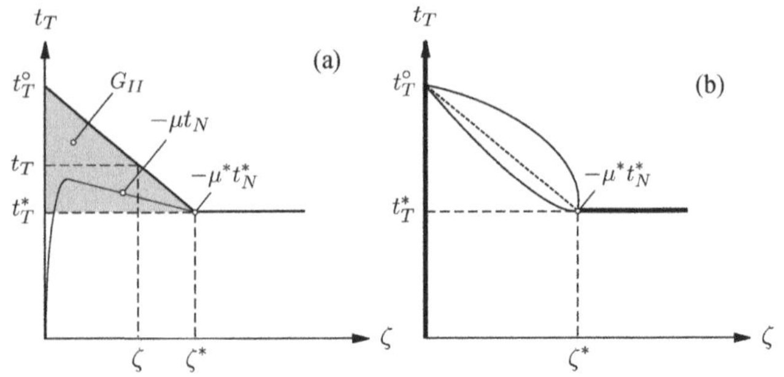

Fig. 5. Slip weakening models for rocks during fault nucleation: (a) linear slip weakening resulting from a combined cohesion softening-friction hardening on an emerging fault; (b) nonlinear slip weakening. Shaded area G_{II} represents shear fracture energy that has been correlated with the magnitude of an earthquake for regional-scale faulting, see [20]

processes, a much lower coefficient of friction may be activated by additional weakening mechanisms such as flash heating [17].

The concept of slip weakening leading to the above rate- and state-dependent residual friction model was motivated by the cohesive zone models for tensile fracture and extended to the shear fracture problem by Ida [20] and Palmer and Rice [21]. To better understand this process we consider a rock through which a shear band (or fault) is nucleating and propagating, as shown in Fig. 4. At the bifurcation point we assume that slip begins to accumulate, as defined by the integral

$$\zeta = \int_t \dot{\zeta} \, \mathrm{d}t \,, \tag{15}$$

where the integration is taken over the slip path, see Fig. 4a.

Figure 4b shows the resolved nominal shear stress t_T decaying to a residual state value away from the tip of a propagating band, where the cumulative slip ζ has exceeded a value $\zeta^* \approx 0.5$ mm for most rocks (Rice, personal communication, see also [22]). Figure 5 further depicts this narrow region of slip weakening [20], in which t_T decreases either linearly or nonlinearly with ζ (Figs. 5a and 5b, respectively). The shaded area G_{II} represents the shear fracture energy that has been correlated by many authors with the magnitude of an earthquake.

Figure 6 portrays the notion of constitutive branching during slip weakening in the context of classical elastoplasticity. Prior to bifurcation, plastic deformation of an intact rock is described by the yield function $F = 0$, which is assumed isotropic on the π-plane and dependent on all three principal Kirchhoff stresses. A deformation band-type bifurcation is detected at point A, and consequently the constitutive theory branches to the slip weakening law

$$\Phi = t_T - c + \mu t_N = 0 \,. \tag{16}$$

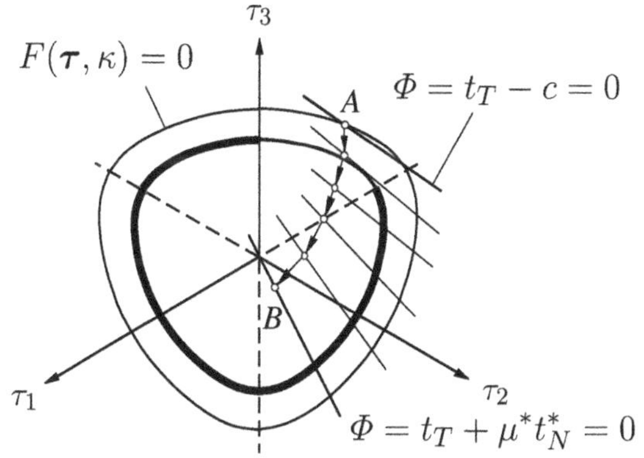

Fig. 6. Slip weakening on the π-plane in principal Kirchhoff stress space. Yield surface for intact rock $F = 0$ branches at the bifurcation stress point A to the Mohr-Coulomb failure criterion $\Phi = 0$ for a planar fault; at residual state B the cohesion decays to zero and the failure criterion passes through the origin

This equation plots as a straight line on the π-plane. However, the slip velocity is zero at the bifurcation point, so $\mu = 0$ initially, and the shear resistance is then purely cohesive. As slip accumulates the coefficient of friction μ quickly increases while the cohesion c decays to zero. During this time t_N may also vary, and the slip weakening law then plots as a fan of straight lines on the π-plane. At the conclusion of slip weakening the cohesion is zero and the shear resistance is purely frictional; hence, the straight line passes through the origin of the π-plane. This interplay between the cohesive and frictional resistances is also shown in Fig. 5a.

Assuming the shear resistance on the band is purely frictional and the slip-weakening constitutive law is linear, we can write an 'effective yield function' of the form

$$\Phi = t_T - \left[t_T^\circ - (t_T^\circ + \mu t_N^*)\frac{\zeta}{\zeta^*}\right] \leq 0\,, \qquad \zeta \in [0, \zeta^*]\,, \tag{17}$$

where $\Phi = 0$ now describes a linear slip weakening. However, the condition $\Phi = 0$ is not a trivial linear constraint to impose since t_N^* is unknown during the slip weakening process. A first-order approximation may be obtained by assuming $t_N^* \approx t_N$ [18], in which case (17) reduces to the form

$$\Phi = t_T - \left[t_T^\circ - (t_T^\circ + \mu t_N)\frac{\zeta}{\zeta^*}\right] \leq 0\,, \qquad \zeta \in [0, \zeta^*]\,. \tag{18}$$

This equation predicts a nonlinear slip weakening similar to Fig. 5b for the case of a variable t_N, and recovers a linear slip weakening for the case of a constant t_N, see [18, 19] for further details.

3 Cataclastic Flow

3.1 E-mode for Cataclastic Flow

The condition for stationary stress rate is given by

$$[\![\dot{\boldsymbol{P}}]\!] = \boldsymbol{\mathcal{A}}^{\mathrm{ep}} : [\![\dot{\boldsymbol{F}}]\!] = \boldsymbol{0}\,. \tag{19}$$

For a non-trivial solution $[\![\dot{\boldsymbol{F}}]\!] \neq \boldsymbol{0}$ to exist, we must have

$$\det(\boldsymbol{\mathcal{A}}^{\mathrm{ep}}) = 0\,. \tag{20}$$

The e-mode in this case is the jump tensor $[\![\dot{\boldsymbol{F}}]\!]$. Borja [6] provides details on how the correct sign of this e-mode may be determined.

To further characterize the e-mode associated with condition (19), we write

$$\boldsymbol{\tau} = \boldsymbol{P} \cdot \boldsymbol{F}^{\mathrm{t}}\,, \tag{21}$$

where $\boldsymbol{\tau}$ is the symmetric Kirchhoff stress tensor. Taking the jumps in the rates gives

$$[\![\dot{\boldsymbol{\tau}}]\!] = \boldsymbol{P} \cdot [\![\dot{\boldsymbol{F}}]\!]^{\mathrm{t}} + [\![\dot{\boldsymbol{P}}]\!] \cdot \boldsymbol{F}^{\mathrm{t}}\,. \tag{22}$$

Noting that $[\![\dot{\boldsymbol{P}}]\!] = \boldsymbol{0}$ and $[\![\dot{\boldsymbol{F}}]\!] = [\![\boldsymbol{l}]\!] \cdot \boldsymbol{F}$ at bifurcation, we get

$$[\![\dot{\boldsymbol{\tau}}]\!] = (\boldsymbol{\tau} \ominus \boldsymbol{1}) : [\![\boldsymbol{l}]\!]\,, \tag{23}$$

where $(\boldsymbol{\tau} \ominus \boldsymbol{1})_{ijkl} = \tau_{il}\delta_{jk}$ and δ_{jk} is the Kronecker delta. Now, consider a rate constitutive equation of the form [10]

$$\dot{\boldsymbol{\tau}} = \boldsymbol{\alpha}^{\mathrm{ep}} : \boldsymbol{l}\,, \tag{24}$$

where $\boldsymbol{\alpha}^{\mathrm{ep}}$ is a rank-four spatial elastoplastic tangential moduli tensor with minor symmetry on its first two indices. This yields an eigenvalue problem

$$\mathbf{a}^{\mathrm{ep}} : [\![\boldsymbol{l}]\!] = \boldsymbol{0}\,, \qquad \mathbf{a}^{\mathrm{ep}} = \boldsymbol{\alpha}^{\mathrm{ep}} - \boldsymbol{\tau} \ominus \boldsymbol{1}\,. \tag{25}$$

Thus, $\boldsymbol{\mathcal{A}}^{\mathrm{ep}}$ is singular whenever $\mathbf{a}^{\mathrm{ep}}$ is singular.

The singularity of $\mathbf{a}^{\mathrm{ep}}$ is of three types [23], and here we shall consider only the one relevant to pore collapse instability. We recall the polar decomposition of the deformation gradient $\boldsymbol{F}$ and spectral decomposition of the left stretch tensor $\boldsymbol{V}$,

$$\boldsymbol{F} = \boldsymbol{V} \cdot \boldsymbol{R}\,, \qquad \boldsymbol{V} = \sum_{A=1}^{3} \lambda_A \boldsymbol{m}^{(A)}\,, \tag{26}$$

where $\boldsymbol{m}^{(A)} = \boldsymbol{n}^{(A)} \otimes \boldsymbol{n}^{(A)}$, $\boldsymbol{R}$ is a proper orthogonal rotation tensor, λ_A are the principal stretches, and $\boldsymbol{n}^{(A)}$ are the principal directions of $\boldsymbol{V}$. Taking the time derivative gives

$$\dot{\boldsymbol{F}} = \dot{\boldsymbol{V}} \cdot \boldsymbol{R} + \boldsymbol{V} \cdot \dot{\boldsymbol{R}}\,, \tag{27}$$

where

$$\dot{V} = \sum_{A=1}^{3} \dot{\lambda}_A m^{(A)} + \sum_{A=1}^{3} \sum_{B \neq A} \omega_{AB}(\lambda_B - \lambda_A) m^{(AB)}, \tag{28}$$

$m^{(AB)} = n^{(A)} \otimes n^{(B)}$, and ω_{AB} is the spin of the (Eulerian) principal axes of V.

The velocity gradient can be expressed in the form

$$l = \dot{F} \cdot F^{-1} = \dot{V} \cdot V^{-1} + V \cdot \Omega \cdot V^{-1}, \qquad \Omega = \dot{R} \cdot R^{\mathrm{t}}. \tag{29}$$

Using the spectral forms and taking the jumps gives

$$[\![l]\!] = \sum_{A=1}^{3} [\![\dot{\varepsilon}_A]\!] m^{(A)} \tag{30}$$

$$+ \sum_{A=1}^{3} \sum_{B \neq A} \Big\{ [\![\omega_{AB}]\!] \Big(1 - \frac{\lambda_A}{\lambda_B}\Big) + [\![w_{AB}]\!] \frac{\lambda_A}{\lambda_B} \Big\} m^{(AB)}, \tag{31}$$

where $w_{AB} = \Omega : m^{(AB)}$ and $\varepsilon_A = \ln \lambda_A$ is the principal logarithmic stretch. Note that w_{AB} arises from the finite rotation of the stretch tensor V whereas ω_{AB} describes the spin of the principal axes of V; the former vanishes in the infinitesimal theory whereas the latter generally does not.

The particular e-mode of interest is characterized by $[\![\omega_{AB}]\!] = [\![w_{AB}]\!] = 0$, where the volume simply implodes at fixed principal directions. In this case the jump in the velocity gradient $[\![l]\!]$ is equal to the jump in the rate of deformation $[\![d]\!]$ (since the spins are 'frozen'), and the relevant jump tensor becomes

$$[\![l]\!] = [\![d]\!] = \sum_{A=1}^{3} [\![\dot{\varepsilon}_A]\!] m^{(A)}. \tag{32}$$

By a simple tensor manipulation the jumps in the principal logarithmic stretch rates, characterizing the e-mode at bifurcation, may be obtained from the eigenvalue problem

$$\sum_{A=1}^{3} a_{AB}^{\mathrm{ep}} [\![\dot{\varepsilon}_B]\!] = 0, \tag{33}$$

where

$$a_{AB}^{\mathrm{ep}} = m^{(A)} : \mathsf{a}^{\mathrm{ep}} : m^{(B)} \tag{34}$$

is a 3×3 matrix of elastoplastic moduli, which also includes the initial stress terms arising from finite deformation effects. Pore collapse is characterized by the condition

$$\sum_{A=1}^{3} [\![\dot{\varepsilon}_A]\!] < 0. \tag{35}$$

Note that the e-mode predicted above is diffuse in the sense that it does not entail the formation of a deformation band.

3.2 Constitutive Branching for Cataclastic Flow

As in classical continuum plasticity we consider yield and plastic potential functions, Φ and Ψ, respectively, representing yield behavior of the compacted material at the conclusion of cataclastic flow. We assume that these functions depend on the Kirchhoff stress tensor $\boldsymbol{\tau}$ and some plastic internal variable κ. By multiplicative plasticity we write the velocity gradient as the sum of elastic and plastic parts,

$$\boldsymbol{l} = \boldsymbol{l}^{\mathrm{e}} + \boldsymbol{l}^{\mathrm{p}}, \tag{36}$$

where

$$\boldsymbol{l}^{\mathrm{e}} = \dot{\boldsymbol{F}}^{\mathrm{e}} \cdot \boldsymbol{F}^{\mathrm{e}-1} \tag{37}$$

and

$$\boldsymbol{l}^{\mathrm{p}} = \boldsymbol{F}^{\mathrm{e}} \cdot \boldsymbol{L}^{\mathrm{p}} \cdot \boldsymbol{F}^{\mathrm{e}-1}, \qquad \boldsymbol{L}^{\mathrm{p}} = \dot{\boldsymbol{F}}^{\mathrm{p}} \cdot \boldsymbol{F}^{\mathrm{p}-1}. \tag{38}$$

Ignoring plastic spin, we have $\boldsymbol{l}^{\mathrm{p}} = \mathrm{sym}(\boldsymbol{l}^{\mathrm{p}}) := \boldsymbol{d}^{\mathrm{p}}$, and the flow rule thus gives

$$\boldsymbol{d}^{\mathrm{p}} = \dot{\lambda}\boldsymbol{q}, \qquad \boldsymbol{q} = \frac{\partial \Psi}{\partial \boldsymbol{\tau}}. \tag{39}$$

Subject to the usual Kuhn-Tucker conditions

$$\Phi(\boldsymbol{\tau}, \kappa) \leq 0, \qquad \dot{\lambda} \geq 0, \qquad \Phi(\boldsymbol{\tau}, \kappa)\dot{\lambda} = 0, \tag{40}$$

along with an appropriate hardening law, we have thus formulated a constitutive model appropriate for the 'new' material at post-cataclasis.

It has been reported [1] that cataclastic flow is only a transient behavior. When the pores have collapsed and the compacted material is loaded further, any subsequent deformation could result in a net dilatancy leading to strain localization. For porous quartzite this occurs at nominal axial strains greater than about 25%. This suggests that constitutive models for a porous but intact rock, initially exhibiting compactive tendencies, should branch to a model exhibiting dilative tendencies at the conclusion of cataclastic flow. During this time the effective cohesion exhibited by the intact rock decays to a near-zero value while the effective frictional strength increases with increased compaction. Interestingly, these features are analogs of those encountered in the development of the slip weakening model of the previous section, with the exception that they now apply to the bulk response of the material.

For a more concrete illustration, we assume the following form of the yield function to which the constitutive response branches at the onset of cataclastic flow:

$$\Phi = (f_1 - 27)(\overline{I}_1/p_{\mathrm{a}})^m - \kappa \leq 0, \qquad f_1 = \overline{I}_1^3/\overline{I}_3, \tag{41}$$

where

$$\overline{I}_1 = \overline{\tau}_1 + \overline{\tau}_2 + \overline{\tau}_3, \qquad \overline{I}_3 = \overline{\tau}_1\overline{\tau}_2\overline{\tau}_3 \tag{42}$$

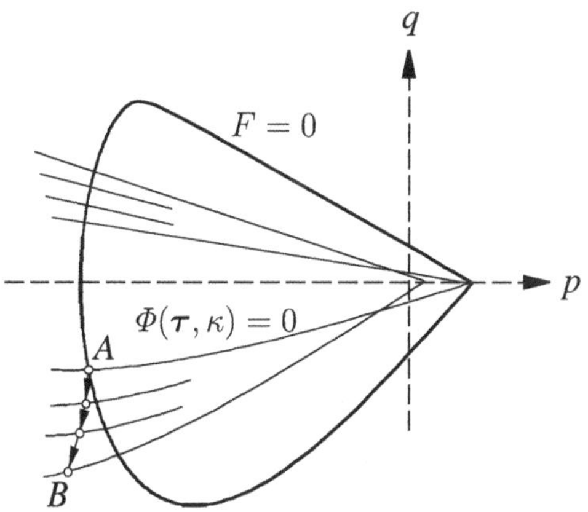

Fig. 7. Yield surface for intact rock $F = 0$ branches at the bifurcation stress point A into yield surface $\Phi(\boldsymbol{\tau}, \kappa) = 0$ signifying onset of cataclastic flow. Post-bifurcation yield surface undergoes cohesion softening-friction hardening. Transient cataclastic flow concludes at stress point B where the compacted continuum can now potentially undergo shear localization with plastic dilatancy

are, respectively, the first and third invariants of the translated principal Kirchhoff stresses

$$\overline{\tau}_1 = \tau_1 - a\,, \qquad \overline{\tau}_2 = \tau_2 - a\,, \qquad \overline{\tau}_3 = \tau_3 - a\,, \tag{43}$$

$a > 0$ is a stress shift on the hydrostatic axis to account for cohesion, and p_{a} is the atmospheric pressure (a normalizing constant). The exponent m is a given parameter describing the curvature of the surface on a meridian plane (straight line for $m = 0$), and κ is the slope that varies with the frictional yield strength. Lade [24] utilizes this function to characterize yielding of dense sands.

The yield function described above has the shape of an asymmetric bullet with the pointed apex initially on the tensile side of the hydrostatic axis (Fig. 7) and initially coinciding with the apex of the yield surface for the intact rock, which is represented in the same figure by the boundary of the region with a compression cap. Cataclastic flow-type bifurcation occurs at point A, and by requiring that $\Phi = 0$ for this given stress state we obtain the initial value κ_0 of the frictional parameter κ. Cohesion softening during cataclastic flow is simulated by a degrading value of a, while friction hardening is captured by an increasing value of κ.

4 Transient Plastic Flow

During slip weakening and cataclastic flow, the directions of plastic flow evolve from those predicted at bifurcation to those attained at residual state. To capture the evolution during these 'transient' conditions we can interpolate these plastic flow directions to satisfy the initial (bifurcation) and final (residual) states.

4.1 Slip Weakening

The region of interest is $0 \leq \zeta \leq \zeta'$, as shown in Fig. 4c. Inside this region the velocity jump $[\![\boldsymbol{v}]\!]$ evolves from being parallel to $\boldsymbol{m}$ at bifurcation to being parallel to $\boldsymbol{\xi}$ when the fault begins to deform by pure tangential sliding. Interpolating linearly with ζ (as a first-order approximation), we get

$$[\![\dot{\boldsymbol{F}}]\!] = \dot{\zeta}\boldsymbol{b} \otimes \boldsymbol{N}\,, \tag{44}$$

where

$$\boldsymbol{b} = \frac{\boldsymbol{\beta}}{\|\boldsymbol{\beta}\|}\,, \qquad \boldsymbol{\beta} = \left(1 - \frac{\zeta}{\zeta'}\right)\boldsymbol{m} + \frac{\zeta}{\zeta'}\boldsymbol{\xi}\,. \tag{45}$$

The corresponding jump in velocity gradient is

$$[\![\boldsymbol{l}]\!] = [\![\dot{\boldsymbol{F}}]\!] \cdot \boldsymbol{F}^{-1} = \dot{\zeta}\frac{h_0}{h}\boldsymbol{b} \otimes \boldsymbol{n}\,. \tag{46}$$

It follows that the jump in the Jacobian rate is

$$[\![\dot{J}]\!] = J\mathrm{tr}[\![\boldsymbol{l}]\!] = \dot{\zeta}\frac{J}{\|\boldsymbol{\beta}\|}\frac{h_0}{h}\left(1 - \frac{\zeta}{\zeta'}\right)\boldsymbol{m} \cdot \boldsymbol{n}\,. \tag{47}$$

This yields $[\![\dot{J}]\!] = 0$ when $\zeta = \zeta'$, so at this level of deformation any initial dilation or compaction of the band will cease and subsequent motion will be characterized by pure tangential sliding.

4.2 Cataclastic Flow

The region of interest is $J_1 \leq J \leq J_0$, where J_0 is the value of the Jacobian at the bifurcation point and J_1 is the value at the conclusion of cataclastic flow. It is fitting that in this case we select J as our parameter since we are concerned with pore collapse instability and J is a measure of volume change.

At the bifurcation stress we take the e-mode as the instantaneous plastic flow direction,

$$\boldsymbol{d}_0^{\mathrm{p}} = \dot{\lambda}\sum_{A=1}^{3}[\![\dot{\varepsilon}_A]\!]\boldsymbol{m}^{(A)}\,, \qquad J = J_0\,. \tag{48}$$

At the conclusion of cataclastic flow the plastic flow direction is given by the flow rule

$$\boldsymbol{d}_1^{\mathrm{p}} = \dot{\lambda}\frac{\partial \Psi}{\partial \boldsymbol{\tau}}\,, \qquad J = J_1\,. \tag{49}$$

A linear interpolation gives

$$\boldsymbol{d}^{\mathrm{p}} = \frac{J - J_1}{J_0 - J_1}\boldsymbol{d}_0^{\mathrm{p}} + \frac{J_0 - J}{J_0 - J_1}\boldsymbol{d}_1^{\mathrm{p}}\,, \qquad J_1 \leq J \leq J_0\,. \tag{50}$$

Of course, other forms of interpolation are possible provided they satisfy the two end conditions.

We emphasize that both slip weakening and cataclastic flow are micromechanical processes. The plastic flow theory presented in this section should thus be interpreted simply as a rough approximation of the micromechanical phenomena and not as a faithful description of the grain-scale flow processes taking place during these transient periods of instability.

5 Closure

We have presented a framework for mathematical modeling of shear localization and cataclastic flow in porous rocks. Classical bifurcation theory provides conditions for the occurrence of these two instability modes, in which shear localization is equated to a singular acoustic tensor whereas cataclastic flow is equated to a singular tangent constitutive operator. For general 3D kinematics, the corresponding e-modes are given by velocity gradient jump tensors of ranks one and three, respectively. We have also proposed a framework for constitutive branching to allow the model to capture post-bifurcation responses. Slip weakening provides a critical link to modeling post-shear localization in rocks. The process is captured by a cohesion softening-friction hardening law applied to the nucleating fault plane. A similar cohesion softening-friction hardening law has been proposed for bifurcation leading to cataclastic flow, except that the law is applied to the bulk constitutive response.

Acknowledgements

The author is grateful to Professor James R. Rice of Harvard University and Professor Teng-fong Wong of State University of New York at Stony Brook for discussions on the basic issues of slip weakening and cataclastic flow addressed in this paper, as well as for providing numerous references relevant to this work. Discussions with Professors David Pollard and Atilla Aydin of Stanford University also helped elucidate the processes and mechanisms of rock deformation addressed in this work. Graduate student Joshua White assisted in the preparation of the manuscript. This work has been supported in part by the U.S. National Science Foundation, Grant No. CMG-0417521 (Collaborations in Mathematical Geosciences) and the U.S. Department of Energy Grant No. DE-FG02-03ER15454.

References

[1] Hirth G, Tullis J (1989) The effects of pressure and porosity on the micromechanics of the brittle-ductile transition in quartzite. Journal of Geophysical Research 94:17825–17838

[2] Menéndez B, Zhu W, Wong TF (1996) Micromechanics of brittle faulting and cataclastic flow in Berea sandstone. Journal of Structural Geology 18:1–16
[3] Tullis J, Yund R (1992) The brittle-ductile transition in feldspar aggregates: An experimental study. In B. Evans and T.F. Wong, editors, Fault mechanics and transport properties of rocks. Cambridge University Press, United Kingdom,
[4] Wong TF, David C, Zhu W (1997). The transition from brittle faulting to cataclastic flow in porous sandstones: Mechanical deformation. Journal of Geophysical Research 102:3009–3025
[5] Rudnicki JW, Rice JR (1975) Conditions for localization of deformation in pressure-sensitive dilatant materials. Journal of the Mechanics and Physics of Solids 23:371394
[6] Borja RI (2006) Conditions for instabilities in collapsible solids including volume implosion and compaction banding. Acta Geotechnica, in press
[7] Chau K, Rudnicki JW (1990) Bifurcations of compressible pressure-sensitive materials in plane strain tension and compression. Journal of the Mechanics and Physics of Solids 38:875–898
[8] Aydin A, Borja RI, Eichnubl P (2006) Geological and mathematical framework for failure modes in granular rock. Journal of Structural Geology 28:83–98
[9] Borja RI, Aydin A (2004) Computational modeling of deformation bands in granular media. I. Geological and mathematical framework. Computer Methods in Applied Mechanics and Engineering 193:26672698
[10] Borja RI (2002) Bifurcation of elastoplastic solids to shear band mode at finite strains. Computer Methods in Applied Mechanics and Engineering 191:52875314
[11] Laursen TA, Simo JC (1993) A continuum-based finite element formulation for the implicit solution of multibody, large deformation frictional contact problems. International Journal for Numerical Methods in Engineering 36:3451-3485
[12] Borja RI (2000) A finite element model for strain localization analysis of strongly discontinuous fields based on standard Galerkin approximations. Computer Methods in Applied Mechanics and Engineering 190:1529–1549
[13] Borja RI (2002) Finite element simulation of strain localization with large deformation: Capturing strong discontinuity using a Petrov-Galerkin multiscale formulation. Computer Methods in Applied Mechanics and Engineering 191:2949–2978
[14] Dieterich JH (1979) Modeling of rock friction 1. Experimental results and constitutive equations. Journal of Geophysical Research 84:2161–2168
[15] Ruina AL (1983) Slip instability and state variable friction laws. Journal of Geophysical Research 88:10359–10370

[16] Dieterich JH, Kilgore BD (1996) Imaging surface contacts: Power law contact distributions and contact stresses in quartz, calcite, glass and acrylic plastic. Tectonophysics 256:219–239

[17] Lapusta N, Rice JR, Ben-Zion Y, Zheng G (2000) Elastodynamic analysis for slow tectonic loading with spontaneous rupture episodes on faults with rate- and state-dependent friction. Journal of Geophysical Research 105:23765–23789

[18] Borja RI, Foster CD (2006) Continuum mathematical modeling of slip weakening in geological systems. Journal of Geophysical Research, in review.

[19] Foster CD, Borja RI, Regueiro RA (2006) Embedded strong discontinuity finite elements for fractured geomaterials with variable friction. International Journal for Numerical Methods in Engineering, in review.

[20] Ida Y (1972) Cohesive force across the tip of a longitudinal shear crack and Griffith's specific surface energy. Journal of Geophysical Research 77:3796–3805

[21] Palmer AC, Rice JR (1973) The growth of slip surfaces in the progressive failure of overconsolidated clay. In Proceedings of the Royal Society of London Ser. A332, pp. 527–548

[22] Wong TF (1982) Shear fracture energy of Westerly granite from post-failure behavior. Journal of Geophysical Research 87:990–1000

[23] Ogden RW (1984) Nonlinear Elastic Deformations. Chichester, Ellis Horwood

[24] Lade PV (1977) Elasto-plastic stress-strain theory for cohesionless soil with curved yield surfaces. International Journal of Solids and Structures 13:1019–1035

Dispersion and Localisation in a Strain–Softening Two–Phase Medium

René de Borst[1,2] and Marie-Angèle Abellan[3]

[1] Delft University of Technology, Faculty of Aerospace Engineering, Kluyverweg 1, Delft, The Netherlands r.deborst@tudelft.nl
[2] LaMCoS, UMR CNRS 5514, INSA de Lyon, 69621 Villeurbanne, France
[3] LTDS-ENISE, UMR CNRS 5513, ENISE, 42023 Saint-Etienne, France

Summary. In fluid–saturated media wave propagation is dispersive, but the associated internal length scale vanishes in the short wave–length limit. Accordingly, upon the introduction of softening, localisation in a zero width will occur and no regularisation is present. This observation is corroborated by numerical analyses of wave propagation in a finite one–dimensional bar.

1 Introduction

Strain softening and the ensuing phenomenon of localisation have been the subject of profound investigations in the past two decades. While, initially, the incorporation of strain softening in constitutive equations was considered to be a straightforward exercise, it soon appeared that the use of strain-softening models led to an excessive dependency of the solution on the discretisation in numerical analyses. At first, deficiencies in the numerical methods were believed to cause this severe mesh dependency. However, it was demonstrated that the underlying cause was the local change of character of the partial differential equations that govern the initial/boundary value problem: from elliptic to hyperbolic for quasi–static problems and from hyperbolic to elliptic in dynamic problems. This local change of character renders the initial/boundary value ill–posed, unless special interface conditions are imposed between both regimes. For ill–posed problems, numerical methods, including finite element methods, still try to capture 'the best possible' solution, but this solution changes for every other discretisation.

To repair this ill–posedness, several proposals have been put forward. Invariably, the aim is to enrich the continuum description to include more of the underlying physical properties of the material, such as grain rotations in granular materials — the Cosserat continuum approach, e.g. [1, 2] —, the incorporation of viscosity or rate–dependency, e.g. [3, 4], or nonlocal approaches which reflect medium and long–range forces which emerge in materials where

Eugenio Oñate and Roger Owen (eds.), Computational Plasticity, 55–66.

the heterogeneity is in the same order of magnitude as the fracture process zone [5, 6, 7, 8]. While these ideas have been suggested and elaborated for single–phase media, they are also effective for multi–phase media, such as fluid–saturated porous solids, e.g. [9]. The question has arisen whether the diffusive character of the movement of the fluid in such a medium already provides a physically based regularisation mechanics. Indeed, it has already been shown by Biot [10], see also Loret and co-workers [11, 12] that wave propagation in such a medium is dispersive, and, accordingly, that an internal length scale must exist. This issue has been debated intensely in recent years [13, 14, 15, 16, 17].

In a previous contribution [17], we have demonstrated that stability in a 'standard' two–phase medium is assured until the tangent modulus ceases to be positive, at least in a one–dimensional medium and for a normal range of material parameters. Thus, the stability condition coincides with that of a single–phase medium. Moreover, it was shown by an analysis of dispersive waves that the length scale associated with wave dispersion vanishes in the short wave–length limit. In this contribution, we supplement the previous analysis by a more comprehensive study in which the momentum balance in the fluid is kept explicitly in the analysis, which enables the identification of the second wave speed in the mixture. The main conclusion of the previous study, namely that the length scale associated with wave dispersion vanishes in the short wave–length limit, so that no regularisation exists, is corroborated by the present analysis, and therefore put on a solid basis.

2 Governing Equations

We consider a two–phase medium subject to the restriction of small displacement gradients and small variations in the concentrations [18]. Furthermore, the assumptions are made that there is no mass transfer between the constituents and that the processes which we consider, occur isothermally. With these assumptions, the balances of linear momentum for the solid and the fluid phases read:

$$\nabla \cdot \boldsymbol{\sigma}_\pi + \hat{\mathbf{p}}_\pi + \rho_\pi \mathbf{g} = \frac{\partial(\rho_\pi \, \mathbf{v}_\pi)}{\partial t} + \nabla(\rho_\pi \mathbf{v}_\pi \otimes \mathbf{v}_\pi) \tag{1}$$

with $\boldsymbol{\sigma}_\pi$ the stress tensor, ρ_π the apparent mass density, and $\mathbf{v}_\pi$ the absolute velocity of constituent π. As in the remainder of this paper, $\pi = s, f$, with s and f denoting the solid and fluid phases, respectively. Further, $\mathbf{g}$ is the gravity acceleration and $\hat{\mathbf{p}}_\pi$ is the source of momentum for constituent π from the other constituent, which takes into account the possible local drag interaction between the solid and the fluid. Evidently, the latter source terms must satisfy the momentum production constraint:

$$\sum_{\pi=s,f} \hat{\mathbf{p}}_\pi = \mathbf{0} \tag{2}$$

We now neglect convective terms and the gravity acceleration, so that the momentum balances reduce to:

$$\nabla \cdot \boldsymbol{\sigma}_\pi + \hat{\mathbf{p}}_\pi = \rho_\pi \frac{\partial \mathbf{v}_\pi}{\partial t} \tag{3}$$

Adding both momentum balances, and taking into account Eq.(2), one obtains the momentum balance for the mixture:

$$\nabla \cdot \boldsymbol{\sigma}_s + \nabla \cdot \boldsymbol{\sigma}_f - \rho_s \frac{\partial \mathbf{v}_s}{\partial t} - \rho_f \frac{\partial \mathbf{v}_f}{\partial t} = \mathbf{0} \tag{4}$$

where

$$\boldsymbol{\sigma}_f = -\alpha p \mathbf{I} \tag{5}$$

with p the fluid pressure, $\mathbf{I}$ the second–order identity tensor, and α the Biot coefficient, cf. [19]. Substitution of Eq.(5) into the momentum balance of the mixture gives:

$$\nabla \cdot \boldsymbol{\sigma}_s - \alpha \nabla p - \rho_s \frac{\partial \mathbf{v}_s}{\partial t} - \rho_f \frac{\partial \mathbf{v}_f}{\partial t} = \mathbf{0} \tag{6}$$

In a similar fashion as for the balances of momentum, one can write the balance of mass for each phase as:

$$\frac{\partial \rho_\pi}{\partial t} + \nabla \cdot (\rho_\pi \mathbf{v}_\pi) = 0 \tag{7}$$

Again neglecting convective terms, the mass balances can be simplified to give:

$$\frac{\partial \rho_\pi}{\partial t} + \rho_\pi \nabla \cdot \mathbf{v}_\pi = 0 \tag{8}$$

We multiply the mass balance for each constituent π by its volumic ratio n_π, add them and utilise the constraint

$$\sum_{\pi=s,f} n_\pi = 1 \tag{9}$$

to give:

$$\nabla \cdot \mathbf{v}_s + n_f \nabla \cdot (\mathbf{v}_f - \mathbf{v}_s) + \frac{n_s}{\rho_s}\frac{\partial \rho_s}{\partial t} + \frac{n_f}{\rho_f}\frac{\partial \rho_f}{\partial t} = 0 \tag{10}$$

The change in the mass density of the solid material is related to its volume change by:

$$\nabla \cdot \mathbf{v}_s = -\frac{K_s}{K_t}\frac{n_f}{\rho_s}\frac{\partial \rho_s}{\partial t} \tag{11}$$

with K_s the bulk modulus of the solid material and K_t the overall bulk modulus of the porous medium. Using the definition of the Biot coefficient, $1 - \alpha = K_t / K_s$ [19], this equation can be rewritten as

$$(\alpha - 1)\nabla \cdot \mathbf{v}_s = \frac{n_f}{\rho_s}\frac{\partial \rho_s}{\partial t} \tag{12}$$

For the fluid phase, a phenomenological relation is assumed between the incremental changes of the apparent fluid mass density and of the fluid pressure [19]:

$$\frac{1}{Q}\mathrm{d}p = \frac{n_f}{\rho_f}\mathrm{d}\rho_f \tag{13}$$

with the overall compressibility, or Biot modulus

$$\frac{1}{Q} = \frac{\alpha - n_f}{K_s} + \frac{n_f}{K_f} \tag{14}$$

where K_f is the bulk modulus of the fluid. Inserting relations (12) and (13) into the balance of mass of the total medium, Eq.(10), gives:

$$\alpha\nabla\cdot\mathbf{v}_s + n_f\nabla\cdot(\mathbf{v}_f - \mathbf{v}_s) + \frac{1}{Q}\frac{\partial p}{\partial t} = 0 \tag{15}$$

The governing equations, i.e. the balance of momentum of the saturated medium, Eq.(6), that of the fluid, Eq.(3) with $\pi = f$, and the balance of mass, Eq.(15), are complemented by the kinematic relation,

$$\boldsymbol{\epsilon}_s = \nabla^s\mathbf{u}_s \tag{16}$$

with $\mathbf{u}_s$, $\boldsymbol{\epsilon}_s$ the displacement and strain fields of the solid, respectively, the superscript s denoting the symmetric part of the gradient operator, and an incrementally linear stress–strain relation for the solid skeleton,

$$\dot{\boldsymbol{\sigma}}_s = \mathbf{D}^{tan} : \dot{\boldsymbol{\epsilon}}_s \tag{17}$$

where $\mathbf{D}^{tan}$ is the fourth–order tangent stiffness tensor of the solid material and the superimposed dot denotes differentiation with respect to a virtual time. For the pore fluid flow, Darcy's relation for isotropic media is assumed to hold,

$$n_f(\mathbf{v}_f - \mathbf{v}_s) = -k_f\nabla p \tag{18}$$

with k_f the permeability coefficient of the porous medium, and defines the drag force of the solid on the fluid:

$$\hat{\mathbf{p}}_f = -n_f k_f^{-1}(\mathbf{v}_f - \mathbf{v}_s) \tag{19}$$

The boundary conditions

$$\mathbf{n}_\Gamma\cdot\boldsymbol{\sigma} = \mathbf{t}_p \quad , \quad \mathbf{v} = \mathbf{v}_p \tag{20}$$

hold on complementary parts of the boundary $\partial\Omega_t$ and $\partial\Omega_v$, with $\Gamma = \partial\Omega = \partial\Omega_t \cup \partial\Omega_v$, $\partial\Omega_t \cap \partial\Omega_v = \emptyset$, $\mathbf{t}_p$ being the prescribed external traction and $\mathbf{v}_p$ the prescribed velocity, and

$$n_f(\mathbf{v}_f - \mathbf{v}_s) = \mathbf{q}_p \quad , \quad p = p_p \tag{21}$$

hold on complementary parts of the boundary $\partial\Omega_q$ and $\partial\Omega_p$, with $\Gamma = \partial\Omega = \partial\Omega_q \cup \partial\Omega_p$ and $\partial\Omega_q \cap \partial\Omega_p = \emptyset$, $\mathbf{q}_p$ and p_p being the prescribed outflow of pore fluid and the prescribed pressure, respectively.

3 Reduction of the Governing Equations

Henceforth, we shall consider the problem of a uniaxially stressed homogeneous bar. Then, $v_{sx} \neq 0, v_{sy} = 0, v_{sz} = 0$ and the momentum balances for the mixture and for the fluid reduce to:

$$\frac{\partial \sigma_s}{\partial x} - \alpha \frac{\partial p}{\partial x} - \rho_s \frac{\partial v_s}{\partial t} - \rho_f \frac{\partial v_f}{\partial t} = 0 \tag{22}$$

where for notational simplicity the subscript x has been dropped and σ_s denotes the axial stress in the solid, and

$$\alpha \frac{\partial p}{\partial x} + n_f k_f^{-1}(v_f - v_s) + \rho_f \frac{\partial v_f}{\partial t} = 0 \tag{23}$$

respectively. From Eq.(23) we observe that Eq.(19) has been used as the source of momentum for the fluid from the solid phase. The mass balance of the mixture, Eq.(15) becomes:

$$\alpha \frac{\partial v_s}{\partial x} + n_f \left(\frac{\partial v_f}{\partial x} - \frac{\partial v_s}{\partial x} \right) + Q^{-1} \frac{\partial p}{\partial t} = 0 \tag{24}$$

To allow for inelastic constitutive equations, we take the incremental format of Eqs.(22)–(24):

$$\frac{\partial \dot{\sigma}_s}{\partial x} - \alpha \frac{\partial \dot{p}}{\partial x} - \rho_s \frac{\partial \dot{v}_s}{\partial t} - \rho_f \frac{\partial \dot{v}_f}{\partial t} = 0 \tag{25}$$

$$\alpha \frac{\partial \dot{p}}{\partial x} + n_f k_f^{-1}(\dot{v}_f - \dot{v}_s) + \rho_f \frac{\partial \dot{v}_f}{\partial t} = 0 \tag{26}$$

and

$$\alpha \frac{\partial \dot{v}_s}{\partial x} + n_f \left(\frac{\partial \dot{v}_f}{\partial x} - \frac{\partial \dot{v}_s}{\partial x} \right) + Q^{-1} \frac{\partial \dot{p}}{\partial t} = 0 \tag{27}$$

We will observe in the next section, where, using an analysis of wave dispersion in this medium, the localisation properties are derived, that the ensuing equations are rather complicated. For this reason, in [17] the pressure p was eliminated from the above equations by inserting Darcy's relation explicitly in the balances of momentum and mass for the mixture. For the momentum balance this results in:

$$\frac{\partial \sigma_s}{\partial x} + \alpha n_f k_f^{-1}(v_f - v_s) - \rho_s \frac{\partial v_s}{\partial t} - \rho_f \frac{\partial v_f}{\partial t} = 0 \tag{28}$$

The mass balance, Eq.(24) is first differentiated with respect to x. Interchanging the order of spatial and temporal differentiation and inserting Darcy's relation then results in:

$$\alpha \frac{\partial^2 v_s}{\partial x^2} + n_f \left(\frac{\partial^2 v_f}{\partial x^2} - \frac{\partial^2 v_s}{\partial x^2} \right) - n_f (k_f Q)^{-1} \left(\frac{\partial v_f}{\partial t} - \frac{\partial v_s}{\partial t} \right) = 0 \tag{29}$$

The above two equations solely have the velocity in the solid, v_s, and that in the fluid, v_f, as unknowns. They are better amenable to analytical manipulations. However, the reduction to two equations makes that the velocity of the wave in the fluid is no longer contained in the set of equations.

4 Dispersion Analysis

In a strain–softening medium the presence of a non-vanishing internal length scale that arises from physical properties of the system, is directly related to the well–posedness of the initial value problem. A method for the quantification of this internal length scale is to investigate the dispersive properties of wave propagation. Wave propagation is called dispersive when harmonics propagate with different velocities [20]. Since a wave is composed of different harmonics, the shape of a dispersive wave can then change upon propagation. The ability to transform the shape of waves is a necessary condition for continua to properly capture localisation phenomena, since it is otherwise impossible that the shape of an arbitrary loading wave is changed into a stationary wave with for instance a sinusoidal shape in the localisation zone. On the other hand, dispersivity of loading waves in a strain–softening medium is not a sufficient condition for localisation to be captured in a zone of finite size, and thus, for the initial value problem to be regularised. As said, such a regularisation will only be present if, in addition to dispersivity, a non-vanishing internal length scale can be identified.

To analyse the characteristics of wave propagation in the two–phase medium, a damped, harmonic wave is considered:

$$\begin{pmatrix} \delta\dot{u}_s \\ \delta\dot{u}_f \\ \delta\dot{p} \end{pmatrix} = \begin{pmatrix} A_s \\ A_f \\ A_p \end{pmatrix} \exp\left(\mathrm{i}kx + \lambda t\right) \tag{30}$$

where A_s, A_f, A_p are the amplitudes of the perturbations for the displacement rates in the solid, $\dot{u}_s$, in the fluid, $\dot{u}_f$, and for the pressure rate, $\dot{p}$, respectively, while k is the wave number. The eigenvalue $\lambda = \lambda_r - \mathrm{i}\omega$ can have a real component λ_r, which characterises the damping properties of the propagating wave, and an imaginary component ω, which is the angular frequency. Substitution of the first of these equations into the one–dimensional versions of the kinematic relation (16) and the incremental stress–strain relation (17) yields after differentiation with respect to x:

$$\frac{\partial \dot{\sigma}_s}{\partial x} = -E^{tan} A_s k^2 \exp\left(\mathrm{i}kx + \lambda t\right) \tag{31}$$

with E^{tan} the tangential stiffness modulus of the solid. Substitution of this relation and the perturbation (30) into Eqs.(25) – (27) yields:

$$\begin{aligned} -E^{tan}k^2 A_s - \mathrm{i}\alpha k A_p - \rho_s \lambda^2 A_s - \rho_f \lambda^2 A_f &= 0 \\ \mathrm{i}\alpha k A_p + n_f k_f^{-1} \lambda A_f - n_f k_f^{-1} \lambda A_s + \rho_f \lambda^2 A_f &= 0 \\ (n_f - \alpha)\lambda k^2 A_s - n_f \lambda k^2 A_f + \mathrm{i}Q^{-1} k \lambda A_p &= 0 \end{aligned} \tag{32}$$

A non–trivial solution to this set of homogeneous equations exists if and only if:

$$\begin{vmatrix} E^{tan}k^2 + \rho_s\lambda^2 & \rho_f\lambda^2 & \mathrm{i}\alpha k \\ -n_f k_f^{-1}\lambda & n_f k_f^{-1}\lambda + \rho_f\lambda^2 & \mathrm{i}\alpha k \\ (n_f - \alpha)k & -n_f k & \mathrm{i}Q^{-1} \end{vmatrix} = 0 \tag{33}$$

from which the characteristic equation for the eigenvalues λ derives in a straightforward manner as:

$$\lambda^4 + a\lambda^3 + bk^2\lambda^2 + ck^2\lambda + dk^4 = 0 \tag{34}$$

with

$$a = \frac{n_f k_f^{-1}(\rho_s + \rho_f)}{\rho_s\rho_f} \tag{35a}$$

$$b = \frac{n_f\rho_s\alpha Q + \rho_f E^{tan}}{\rho_s\rho_f} \tag{35b}$$

$$c = \frac{n_f k_f^{-1}(E^{tan} + \alpha^2 Q)}{\rho_s\rho_f} \tag{35c}$$

$$d = \frac{n_f\alpha Q E^{tan}}{\rho_s\rho_f} \tag{35d}$$

Decomposing Eq.(34) into real and imaginary parts leads to:

$$\lambda_r^4 + a\lambda_r^3 + (bk^2 - 6\omega^2)\lambda_r^2 + (ck^2 - 3a\omega^2)\lambda_r + dk^4 - b\omega^2k^2 + \omega^4 = 0 \tag{36}$$

and

$$(a + 4\lambda_r)\omega^2 - 4\lambda_r^3 - 3a\lambda_r^2 - 2bk^2\lambda_r - ck^2 = 0 \tag{37}$$

From the latter equation the phase velocity can formally be deduced as:

$$c_f = \frac{\omega}{k} = \sqrt{\frac{1}{a + 4\lambda_r}(4k^{-2}\lambda_r^3 + 3ak^{-2}\lambda_r^2 + 2b\lambda_r + c)} \tag{38}$$

Evidently, wave propagation is dispersive, since Eq.(38) is such that the phase velocity c_f is dependent on the wave number k, cf. [10, 11, 12, 13, 14, 15, 16].

Taking the long wave–length limit in Eqs.(36) and (37) by letting $k \to 0$, and eliminating ω yields the following sixth-order equation in λ_r:

$$\lambda_r^3(8\lambda_r^3 + 12a\lambda_r^2 + 6a^2\lambda_r + a^3) = 0 \tag{39}$$

which has two triple roots: $\lambda_r = 0$ and $\lambda_r = -\frac{1}{2}a$. Substitution of the first root in Eq.(37) gives for the long–wave limit $a\omega^2 = ck^2$, so that with Eqs.(35–a) and (35–c), the phase velocity in the mixture is obtained as:

$$c_f = \sqrt{\frac{E^{tan} + \alpha^2 Q}{\rho_s + \rho_f}} \tag{40}$$

This expression is identical to that which has been found by an analysis of the reduced equations (28)–(29). The second wave speed is obtained by substituting the second independent root $\lambda_r = -\frac{1}{2}a$ into Eq.(37), which results in $c_f = \sqrt{(b - c/a) - (a/2k)^2}$. For $k \rightarrow 0$ this expression becomes imaginary, and harmonics with small wave numbers cannot propagate. The cut-off wave number below which harmonics cannot propagate is given by $k = \sqrt{a^3/(ab - c)}$. This situation is somewhat reminiscent of some gradient–enhanced plasticity models [21].

For the short wave-length limit, i.e. when $k \rightarrow \infty$, we assume, inspired by the closed-form solution of the reduced equations (28)–(29), a general form for the damping coefficient as $\lambda_r \sim -k^n, n > 1$. Substitution of this identity into Eq.(38) and taking $k \rightarrow \infty$ yields that $c_f \rightarrow k^{n-1}$. In analogy with a single–phase, rate–dependent medium [4], an internal length scale can be defined as:

$$l = \lim_{k \rightarrow \infty} \left(-\frac{c_f}{\lambda_r} \right) \sim \lim_{k \rightarrow \infty} k^{-1} = 0 \tag{41}$$

which indicates that the internal length scale l vanishes in the short wave–length limit. Again, this result is in agreement with earlier analyses using the reduced set of equations [17].

The observation that in a fluid–saturated medium a non–vanishing physical internal length scale cannot be identified for the short–wave length limit, is different from the situation in a rate–dependent single–phase medium [4]. The lack of a non–vanishing physical internal length scale in the present case causes that in numerical analyses the grid spacing takes the role of the internal length scale and localisation necessarily occurs between two neighbouring grid points. Evidently, this leads to a dependence of the solution on the discretisation, as is the case for localisation in the underlying strain–softening, single–phase continuum.

5 Numerical Examples

To verify and elucidate the theoretical results of the preceding section, a finite difference analysis has been carried out. The spatial derivatives in Eqs.(28) and (29) have been approximated with a second–order accurate finite difference scheme. Explicit forward finite differences have been used to approximate the temporal derivatives, which is first-order accurate. The choice for a fully explicit time integration scheme was motivated by the analysis of Benallal and Comi [16], in which they showed that in this case no numerical length scale was introduced in the analysis, apart from the grid spacing. As implied in Eqs.(28) and (29) the velocities v_s and v_f of the solid skeleton and the fluid have been taken as fundamental unknowns and the displacements have been obtained by integration. This scheme may not be the most accurate, but it suffices to provide the numerical evidence needed to support the analytical findings of the preceding section.

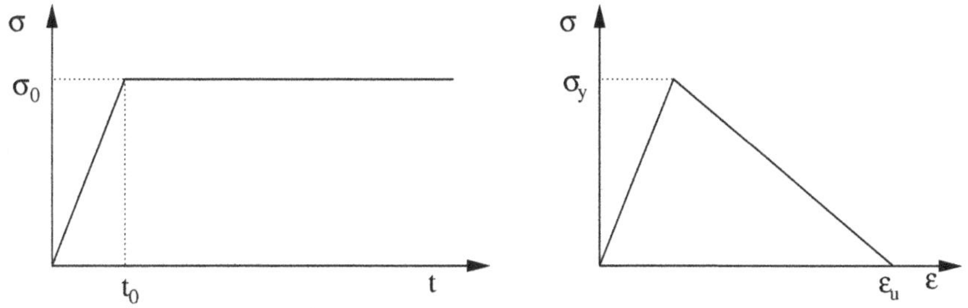

Fig. 1. Applied stress as function of time (left) and local stress–strain diagram (right)

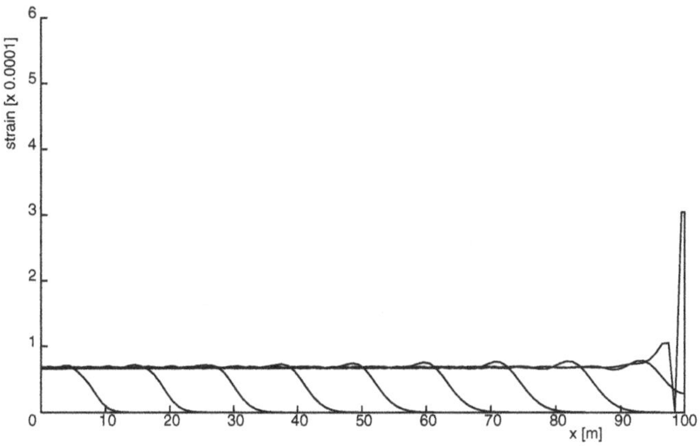

Fig. 2. Strain profiles along the bar for 101 grid points and time step $\Delta t = 0.5{\cdot}10^{-3}$ s

All calculations have been carried out for a bar with a length $L = 100$ m. For the solid material, a Young's modulus $E = 20$ GPa and an absolute mass density $\rho_s' = \rho_s/n_s = 2000$ kg/m^3 have been assumed. For the fluid, an absolute mass density $\rho_f' = \rho_f/n_f = 1000$ kg/m^3 was adopted and a compressibility modulus $Q = 5$ GPa was assumed. As regards the porosity, a value $n_f = 0.3$ was adopted and in the reference calculations $\alpha = 0.6$ and the permeability $k_f = 10^{-10}$ m^3/Ns. In all cases, the external compressive stress was applied according to the scheme shown in Fig. 1, with a rise time $t_0 = 0.05$ s to reach the peak level $\sigma_0 = 1.5$ MPa. In the reference calculations a time step $\Delta t = 0.5 \cdot 10^{-3}$ s was adopted, which is about half the critical time step for this explicit scheme.

The results of the reference calculation are shown in Fig. 2 in terms of the strain profile along the bar for $t = 0.1, 0.2, 0.3, 0.4, 0.5, 0.6, 0.7, 0.8, 0.9, 0.95$ s. For the above set of parameters, the phase velocity for the long wave–length limit is captured exactly. In line with this expression, a variation of the permeability k_f does not influence the phase velocity. Also, the influence of α according to Eq.(40) was correctly reproduced, as was verified by varying

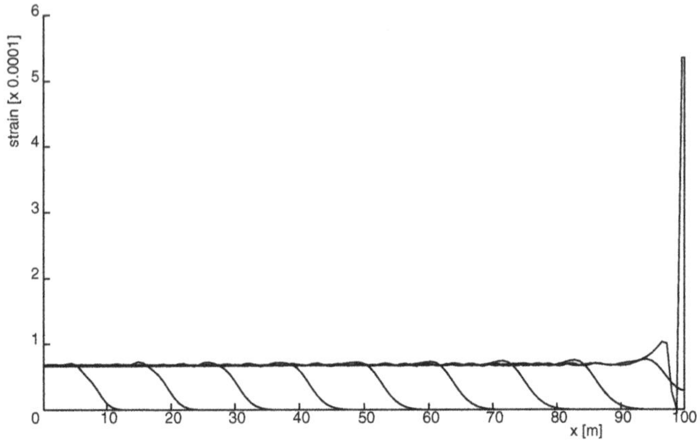

Fig. 3. Strain profiles along the bar for 126 grid points and time step $\Delta t = 0.5 \cdot 10^{-3}$ s

α between 0 and 1. In all cases the maximum error with respect to the phase velocity remained within 3%.

Upon reflection at the right boundary, the stress intensity doubles and the stress in the solid exceeds the yield strength $\sigma_y = 2.5$ MPa and enters a linear descending branch with an ultimate strain $\epsilon_u = 1.125 \cdot 10^{-3}$, see Fig. 1. Figure 2 shows that a Dirac–like strain distribution develops immediately upon wave reflection. This is logical, since a two–phase medium with neither constituent being equipped with viscosity in its constitutive model, does not have regularising properties. To further strengthen this observation the analysis was repeated with a slightly refined mesh (126 grid points), which resulted in a marked increase of the localised strain (Fig. 3, which has been plotted on the same scale as the results of the original discretisation in Fig. 2). In dynamic calculations of softening media without regularisation, not only the spatial discretisation strongly influences the results, but also the time discretisation [16]. This is exemplified in Fig. 4 for a time step that is only 20% smaller than the time step used in the reference calculation.

6 Concluding Remarks

In two–phase media waves propagation is dispersive. However, in the short wave–length limit, the *physical* internal length scale disappears. Since this limiting case governs the development of localisation in a zone of finite width after the onset of strain softening, regularising properties, which are directly related to the existence of a non–vanishing internal length scale, are absent. This conclusion is corroborated by the results of numerical analyses of wave propagation in a finite one–dimensional bar, which show that, upon the introduction of a softening stress–strain relation for the solid constituent, strain

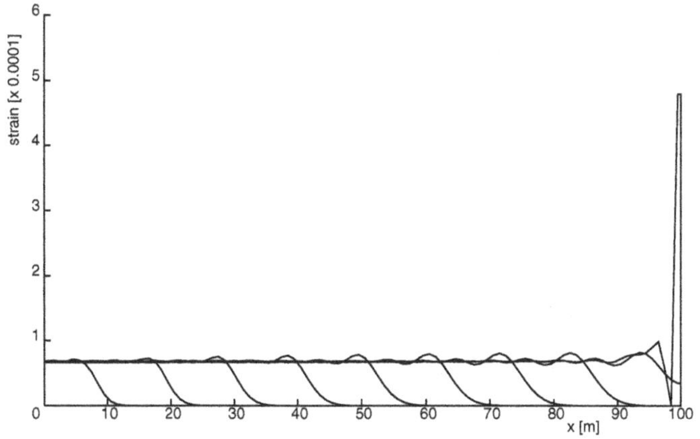

Fig. 4. Strain profiles along the bar for 101 grid points and time step $\Delta t = 0.4{\cdot}10^{-3}$ s

localisation develops in the smallest possible size, i.e. between two neighbouring grid points. Additional computations with a different spatial resolution and with a different time step confirm this observation. Regularisation can be introduced in a two–phase medium, but this necessitates the introduction of rate or gradient dependence in the solid constituent, or explicitly taking into account the effect of grain rotations.

References

1. Mühlhaus HB, Vardoulakis I (1987) The thickness of shear bands in granular materials. Géotechnique 37: 271–283
2. de Borst R, Sluys LJ (1991) Localisation in a Cosserat continuum under static and dynamic loading conditions. Comp Meth Appl Mech Eng 90: 805–827
3. Needleman A (1987) Material rate dependence and mesh sensitivity in localization problems. Comp Meth Appl Mech Eng 67: 68–85
4. Sluys LJ, de Borst R (1992) Wave propagation and localisation in a rate-dependent cracked medium — Model formulation and one-dimensional examples. Int J Solids Struct 29: 2945–2958
5. Aifantis EC (1984) On the microstructural origin of certain inelastic models. ASME J Eng Mater Technol 106: 326–330
6. Pijaudier-Cabot G, Bazant ZP (1987) Nonlocal damage theory. ASCE J Eng Mech 113: 1512–1533
7. de Borst R, Mühlhaus HB (1992) Gradient-dependent plasticity: Formulation and algorithmic aspects. Int J Num Meth Eng 35: 521–539
8. Peerlings RHJ, de Borst R, Brekelmans WAM, de Vree JHP (1996) Gradient enhanced damage modelling for quasi–brittle materials. Int J Num Meth Eng 39: 3391–3403
9. Ehlers W, Volk W (1997) On shear band localization phenomena in liquid–saturated granular elastoplastic porous solid materials accounting for fluid viscosity and micropolar solid rotations. Mech Coh-frict Mat 2: 301–330

10. Biot MA (1956) Theory of propagation of elastic waves in a fluid-saturated porous solid. I. Low-frequency range. J Acoust Soc America 28: 168–178
11. Rizzi E, Loret B (1999) Strain localization in fluid–saturated anisotropic elastic–plastic porous media. Int J Eng Sci f37: 235–251
12. Simoes FMF, Martins JAC, Loret B (1999) Instabilities in elastic-plastic saturated porous media: harmonic wave versus acceleration wave analysis. Int J Solids Struct 36: 1277–1295
13. Zhang HW, Sanavia L, Schrefler BA (1999) An internal length scale in dynamic strain localization of multiphase porous media. Mech Coh-frict Mat 4: 445–460
14. Zhang HW, Schrefler BA (2004) Particular aspects of internal length scales in strain localisation analysis of multiphase porous materials. Comp Meth Appl Mech Eng 193: 2867–2884
15. Benallal A, Comi C (2002) Material instabilities in inelastic saturated porous media under dynamic loadings. Int J Sol Struct 39: 3693–3716
16. Benallal A, Comi C (2003) On numerical analyses in the presence of unstable saturated porous materials. Int J Num Meth Eng 56: 883–910
17. Abellan MA, de Borst R (2006) Wave propagation and localisation in a softening two–phase medium. Comp Meth Appl Mech Eng, doi: 10.1016/j.cma.2005.05.056
18. Jouanna P, Abellan MA (1995) Generalized approach to heterogeneous media. In: Gens A, Jouanna P, Schrefler B (eds) Modern Issues in Non-Saturated Soils, 1–128. Springer, Berlin Heidelberg New York
19. Lewis RW, Schrefler BA (1998) The Finite Element Method in the Static and Dynamic Deformation and Consolidation of Porous Media, Second Edition. John Wiley & Sons, Chichester
20. Whitham GB (1974) Linear and Nonlinear Waves. Wiley-Interscience, New York
21. Sluys LJ, de Borst R, Mühlhaus HB (1993) Wave propagation and dispersion in a gradient-dependent medium. Int J Solids Struct 30: 1153–1171

New Developments in Surface-to-Surface Discretization Strategies for Analysis of Interface Mechanics

Tod A. Laursen and Bin Yang

Computational Mechanics Laboratory
Department of Civil and Environmental Engineering
Duke University, Durham, NC 27708-0287, USA
laursen@duke.edu

Summary. This article summarizes recent results pertaining to the implementation of mortar-based contact formulations in nonlinear computational solid mechanics. In particular, the authors discuss extension of the mortar framework to encompass large sliding, searching algorithms, treatment of self-contact phenomena, and use of the mortar framework to treat problems of lubricated contact.

1 Introduction

This paper describes recent progress in the development of mortar-based methods for analysis of contact mechanics problems using finite elements. The mortar element method was first conceived as a domain decomposition method, whereby possibly distinct meshes may be joined across interfaces such that optimal convergence results expected from the underlying finite element method can be obtained (the reader may consult [4, 2, 3, 15] for early descriptions of the idea, as well as useful overviews). In the context of nonlinear mechanics, one of the appealing applications of this idea is "mesh tying." In such an application, separate meshes can be generated for subcomponents of an assembly to be analyzed, and then the mortar formulation can be used to "join" these parts by enforcing compatibility of the solution across the interfaces where they meet. Using this approach, accurate results can be obtained even for nonconforming meshes, as long is there is not too large a discrepancy in mesh fineness across each interface. In mortar methods in general, the key idea is that the kinematic constraints between neighboring domains are enforced in an integral sense, rather than locally at each of a finite set of collocation points.

Since contact problems frequently feature nonconforming meshes in contact regions, the potential applicability of mortar techniques to such problems

Eugenio Oñate and Roger Owen (eds.), Computational Plasticity, 67–86.

is not hard to imagine. Several early authors recognized this fact and proposed mortar-based formulations for kinematically linear applications with considerable success (see [1, 9, 10, 6]). In extending such methods to the large deformation, large sliding regime, however, one is confronted with the need to not only reformulate the constraint definitions to reflect changing connectivity during relative sliding, but also to provide robust algorithmic implementations of these definitions such that (for example) Newton-Raphson or similar iterative schemes for solution of the nonlinear equations will be effective. In a set of recent papers [12, 13, 17], such large deformation formulations have been thoroughly described, and have been shown to be robust for large deformation applications in both two and three dimensions, and under either dynamic or quasistatic circumstances.

In this paper, we review some of the key constructs of mortar-based contact algorithms, starting by looking at the mesh tying problem as a template and briefly discussing the extension to truly large sliding contact. After this review, the emphasis will be on recent extensions of the mortar methodology, which allow it to encompass a much broader set of applications. Specifically, the self-contact problem and the lubricated contact problem will be examined. Generally speaking, the mortar framework allows us to put the spatial discretization of contact phenomena on a much firmer theoretical basis, opening up a large range of interfacial phenomena that can be reliably simulated. In addition to the accuracy exhibited by these approaches, they also lend considerable robustness to a nonlinear equation solving strategy as compared to traditional contact descriptions, due to the nonlocal nature of the constraints produced by the formulation.

2 Background: Mortar Projection in Contact Mechanics

2.1 The Mesh Tying Problem

The key conceptual concept behind mortar projection is illuminated by taking the mesh tying application as a first example. To fix ideas, we consider the large deformation response of two bodies (i), with an eye toward approximately computing the deformation mappings $\boldsymbol{\varphi}^{(i)} \in H^1(\Omega^{(i)})$, describing the current positions of reference configurations $\Omega^{(i)}$. As per a common procedure, the surfaces of the bodies $\partial\Omega^{(i)}$ are assumed to be decomposed into prescribed traction, displacement and contact regions via

$$\begin{aligned} \partial\Omega^{(i)} &= \Gamma_\sigma^{(i)} \cup \Gamma_u^{(i)} \cup \Gamma_c^{(i)} \\ \Gamma_\sigma^{(i)} \cap \Gamma_u^{(i)} = \Gamma_u^{(i)} \cap \Gamma_c^{(i)} &= \Gamma_\sigma^{(i)} \cap \Gamma_c^{(i)} = \emptyset \end{aligned} \tag{1}$$

and boundary conditions are summarized as

$$\begin{aligned} P_{iJ}^{(J)} N_J^{(i)} &= \bar{t}_i^{(i)} \text{ on } \Gamma_\sigma^{(i)} \\ \varphi_i^{(i)} &= \bar{\varphi}_i^{(i)} \text{ on } \Gamma_u^{(i)} \end{aligned} \tag{2}$$

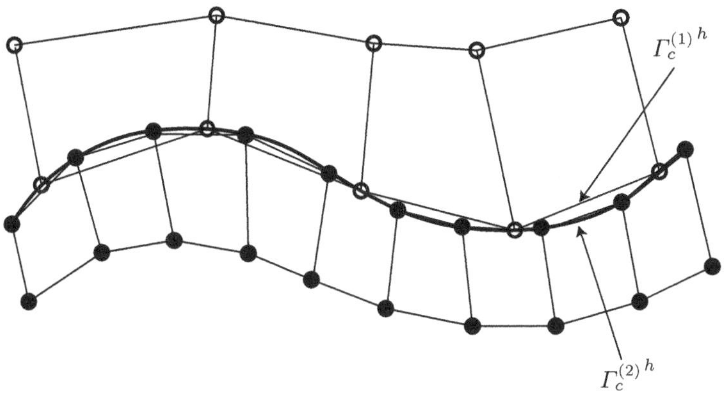

Fig. 1. Simple two dimensional illustration of how dissimilar polynomial interpolation of the same surface causes $\Gamma_c^{(1)^h} \neq \Gamma_c^{(2)^h}$. The dark curve indicates the surface to be approximated

In (2), P_{iJ} denotes the components of the first Piola-Kirchhoff stress, N_J represents the components of the outward normal to the *reference* configuration, and $\bar{t}_i^{(i)}$ and $\bar{\varphi}_i^{(i)}$ are the prescribed tractions and displacements, respectively. The summation convention is applied, where uppercase Roman indices refer to reference quantities, while lowercase counterparts are spatial. We leave unspecified the specific constitutive law governing P_{iJ}; in fact, the formulations we propose in this paper for treatment of interfaces are conceived independently of the material laws used to describe the contacting continua.

The mortar method is to be applied over the tied regions, which are denoted here by $\Gamma_c^{(i)}$, $i = 1, 2$, such that the mesh tying constraint becomes

$$\boldsymbol{\varphi}^{(1)} = \boldsymbol{\varphi}^{(2)} \text{ on } \Gamma_c^{(1)} = \Gamma_c^{(2)}. \tag{3}$$

In the continuum setting, the reference positions of the tied regions coincide, so that one could equivalently require

$$\boldsymbol{U}^{(1)} = \boldsymbol{U}^{(2)} \text{ on } \Gamma_c^{(1)} = \Gamma_c^{(2)}, \tag{4}$$

where $\boldsymbol{U}^{(i)} = \boldsymbol{\varphi}^{(i)} - \boldsymbol{X}^{(i)}$ is the Lagrangian displacement. In a numerical approximation, the discretized versions of the surfaces (denoted $\Gamma_c^{(i)^h}$, $i = 1, 2$) are no longer coincident; indeed, it is easy to show that discretizations of the same (curved) surface will in general produce areas of unintended overlap and void (see Fig. 1). This fact is important in practice (see for example [11]), and means that quite often reference geometries must be corrected after mesh generation so that the initial configuration obeys the mortar-based constraints. This is particularly true if the continuity constraints are written in terms of (3) rather than (4), which are *not* equivalent in the spatially discrete case. To simplify our discussion here, however, we neglect this subtlety.

By convention, we introduce a mortar multiplier field $\boldsymbol{\lambda}^h$, defined over the non-mortar (sometimes termed *slave*) surface $\Gamma_c^{(1)^h}$. This field is written in terms of the shape functions N_A describing $\Gamma_c^{(1)^h}$, such that

$$\boldsymbol{\lambda}^h = \sum_{A=1}^{ns} \boldsymbol{\lambda}_A N_A^{(1)}, \tag{5}$$

where ns is the set of all nodes comprising $\Gamma_c^{(1)}$. A mortar method enforcing compatibility of the two bodies' responses over the tying region is defined by enforcing:

$$0 = G^{int,ext}(\boldsymbol{\varphi}^h, \overset{*}{\boldsymbol{\varphi}}{}^h) + G^c(\boldsymbol{\varphi}^h, \overset{*}{\boldsymbol{\varphi}}{}^h) \tag{6}$$

which must hold for all admissible variations $\{\overset{*}{\boldsymbol{\varphi}}{}^{(1)^h}, \overset{*}{\boldsymbol{\varphi}}{}^{(2)^h}\}$, and

$$0 = \int_{\Gamma_c^{(1)^h}} \boldsymbol{q}^h \cdot (\boldsymbol{U}^{(1)^h} - \boldsymbol{U}^{(2)^h})\, d\Gamma \tag{7}$$

which must hold for all $\boldsymbol{q}^h \in \mathcal{M}^h$, where $\mathcal{M}^h$ is the space of admissible variations of the multiplier field $\boldsymbol{\lambda}^h$. In (6), $G^{int,ext}$ is the usual virtual work of the internal stresses and applied loads, while G^c denotes the contact virtual work, given here as

$$G^c(\boldsymbol{\varphi}^h, \overset{*}{\boldsymbol{\varphi}}{}^h) = -\int_{\Gamma_c^{(1)^h}} \boldsymbol{\lambda}^h \cdot (\overset{*}{\boldsymbol{\varphi}}{}^{(1)^h} - \overset{*}{\boldsymbol{\varphi}}{}^{(2)^h})\, d\Gamma. \tag{8}$$

Equation (6) is recognized as the usual expression of global equilibrium, while the constraints implied by (7) weakly enforce compatibility between $\boldsymbol{U}^{(1)^h}$ and $\boldsymbol{U}^{(2)^h}$. Weighting functions associated with the constraint equation are given as

$$\boldsymbol{q}^h = \sum_{A=1}^{ns} \boldsymbol{q}_A N_A^{(1)}. \tag{9}$$

Writing the Lagrangian displacements on both surfaces as

$$\boldsymbol{U}^{(1)^h} = \sum_{A=1}^{ns} \boldsymbol{U}_A N_A^{(1)},\ \boldsymbol{U}^{(2)^h} = \sum_{C=1}^{nm} \boldsymbol{U}_C N_C^{(2)}, \tag{10}$$

where nm is the number of nodes C that comprise $\Gamma_c^{(2)^h}$ (the mortar surface), we find that substitution into (7) renders

$$0 = \sum_{A=1}^{ns} \boldsymbol{q}_A \cdot \left\{ \int_{\Gamma_c^{(1)^h}} N_A^{(1)} \left(\sum_{B=1}^{ns} \boldsymbol{U}_B N_B - \sum_{C=1}^{nm} \boldsymbol{U}_C N_C^{(2)} \right) d\Gamma \right\}. \tag{11}$$

Equation (11) must hold for arbitrary $\boldsymbol{q}_A$, so that a vector-valued constraint is implied at each node A in $\Gamma_c^{(1)^h}$:

$$\begin{aligned} 0 = \boldsymbol{c}_A &= \int_{\Gamma_c^{(1)^h}} N_A^{(1)} \left(\sum_{B=1}^{ns} \boldsymbol{U}_B N_B - \sum_{C=1}^{nm} \boldsymbol{U}_C N_C^{(2)} \right) d\Gamma \\ &= \sum_{B=1}^{ns} n_{AB}^{(1)} \boldsymbol{U}_B - \sum_{\hat{B}=1}^{nm} n_{AC}^{(2)} \boldsymbol{U}_C . \end{aligned} \tag{12}$$

In (12), $n_{AB}^{(1)}$ and $n_{AC}^{(2)}$ are given by

$$n_{AB}^{(1)} = \int_{\Gamma_c^{(1)^h}} N_A^{(1)} N_B^{(1)} \, d\Gamma , \quad \text{and} \quad n_{AC}^{(2)} = \int_{\Gamma_c^{(1)^h}} N_A^{(1)} N_C^{(2)} \, d\Gamma . \tag{13}$$

One may readily see that (12) forces $\boldsymbol{U}^{(1)^h}$ to equal a *least squares projection* of $\boldsymbol{U}^{(2)^h}$ onto the discretization $\Gamma_c^{(1)^h}$ of $\Gamma_c^{(1)}$. The multiplier field $\boldsymbol{\lambda}^h$ defined in (5) enforces these discrete constraints, and may be understood as the surface tractions distribution appearing in the contact virtual work, as given in (8).

2.2 Generalization to Sliding Contact

In extending the mortar idea to true contact-impact interaction, one must recognize that the definition of the mortar operators (as summarized above in (13)) is now affected by changing connectivity as sliding occurs. This suggests that it is now more natural (in contrast to (8)) to represent the contact virtual work G^c in terms of spatial configuration quantities, via

$$G^c(\boldsymbol{\varphi}, \overset{*}{\boldsymbol{\varphi}}) = - \int_{\gamma_c^{(1)}} \boldsymbol{\lambda} \cdot \left(\overset{*}{\boldsymbol{\varphi}}^{(1)}(\boldsymbol{X}) - \overset{*}{\boldsymbol{\varphi}}^{(2)}(\bar{\boldsymbol{Y}}) \right) d\gamma , \tag{14}$$

where $\gamma_c^{(1)}$ is the current configuration of the (non-mortar) contact surface on $\Gamma_c^{(1)}$, the mortar multiplier $\boldsymbol{\lambda}$ now denotes the Cauchy contact traction, and where $\boldsymbol{\varphi}^{(2)}(\bar{\boldsymbol{Y}})$ is the current position of the contact point for $\boldsymbol{X}$. Superscript h's have been omitted in (14) to simplify notation. The notation $\bar{\boldsymbol{Y}}$ has been used to denote the contact point on $\Gamma_c^{(2)}$ corresponding to each point $\boldsymbol{X} \in \Gamma_c^{(1)}$. In traditional contact implementations, this point is typically explicitly determined for each point at which which contact constraints are to be enforced. Here we will employ a spatial integration procedure which will only indirectly define pairings between points $\boldsymbol{X}$ and $\bar{\boldsymbol{Y}}$.

Another important distinction between the contact problem and the mesh tying problem is that inequality constraints now govern the Lagrange multipliers (tractions) and associated kinematics. We may define a gap function $g(\boldsymbol{X}, t)$ as

$$g(\boldsymbol{X}, t) = \boldsymbol{n} \cdot \left[\boldsymbol{\varphi}^{(1)} (\boldsymbol{X}, t) - \boldsymbol{\varphi}^{(2)} (\bar{\boldsymbol{Y}}, t) \right] \tag{15}$$

where $\boldsymbol{n}$ denotes the outward unit normal to $\gamma_c^{(1)}$ at $\boldsymbol{x} = \boldsymbol{\varphi}^{(1)}(\boldsymbol{X}, t)$. The Kuhn-Tucker conditions governing the normal part of the contact interaction are then written as

$$\begin{aligned} g(\boldsymbol{X}, t) &\leq 0 \\ \lambda_N(\boldsymbol{X}, t) &\geq 0 \\ \lambda_N(\boldsymbol{X}, t) g(\boldsymbol{X}, t) &= 0 \end{aligned} \tag{16}$$

where $\lambda_N(\boldsymbol{X}, t) = -\boldsymbol{\lambda} \cdot \boldsymbol{n}$ is the Cauchy contact pressure at material point $\boldsymbol{X}$ on the slave surface. Equation (16_2) implies that only compressive interaction is allowed, while (16_3) allows the contact pressure to be non-zero only when $g(\boldsymbol{X}, t) = 0$.

The frictional response can be characterized by isolating the relative tangential velocity $\boldsymbol{v}_T$ via

$$\begin{aligned} \boldsymbol{v} &= \dot{\boldsymbol{\varphi}}^{(2)}\left(\bar{\boldsymbol{Y}}, t\right) - \dot{\boldsymbol{\varphi}}^{(1)}(\boldsymbol{X}, t) \\ &= \boldsymbol{v}_N + \boldsymbol{v}_T \end{aligned} \tag{17}$$

where $\boldsymbol{v}_T$ is that part of the relative velocity in the tangent plane associated with $\boldsymbol{n}$. The contact traction $\boldsymbol{\lambda}$ is similarly resolved,

$$\boldsymbol{\lambda} = \boldsymbol{\lambda}_N + \boldsymbol{\lambda}_T. \tag{18}$$

With these notions in hand, the conditions for Coulomb friction can be written as:

$$\begin{aligned} \boldsymbol{v}_T - \dot{\gamma} \frac{\boldsymbol{\lambda}_T}{\left\|\boldsymbol{\lambda}_T\right\|} &= \boldsymbol{0} \\ \Phi := \left\|\boldsymbol{\lambda}_T\right\| - \mu \left\|\boldsymbol{\lambda}_N\right\| &\leq 0 \\ \dot{\gamma} &\geq 0 \\ \Phi \dot{\gamma} &= 0, \end{aligned} \tag{19}$$

where, as can be seen, the frictional traction $\boldsymbol{\lambda}_T$ is forced to oppose the tangential velocity $\boldsymbol{v}_T$.

A discrete version of the contact problem is achieved through substitution of finite dimensional approximations for deformation mappings, variations, and Lagrange multipliers. This results in the following mortar approximation to the contact virtual work, denoted as G^{cm}:

$$G^{cm}(\boldsymbol{\varphi}^h, \overset{*}{\boldsymbol{\varphi}}{}^h) = -\sum_A^{ns} \sum_B^{ns} \sum_C^{nm} \boldsymbol{\lambda}_A \cdot \left[n_{AB}^{(1)} \overset{*}{\boldsymbol{\varphi}}{}_B^{(1)} - n_{AC}^{(2)} \overset{*}{\boldsymbol{\varphi}}{}_C^{(2)} \right]. \tag{20}$$

The expressions for $n_{AB}^{(1)}$ and $n_{AC}^{(2)}$ are similar to those given in the tying problem (see (13), but now live in the spatial configuration:

$$\begin{aligned} n_{AB}^{(1)} &= \int_{\gamma^{(1)^h}} N_A^{(1)}\left(\xi^{(1)}(\boldsymbol{X})\right) N_B^{(1)}\left(\xi^{(1)}(\boldsymbol{X})\right) d\gamma, \\ n_{AC}^{(2)} &= \int_{\gamma^{(1)^h}} N_A^{(1)}\left(\xi^{(1)}(\boldsymbol{X})\right) N_C^{(2)}\left(\xi^{(2)}\left(\bar{\boldsymbol{Y}}(\boldsymbol{X})\right)\right) d\gamma. \end{aligned} \tag{21}$$

The normal and tangential portions of the contact operator are now exposed by splitting each nodal $\boldsymbol{\lambda}_A$ into normal and frictional parts:

$$\boldsymbol{\lambda}_A = \boldsymbol{\lambda}_{N_A} + \boldsymbol{\lambda}_{T_A}. \tag{22}$$

The normal part of the contact traction may be represented as (see [12])

$$\boldsymbol{\lambda}_{N_A} = -\lambda_{N_A}\boldsymbol{n}_A \text{ (no sum)} \tag{23}$$

where λ_{N_A} represents the contact pressure at node A. It is subject to Kuhn-Tucker conditions via

$$\begin{aligned} \lambda_{N_A} &\geq 0 \\ g_A &\leq 0 \\ \lambda_{N_A} g_A &= 0 \end{aligned} \tag{24}$$

where the mortar projected gap g_A at slave node A is defined as

$$\begin{aligned} g_A &= \boldsymbol{n}_A \cdot \boldsymbol{g}_A, \\ \boldsymbol{g}_A &= \kappa_A \left[\sum_B^{ns} n_{AB}^{(1)} \boldsymbol{\varphi}_B^{(1)} - \sum_C^{nm} n_{AC}^{(2)} \boldsymbol{\varphi}_C^{(2)} \right], \end{aligned} \tag{25}$$

where κ_A is a scale factor defined as

$$\begin{aligned} \kappa_A &= \frac{1}{\sum_D n_{AD}^{(1,ref)}}, \\ n_{AD}^{(1,ref)} &= \int_{\Gamma^{(1)^h}} N_A^{(1)}\left(\xi^{(1)}(\boldsymbol{X})\right) N_D^{(1)}\left(\xi^{(1)}(\boldsymbol{X})\right) d\Gamma. \end{aligned} \tag{26}$$

The scaling defined in (26), while not by any means unique, is performed so that the gap function $\boldsymbol{g}_A$ is dimensionally correct; this feature is of crucial importance when implementing penalty methods in particular. Equation (25) is written in terms of a nodal normal $\boldsymbol{n}_A$ associated with each slave node A.

Although other implementations of mortar-based frictional contact are possible, we consider here a penalty regularization of conditions (19). This may be expressed via

$$\begin{aligned} \mathcal{L}_v \boldsymbol{\lambda}_T &= \epsilon_T \left[\boldsymbol{v}_T - \dot{\gamma} \frac{\boldsymbol{\lambda}_T}{\left\| \boldsymbol{\lambda}_T \right\|} \right] \\ \Phi := \left\| \boldsymbol{\lambda}_T \right\| - \mu \left\| \boldsymbol{\lambda}_N \right\| &\leq 0 \\ \dot{\gamma} &\geq 0 \\ \Phi \dot{\gamma} &= 0 \end{aligned} \tag{27}$$

where ϵ_T is the frictional penalty parameter. The frictional conditions in (19) are recovered in the limit as $\epsilon_T \to \infty$. $\mathcal{L}_v \boldsymbol{\lambda}_T$ is the *Lie derivative* of the frictional traction, and may be defined (for example) in a two dimensional context via

$$\mathcal{L}_v \boldsymbol{\lambda}_T = \dot{\lambda}_T \boldsymbol{\tau} \tag{28}$$

where $\boldsymbol{\tau}$ is a tangential base vector. Equation (28) contains material time derivatives of the components of $\boldsymbol{\lambda}_T$ only; the absence of derivatives of base vectors assures its frame indifference.

Correspondingly, as shown in [17], the appropriate notion of tangential velocity to use in a mortar projected framework is

$$\boldsymbol{v}_{T_A} = -\kappa_A \left[\sum_{C}^{nm} \dot{n}_{AC}^{(2)} \boldsymbol{\varphi}_C^{(2)} - \sum_{B}^{ns} \dot{n}_{AB}^{(1)} \boldsymbol{\varphi}_B^{(1)} \right] \cdot (\boldsymbol{I} - \boldsymbol{n} \otimes \boldsymbol{n}) \tag{29}$$

where $\dot{n}_{AC}^{(2)}$ and $\dot{n}_{AB}^{(1)}$ are time derivatives of the mortar integrals (holding node A constant). Note that the scaling factors κ_A have again been introduced to retain dimensional consistency. Since the mortar integral time derivatives are invariant with respect to any rigid body motion relative to the original spatial frame, $\boldsymbol{v}_{T_A}$ is frame indifferent.

With these definitions in hand, a trial state-return map strategy is employed to determine the Coulomb frictional tractions in an algorithmic, time stepping procedure. The algorithm begins by computation of a trial state, assuming no slip during the increment:

$$\begin{aligned} \boldsymbol{\lambda}_{T_{A_{n+1}}}^{trial} = \boldsymbol{\lambda}_{T_{A_n}} - \epsilon_T \kappa_A \Bigg[& \sum_{C}^{nm} \left(n_{AC_{n+1}}^{(2)} - n_{AC_n}^{(2)} \right) \boldsymbol{\varphi}_C^{(2)} \\ & - \sum_{B}^{ns} \left(n_{AB_{n+1}}^{(1)} - n_{AB_n}^{(1)} \right) \boldsymbol{\varphi}_B^{(1)} \Bigg] \cdot (\boldsymbol{I} - \boldsymbol{n}_A \otimes \boldsymbol{n}_A), \end{aligned} \tag{30}$$

with a corresponding trial value for the slip function

$$\Phi_{A_{n+1}}^{trial} = \left\| \boldsymbol{\lambda}_{T_{A_{n+1}}}^{trial} \right\| - \mu |\lambda_{N_{A_n}}|. \tag{31}$$

A return map is then used to define the final frictional traction via

$$\boldsymbol{\lambda}_{T_{A_{n+1}}} = \begin{cases} \boldsymbol{\lambda}_{T_{A_{n+1}}}^{trial} & \text{if } \Phi_{A_{n+1}}^{trial} \leq 0 \text{ (stick)}, \\ \mu |\lambda_{N_{A_n}}| \dfrac{\boldsymbol{\lambda}_{T_{A_{n+1}}}^{trial}}{\|\boldsymbol{\lambda}_{T_{A_{n+1}}}^{trial}\|} & \text{otherwise (slip)}. \end{cases} \tag{32}$$

In these expressions, the subscript $n+1$ means a state associated with the current iteration for the unknown solution at t_{n+1}, while n is associated with the last (converged) time level.

3 An Extension of the Framework: Mortar-Based Self-Contact

In many applications, such as are encountered in post-buckling analysis, a surface may fold upon itself such that so-called *self contact* takes place. A two

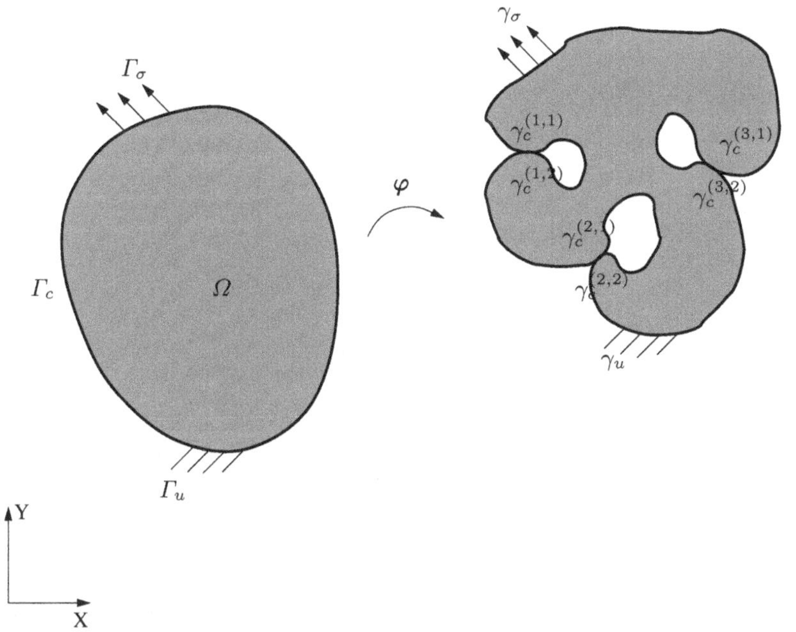

Fig. 2. Notation for a large deformation self-contact problem

dimensional self-contact problem, with large deformation and large sliding, is shown schematically in Fig. 2. Recent work shows that the mortar framework can be extended to such settings, and we highlight some of the key steps here in accomplishing that extension.

In writing the virtual work for such a system, the expression given originally in (8) must be generalized not only to account for the possibility of sliding, but also to account for the possibility of multiple contact patches which can be generated as the surface folds over on itself. We therefore replace (8) by

$$G^c(\boldsymbol{\varphi}, \overset{*}{\boldsymbol{\varphi}}) := -\sum_{i=1}^{k} \int_{\gamma_c^{(i,1)}} \boldsymbol{\lambda}^{(i,1)}(\boldsymbol{X}, t) \cdot \left(\overset{*}{\boldsymbol{\varphi}}{}^{(i,1)}(\boldsymbol{X}, t) - \overset{*}{\boldsymbol{\varphi}}{}^{(i,2)}(\bar{\boldsymbol{Y}}, t) \right) d\gamma, \tag{33}$$

where $\boldsymbol{\varphi}^{(i,2)}(\bar{\boldsymbol{Y}})$ is the current position of the contact point for $\boldsymbol{X}$. Indices i refer to the multiple contact patches which might be identified at any instant in the simulation, with k being the number of these patches. As before, the surface $\gamma_c^{(i,1)}$ is the *nonmortar* surface where the Lagrange multipliers will be interpolated, and $\gamma_c^{(i,2)}$ is the *mortar* surface. As summarized earlier in (22), the contact traction $\boldsymbol{\lambda}$ can be resolved into normal and tangential components, and the conditions of normal contact are given by (24), and the frictional conditions by (19), just as in the frictional contact case. The contact searching

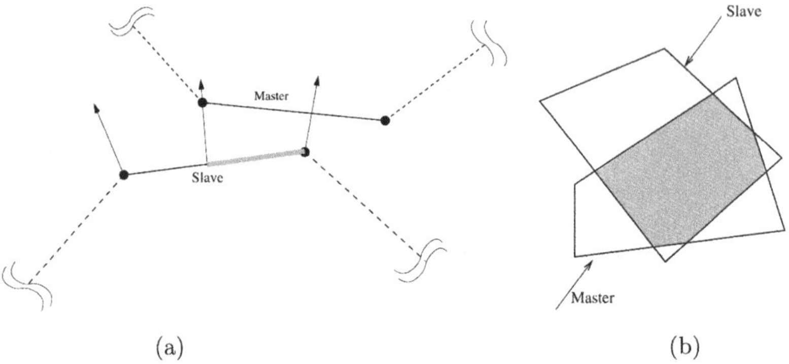

Fig. 3. Mortar segments in (a) two dimensions; (b) three dimensions. A segment is shown in green

algorithm for mortar contact, however, must be considerably generalized to deal with the phenomenon of self-contact.

As discussed in [17] and [12] for general mortar contact formulations, we need to find all the mortar segments to compute the mortar integrals, $n_{AB}^{(1)}$ and $n_{AC}^{(2)}$, defined in (21). For example, in three dimensional problems, the mortar integrals are computed as:

$$\begin{aligned} n_{AB}^{(i,1)} &= \sum_{seg} n_{AB}^{(i,1)^{seg}} = \sum_{seg} \sum_{tri} n_{AB}^{(i,1)^{tri,seg}}, \quad i = 1, ..., k, \\ n_{AB}^{(i,1)^{tri,seg}} &= \int_{\gamma_c^{tri,seg,i}} N_A^{(1)} \left(\boldsymbol{\xi}^{(1)}(\boldsymbol{X}) \right) N_B^{(1)} \left(\boldsymbol{\xi}^{(1)}(\boldsymbol{X}) \right) d\gamma \end{aligned} \tag{34}$$

$$\begin{aligned} n_{AC}^{(i,2)} &= \sum_{seg} n_{AC}^{(i,2)^{seg}} = \sum_{seg} \sum_{tri} n_{AC}^{(i,2)^{tri,seg}}, \quad i = 1, ..., k, \\ n_{AC}^{(i,2)^{tri,seg}} &= \int_{\gamma_c^{tri,seg,i}} N_A^{(1)} \left(\boldsymbol{\xi}^{(1)}(\boldsymbol{X}) \right) N_C^{(2)} \left(\boldsymbol{\xi}^{(2)} \left(\bar{\boldsymbol{Y}} (\boldsymbol{X}) \right) \right) d\gamma \end{aligned} \tag{35}$$

In these equations, the superscript *seg* refers to a *mortar segment*, which is constructed such that each segment is defined by the mutual projection of a single slave element surface with a single master element surface, as shown in Fig. 3. The polygonal intersections of element surfaces, shown in green in Fig. 3(b), are in turn divided into triangles for numerical evaluation of the integrals, giving rise to the superscripts *tri* above. The segments change dynamically during a simulation of sliding contact, and the dependence of the method on the determination of the segments is why we refer to the contact algorithms in this work as being *surface to surface* algorithms.

Given a discretized contact surface, an efficient contact algorithm depends crucially on efficient determination of these mortar segments. In our implementation, we recursively split the surface into subsurfaces to build a tree

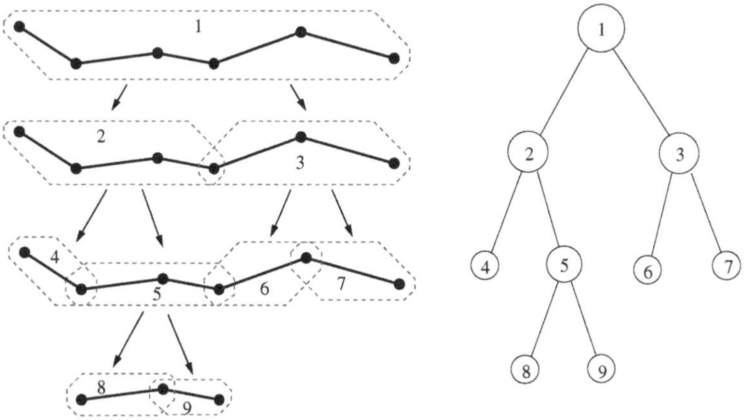

Fig. 4. A bounding volume tree for a 2D surface

structure; the approach has been extensively described in [16] and will only be briefly recounted here. The subsurfaces related to leaf tree nodes are the elements of the surface. Each node of the tree will be termed a T-node, to differentiate from a finite element node. Each subsurface is defined and stored in terms of an inflated bounding volume (specifically, a k-DOP), which is saved in its corresponding T-nodes. Binary trees are used in this work so each non-leaf T-node has exactly two children, a left child and right child. A typical bounding volume hierarchy is shown in Fig. 4. If there are n elements for a contact surface, the number of T-nodes in bounding volume tree is $2n - 1$. The storage space for a bounding volume tree is then $O(n)$.

In [16], two methods for building bounding volume trees are discussed: *top-down* and *bottom-up* approaches. A top-down approach, which is used to build the bounding volume trees for multi-body contact problems, starts with one T-node (a root T-node) which is related to the entire surface. The surface is then recursively split into subsurfaces and each of them is saved in a newly created T-node. The bottom-up approach, by contrast, begins with the input elements as the leaves of the bounding volume tree and applies some algorithms to group elements recursively. Usually, by this approach, the subsurface corresponding to a T-node is a connected surface so it is a more appropriate way than the top-down approach to build the bounding volume trees for self-contact problems. In this work, we apply the surface clustering algorithm introduced in [14, 7] to build bounding volume trees.

In the face clustering algorithm, one first creates a dual graph for the mesh of the contact surface. A dual graph is constructed by mapping each element of the surface mesh to a dual node and connecting every two dual nodes by a dual edge if the corresponding elements are adjacent and share an element edge. An example of a dual graph is shown in Fig. 5, where dashed lines are

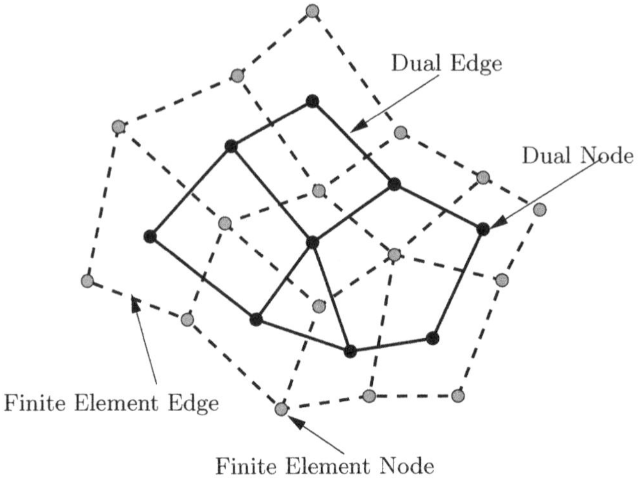

Fig. 5. Dual graph for a surface mesh. Each dual node represents an element or subsurface in the finite element mesh

finite element edges, solid lines are dual edges, grey dots are finite element nodes, and black dots are dual nodes.

By contracting a dual edge (i.e., removing the dual edge and collapsing its two dual nodes into a new dual node) one can group two elements into an element cluster (or a subsurface) which corresponds to a new dual node in the dual graph, as seen in Fig. 6. Basically, this contraction will group two contiguous elements or subsurfaces into a new subsurface, which will be their parent in the bounding volume hierarchy. One may also notice that each dual node in the dual graph corresponds to a T-node in the bounding volume tree. By doing this contraction iteratively, one can finally build a hierarchical structure of the surface mesh, as shown in Fig. 7.

To control the quality of the bounding volume hierarchy, i.e. the balance of the bounding volume tree and the shape and tightness of the bounding volumes, one has to follow a carefully designed rule to contract the dual edges. This rule identifies which edge should be contracted first and which should be contracted later. To do this, a cost function is assigned to each dual edge; the cost function applied in this work is similar with that was used by [14]. All dual edges are saved in a heap data structure (see [5] for an introduction) and the key values of the heap entries are the contraction costs of the corresponding dual edges. Thus, the dual edge with the least cost is saved on the top of the heap. The procedure to construct a bounding volume hierarchy with the face clustering algorithm can be summarized as:

1. Build the initial dual graph for the finite element mesh, compute cost functions of all dual edges, and save all the dual edges in a heap.

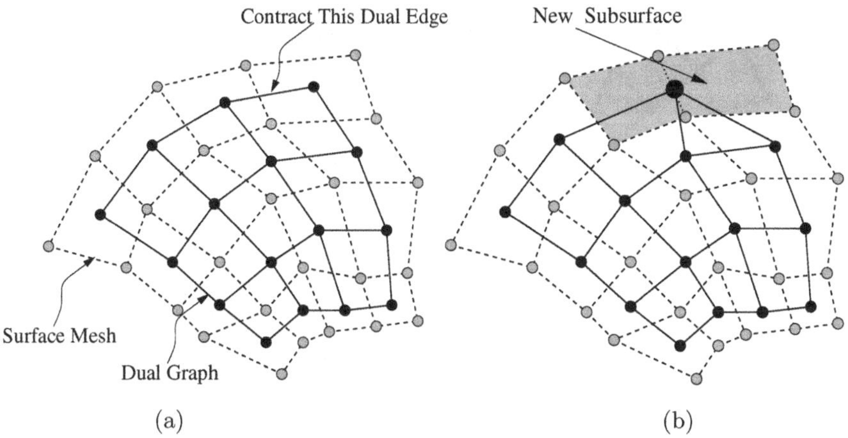

Fig. 6. Contraction of a dual edge to construct a new subsurface (i.e. removal of a dual edge and collapsing of its two dual nodes): (a) initial dual graph, and (b) new dual graph. The bigger dot on (b) denotes the new generated dual node by this contraction, and the new dual node corresponds to a new constructed subsurface denoted by the shaded area

2. Remove a dual edge from the top of the heap and contract it to construct a new subsurface.
3. Generate a new T-node for the new subsurface. The two children of the new T-node are the two T-nodes corresponding to two dual nodes (subsurfaces or elements) connected by the dual edge.
4. Update all cost functions of dual edges that are connected with the two dual nodes of the contracted dual edge.
5. Update the heap.
6. Repeat 2 through 5 iteratively until the whole tree is constructed.

From this procedure, one can see that the bounding volume tree is built in a bottom-up manner. For each node of the tree, the bounding volume of its corresponding subsurface is computed. For large deformation contact problems, the updating of the bounding volume trees is required after each iteration since the bounding volumes are dependent on deformation.

Finally, after the contact detection has been accomplished, it is necessary to do some post processing after the searching, to avoid the situation where the contact traction fields on each contact patch could be defined on discontinuous surfaces, as shown in Fig. 8. As can be seen, without some type of intervention, an element can be a slave element in one contact pair and a master element in another, and there may be a mixture of slave and master elements on a continuous contact element patch (the lower or upper solid line in Fig. 8). A

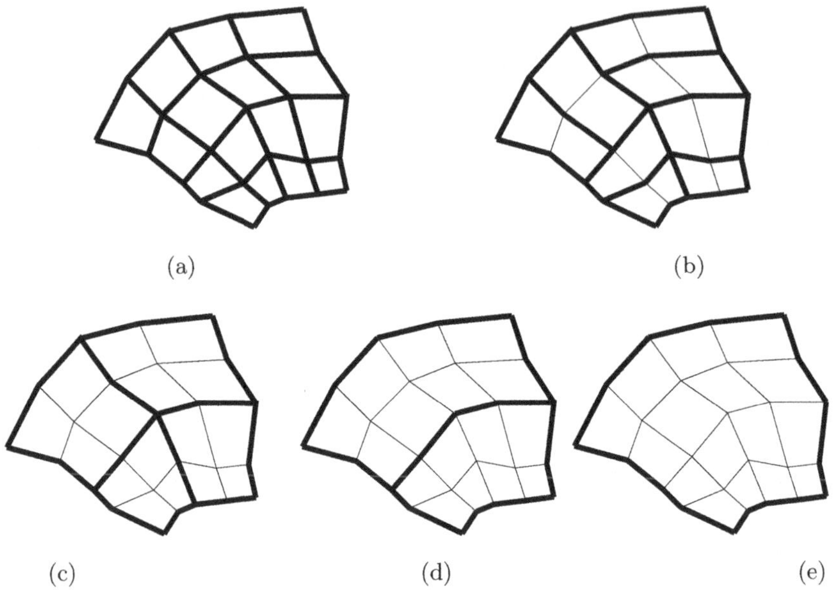

Fig. 7. A hierarchy (tree) structure from the bottom level to the top level

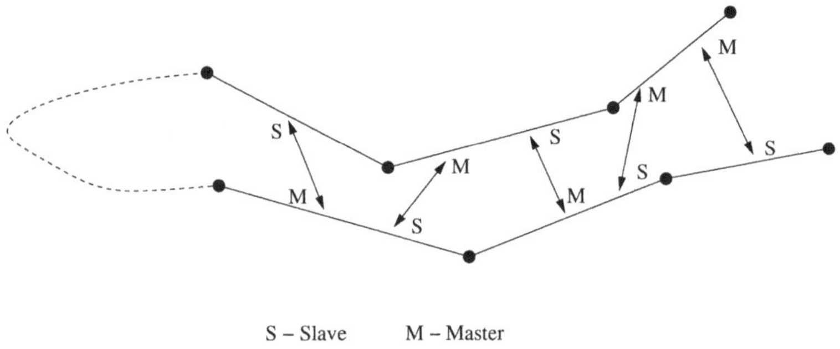

Fig. 8. Discontinuous definition of slave and master elements (a double arrow denotes a contact element pair) which should be avoided

simple sorting procedure is applied so that each side of each contact patch retains its own identity as master or slave.

As an example of the type of problem to which this self-contact approach is applicable, we consider the rolling of a highly loaded tire which is nearly flat. In this problem, the part of the surface in self-contact (inside the tire) continually changes due to the rolling, placing a significant demand for robustness on the searching algorithm and Newton-Raphson iterative scheme. Deformed configurations of the nearly flat tire at different load steps are shown in Fig. 9, where (c) and (d) only show a quarter of the tire to demonstrate self-contact

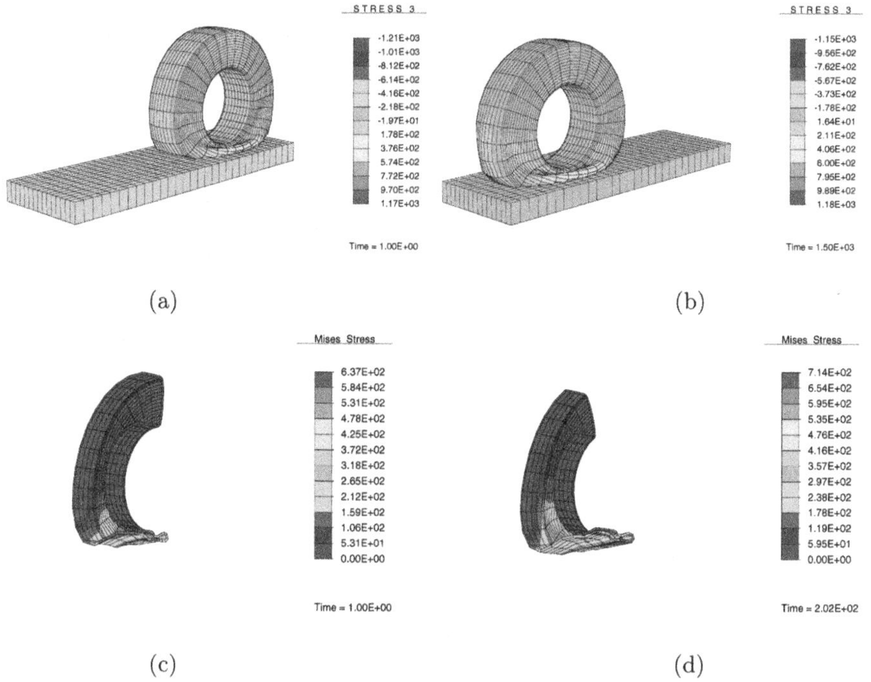

Fig. 9. Deformed configurations of the tire rolling problem at different load steps; parts (c) and (d) only show a quarter of the tire

inside the tire. Significant deformation and sliding are seen to occur in this problem. Notably, it is difficult if not impossible to perform this situation using node-to-surface contact discretization strategies.

4 Mortar Formulation of Lubricated Contact

Another recent extension of the mortar formalism involves its use to describe lubricated contact, such as might occur for example in many metal forming applications, or underneath a tire rolling on wet pavement. Such a problem, with two deformable solid bodies and a thin fluid film between them, is shown in Fig. 10. The two solid bodies occupy the open sets $\Omega_s^{(1)}$ and $\Omega_s^{(2)}$, where a subscript s represents the solid phase.

The solid phase of a lubricated contact problem is governed by the usual solid continuum mechanics equations. Here we assume infinitesimal deformations for the solid phase for simplicity, although it is possible to extend the formulation to large deformation cases with proper treatments of geometric nonlinearities. No assumptions are made for the constitutive laws: the material response may be either elastic or inelastic, rate dependent or independent.

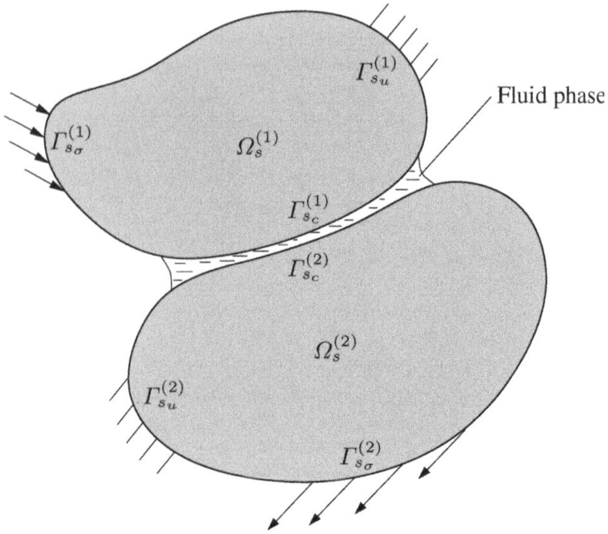

Fig. 10. Notation for two body lubricated contact problems

In the problems considered here, the through-the-thickness dimension of the fluid film is assumed to be much smaller than the size of the contact patch in surface coordinates. Under such conditions, the fluid phase of a lubricated contact problem is governed by the Reynolds equation, which can be derived from the Navier-Stokes and continuity equations based on the thin-film assumption (see, e.g., [8]). The Reynolds equation is defined on a surface domain, which considered to be parameterized by the slave side of a contact surface in this formulation. As shown in Fig. 11, the contact domain Ω_f, corresponding to the slave surface $\Gamma_{s_c}^{(1)}$ less the dry contact region Ω_d, is divided into two different subdomains, Ω_{f_l} and Ω_{f_c}. Ω_{f_l} is the subdomain where the lubricant is continuous and has positive pressure, while Ω_{f_c} is the subdomain where the lubricant has cavitations and is ruptured (the fluid film is assumed to support only a negligible amount of tension before rupture). Γ_{f_D} is the surface where the Dirichlet boundary conditions are prescribed for the Reynolds equation, while Γ_{f_N} is the boundary between the dry contact domain and the lubrication domain and where Neumann boundary conditions are prescribed. Γ_{f_R}, which is called the Reynolds boundary, is the boundary between Ω_{f_l} and Ω_{f_c}. Based on these definitions,

$$\Omega_f = \Gamma_{s_c}^{(1)} \setminus \Omega_d, \tag{36}$$

and

$$\Omega_f = \Omega_{f_l} \cup \Omega_{f_c}. \tag{37}$$

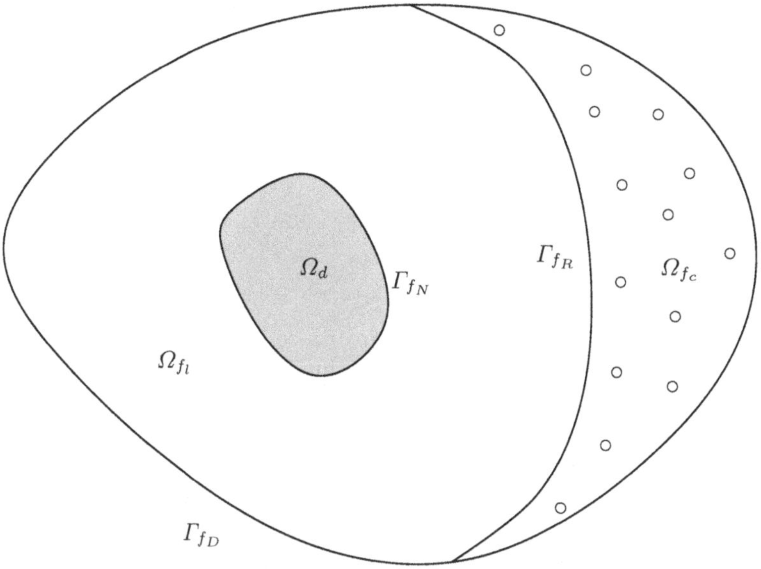

Fig. 11. Notation for the lubrication domain

The virtual work for the solid phase can be written in the usual way via

$$G^c(\boldsymbol{u}, p, \overset{*}{\boldsymbol{u}}) := -\int_{\Gamma_{sc}^{(1)}} \boldsymbol{t}^{(1)}(\boldsymbol{X}) \cdot \left(\overset{*}{\boldsymbol{u}}^{(1)}(\boldsymbol{X}) - \overset{*}{\boldsymbol{u}}^{(2)}(\bar{\boldsymbol{Y}}) \right) d\Gamma, \tag{38}$$

where $\bar{\boldsymbol{Y}}$ is the position of the contact point for $\boldsymbol{X}$, with the understanding that the contact traction $\boldsymbol{t}$ is now inherited from the traction field prevailing in the intervening fluid, consisting in general of a fluid pressure p in the normal direction and a viscous tangential component.

It is the fluid part of the formulation that makes the mortar approach attractive in this case, since the pressure distribution in the thin film is subject to a governing differential equation (the Reynolds equation). The fact that the mortar approach produces a consistent surface discretization gives an ideal way to not only approximate this differential equation, but also to couple it to the solid response. As presented by [8], the general Reynolds equation which governs the fluid pressures within Ω_{f_l} is

$$\frac{\partial(\rho h)}{\partial t} + \tilde{\nabla} \cdot \left(-\frac{\rho h^3}{12\mu} \tilde{\nabla} p + \frac{\rho(\tilde{\boldsymbol{V}}^{(1)} + \tilde{\boldsymbol{V}}^{(2)})}{2} h \right) = 0, \tag{39}$$

where ρ is the current mass density of the lubricant, h is the fluid film thickness (negative of the gap function g), μ is the viscosity of the lubricant, $\tilde{\boldsymbol{V}}^{(1)}$ and

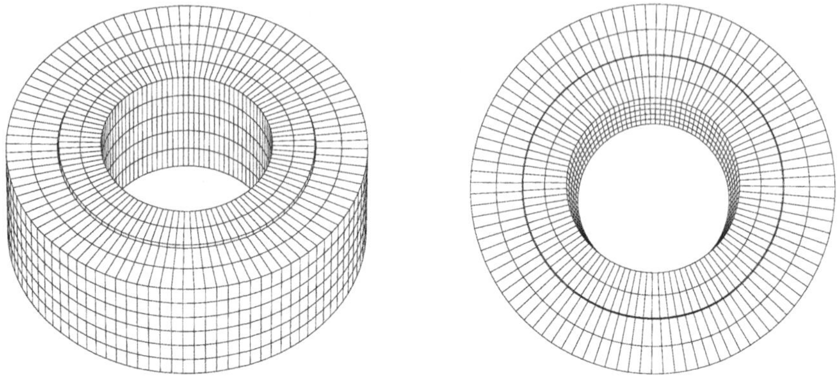

Fig. 12. Finite element mesh of the three dimensional lubricated contact problem with two different perspectives

$\tilde{\boldsymbol{V}}^{(2)}$ are projections of slave and master surface velocity vectors onto the slave surface, and p is the fluid pressure (the primary unknown of Equation (39)). Operators $\tilde{\nabla}$ indicate divergence and gradient operators in reduced (surface) coordinates. The fluid viscosity may be dependent on the fluid pressure; both constant viscosity cases and exponential pressure dependence of viscosity have been tested in our early work.

We have implemented a mortar-based monolithic strategy to solve the coupled Reynolds equations and global equilibrium equations, with the solid phase displacement coupling to the fluid equations through the film thickness h and the fluid equations coupling to the solid mechanics equations through generation of the pressure field p and the viscous shear stresses. Although the full numerical formulation is too involved to recount in detail here, we present a simple three dimensional example to demonstrate the type of simulation for which the technique has been tested. The problem is depicted in Fig. 12, where the relative rotation of two cylinders with a lubricant-filled interface is considered. A quasi-static rotation is applied to the inner surface of the inside cylinder, corresponding to an angular speed of $\omega = 500$. Figure 12 presents the finite element mesh for the two cylinders. The inside surface of the outer cylinder is chosen as the slave surface and the outside surface of the inner cylinder is chosen as the master surface.

Only one load step is applied for this problem. The computed pressure distribution in the fluid film is plotted in Fig. 13.

5 Conclusion

This paper has discussed the mortar method as an underlying spatial discretization technique for large deformation contact problems, and has empha-

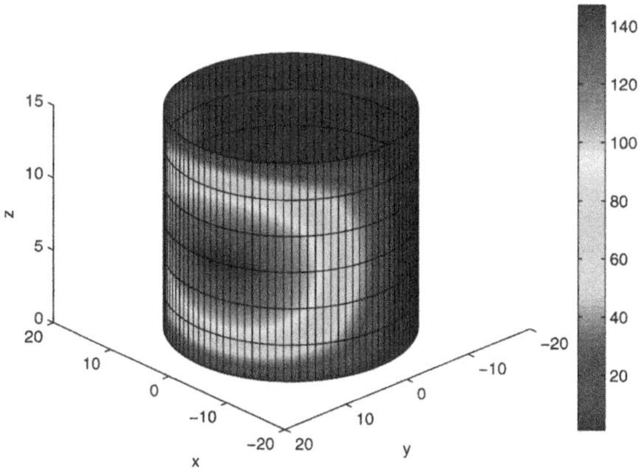

Fig. 13. Computed pressure distribution for the three dimensional lubricated contact problem

sized two promising recent extensions of this idea: incorporation of self-contact phenomena, and consideration of lubricated contact problems. Contact formulations based on mortar concepts have been seen not only to be extremely accurate in comparison with more traditional node-to-surface strategies, but are also extremely robust numerically, particularly within the implicit dynamic and quasistatic applications of primary interest here.

Acknowledgment

This work was supported through a research contract by Michelin Americas Research Corporation. This support, as well as the collaboration of Drs. Stephane Cohade, Ali Rezgui, and Mike Andrews, is greatly appreciated.

References

1. Belgacem FB, Hild P, Laborde P (1997) Approximation of the unilateral contact problem by the mortar finite element method. Comptes Rendus De L'Academie Des Sciences 324:123–127
2. Belgacem FB, Maday Y (1994) A spectral element methodology tuned to parallel implementations. Computer Methods in Applied Mechanics and Engineering 116:59–67
3. Belhachmi Z, Bernardi C (1994) Resolution of fourth order problems by the mortar element method. Computer Methods in Applied Mechanics and Engineering 116:53–58

4. Bernardi C, Maday Y, Patera AT (1992) A new nonconforming approach to domain decomposition: The mortar element method. In H. Brezia and J.L. Lions (Eds.) Nonlinear Partial Differential Equations and Their Applications pp 13–51. Pitman and Wiley
5. Cormen TH, Leiserson CE, Rivest RL, Stein C (2001) Introduction to Algorithms, Second Edition. The MIT Press
6. El-Abbasi N, Bathe K-J (2001) Stability and patch test performance of contact discretizations and a new solution algorithm. Computers and Structures 79:1473–1486
7. Garland M, Willmott AJ, Heckbert PS (2001) Hierarchical face clustering on polygonal surfaces. In SI3D, pp 49–58
8. Hamrock BJ (1991) Fundamentals of Fluid Film Lubrication. NASA, Washington, D.C.
9. Hild P (2000) Numerical implementation of two nonconforming finite element methods for unilateral contact. Computer Methods in Applied Mechanics and Engineering 184:99–123
10. McDevitt TW, Laursen TA (2000) A mortar-finite element formulation for frictional contact problems. International Journal for Numerical Methods in Engineering 48:1525–1547
11. Puso MA (2004) A 3d mortar method for solid mechanics. International Journal for Numerical Methods in Engineering 59:315–336
12. Puso MA, Laursen TA (2004) A mortar segment-to-segment contact method for large deformation solid mechanics. Computer Methods in Applied Mechanics and Engineering 193:601–629
13. Puso MA, Laursen TA (2004) A mortar segment-to-segment frictional contact method for large deformations. Computer Methods in Applied Mechanics and Engineering 193:4891–4913
14. Willmott AJ (2000) Hierarchical Radiosity with Multiresolution Meshes. PhD thesis, Carnegie Mellon University
15. Wohlmuth BI (2001) Discretization Methods and Iterative Solvers Based on Domain Decomposition. Springer-Verlag, Heidelberg
16. Yang B, Laursen TA (2006) A contact searching algorithm for large deformation mortar formulation. Computational Mechanics, (submitted)
17. Yang B, Laursen TA, Meng XN (2005) Two dimensional mortar contact methods for large deformation frictional sliding. International Journal for Numerical Methods in Engineering 62:1183–1225

Particle Finite Element Methods in Solid Mechanics Problems

J. Oliver, J.C. Cante, R. Weyler, C. González and J. Hernandez

E.T.S. Enginyers de Camins, Canals I Ports de Barcelona
E.T.S. d'Enginyería Industrial i Aeronáutica de Terrassa
Universitat Politécnica de Catalunya (UPC)
Campus Nord UPC, Edifici C-1, c/ Jordi Girona 1-3, 08034 Barcelona, Spain
xavier.oliver@upc.edu

Summary. The paper examines the possibilities of extending the Particle finite element methods (PFEM), which have been successfully applied in fluid mechanics, to solid mechanics problems. After a review of the fundamentals of the method, their specific features in solid mechanics are presented. A methodology to face contact problems, the anticipating contact interface mesh, is presented on the basis of a penalty-like constitutive models for imposing the contact and friction conditions. Finally, the PFEM is applied to same representative solid mechanics problems to display the capabilities of the method and some final conclusions are obtained.

1 Introduction

Particle finite element methods (PFEM) have received considerable attention in the recent years due to their modeling capabilities for same specific problems. So far, most of the research and applications of PFEM can be found in the context of computational fluid dynamics (CFD) to tackle fluid mechanics problems in typical solid mechanics settings: i.e. using Lagrangean descriptions of the motion of the continuum medium (Onate et al., 1996; Lohner et al., 2002; Idelsohn et al., 2003a, 2003b, 2004). Their main advantages are found in modeling confined fluids exhibiting moving free surfaces. There, the limited character of the particle displacement makes suitable a Lagrangean description which, in turn, facilitates the tracking and modeling of the existing free surfaces.

On the other hand, Lagrangean descriptions are the natural way of describing motion of solids, and, therefore, there is a long solid mechanics tradition in this sense. However in certain processes, of considerable practical interest, the material undergoes very large deformations, rapidly changing boundaries are involved and the motion resembles that of fluids. Metal forming and machining processes, or manufacturing processes involving powder and granular

Eugenio Oñate and Roger Owen (eds.), Computational Plasticity, 87–103.

materials are typical cases where the border between solid and fluid behaviors becomes fuzzy.

Therefore, the application of PFEM to solid mechanics problems appears as a new research field deserving to be explored. The purpose of this work is precisely investigating the possibilities offered by the particle finite element methods in some representative solid mechanics problems involving large deformations, multiple contacts, new boundaries generation, etc., thus providing some insights on the future developments in that field.

2 Fundamentals: The Particle Finite Element Method

Particle finite element methods emerged as a natural result of previous explorations in the context of the meshless methods (Belytschko et al., 1994, 1996; Onate et al., 1998). They can be characterized by the following ingredients:

1. The use of a Lagrangean format for describing the motion (Malvern, 1969). A selected cloud of particles of infinitesimal size (material points) are tracked along the motion to describe the continuum medium properties evolution (position, displacement, velocities, strain, stresses, internal variables, etc.). When necessary, the properties of the remaining particles of the continuum medium are obtained by interpolation of the properties at points of that cloud. Numerical computations are done on the basis of a finite element mesh that is constructed at every time step on the basis of the particle positions. Then, Delaunay triangulations, allowing the construction of a finite element mesh for a given sets of nodes, emerge as a suitable meshing procedure (George, 1991; Calvo et al., 2003).
2. The use of a boundary recognition procedure to identify what particles of the cloud define an external (or internal) boundary. The so-called alpha-shape method (Calvo et al., 2003; Xu et al., 2003) constitutes a suitable strategy for this purpose (see Figure 1). It essentially consists of identifying those sides/segments of the cloud that can be inserted into an empty circle/ball (not including other particles of the cloud) of size larger than a given parameter (the alpha-shape parameter). The vertices/particles of those segments are then identified as boundary particles. Large values of the alpha-shape parameter result in a boundary which is the convex hull of the cloud. Small values of the alpha-shape parameter return a boundary constituted of all the particles of the cloud.[1]

2.1 Equations of Motion. Boundary Value Problem

Let us consider a solid body B experiencing a deformation $\varphi(\mathbf{X},t) : \mathsf{B} \times [0,T] \to \mathsf{R}^2$, where $[0,T]$ is the time interval of interest. Let us also consider a

[1]For a uniformly distributed cloud of particles (with typical separation h) alpha-shape values of $1.1h - 1.5h$ are recommended for a good estimation of the actual boundary.

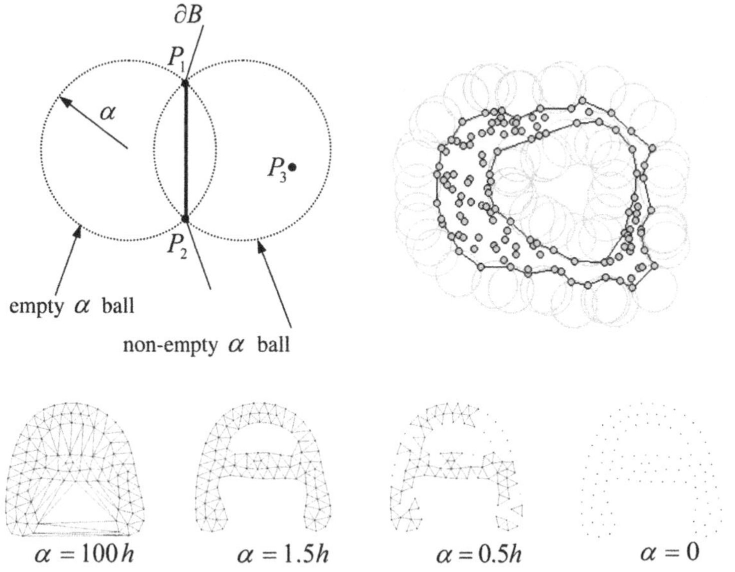

Fig. 1. The alpha-shape method for recognizing boundaries of a cloud of particles

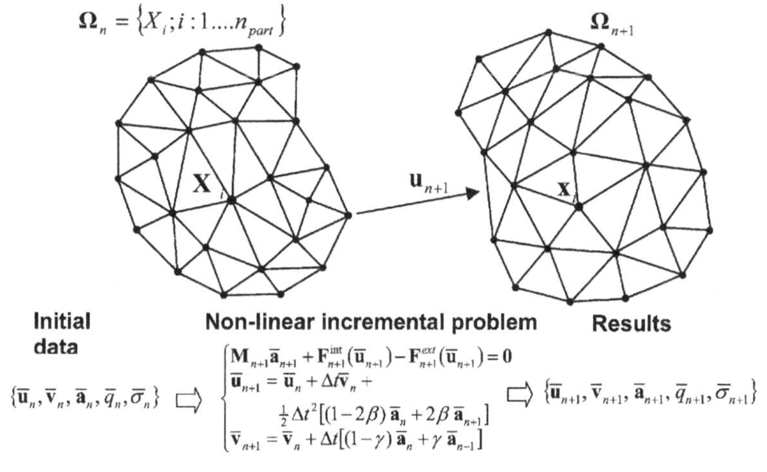

Fig. 2. Incremental non-linear problem at time step $[t_n, t_{n+1}]$

finite set of particles of $\mathsf{B} \supset \mathsf{P} := \{\mathsf{P}_1, \mathsf{P}_2, \ldots, \mathsf{P}_{n_{part}}\}$, occupying, at a specific time interval $t \in [0, T]$, spatial points with coordinates $\bar{\mathbf{X}}_i = [X_1, X_2]^T$, which define the particle reference configuration $\Omega_t := \{\bar{\mathbf{X}}_{\mathsf{P}_1}, \bar{\mathbf{X}}_{\mathsf{P}_2}, \ldots, \bar{\mathbf{X}}_{\mathsf{P}_{n_{part}}}\}$ (see Figure 2).

Let us assume, at time t, a Delaunay triangulation, with all its vertices, i, placed at corresponding positions, $\bar{\mathbf{X}}_i$, of the particle configuration Ω_t, thus defining an open set, V_t, at the Euclidean 2D space, with boundary ∂V_t. The particles of B belonging to the material volume V_t define a body $\tilde{\mathsf{B}} \subset \mathsf{B}$ approaching B as $n_{part} \to \infty$. Let us also consider a time discretization of $[0, T]$ in time intervals and a specific time slice $[t_n, t_{n+1}] \subset [0, T]$ characterizing the time interval/step $n+1$ of length $\Delta t = t_{n+1} - t_n$. The boundary value problem at the material domain during the time interval can be written as:

$$
\begin{aligned}
&\textit{Given } \mathbf{v}_n, \mathbf{a}_n, \boldsymbol{\sigma}_n, \mathbf{q}_n, \mathbf{u}^*_{n+1}, \mathbf{t}^*_{n+1} \\
&\textit{Find } \mathbf{u}_{n+1} \equiv \varphi_{n+1}(\mathbf{X}_n) \quad \textit{such that:} \\
&\nabla \cdot \boldsymbol{\sigma}_{n+1} + \mathbf{b}_{n+1} = \rho_{n+1} \mathbf{a}_{n+1} \quad (\textit{momentum equation}) \\
&\left.\begin{aligned}
\boldsymbol{\sigma}_{n+1} &= \boldsymbol{\sigma}_n + \underbrace{\Sigma(\mathbf{e}_{n+1}, \boldsymbol{\sigma}_{n+1}, \mathbf{q}_{n+1})}_{\Delta \boldsymbol{\sigma}_{n+1}} \\
\mathbf{q}_{n+1} &= \mathbf{q}_n + \underbrace{\mathsf{Q}(\mathbf{e}_{n+1}, \boldsymbol{\sigma}_{n+1}, \mathbf{q}_{n+1})}_{\Delta \mathbf{q}_{n+1}}
\end{aligned}\right\} \quad (\textit{constitutive equations}) \qquad (1) \\
&\mathbf{e}_{n+1} = \mathsf{E}(\mathbf{u}_{n+1}) \quad (\textit{kinematic equation}) \\
&\boldsymbol{\sigma}_{n+1}, \mathbf{n}_{n+1} = \mathbf{t}^*_{n+1} \quad (\textit{natural boundary condition}) \quad \textit{at } \partial_\sigma V_{n+1} \\
&\mathbf{u}_{n+1} = \mathbf{u}^*_{n+1} \quad (\textit{essential boundary condition}) \quad \textit{at } \partial_u V_{n+1}
\end{aligned}
$$

where $\mathbf{v}_n$, $\mathbf{a}_n$, $\boldsymbol{\sigma}_n$, $\mathbf{q}_n$ are, respectively, the velocities, accelerations, stresses and internal variable values at the beginning of the interval, $\mathbf{u}^*_{n+1}$, $\mathbf{t}^*_{n+1}$ are the prescribed values of displacements and tractions, respectively, at the boundary ∂V_{n+1} with outward normal $\mathbf{n}_{n+1}$, $\mathbf{u}_{n+1} = \mathbf{X}_{n+1} - \mathbf{X}_n$ are the interval displacements and $\mathbf{e}_{n+1}$ is a suitable measure of the interval strains.

2.2 Time Marching Scheme

The problem described in Section 2.1 allows considering the motion of the approximating body $\tilde{\mathsf{B}}$ as a sequence of discrete (in time) boundary value problems ruled by equations (1). Then, at every time step the corresponding problem is solved according to the following strategy:

Step I: Finite element discretization: Spatial discrete problem
Let us now consider a finite element discretization of V_n, on the basis of the existing triangularization, so that the nodes match the vertices (therefore, $n_{node} = n_{part}$), and that every property, μ, of the particles of $\tilde{\mathsf{B}}$ is evaluated via interpolation of the corresponding property at the vertices/nodes as:

$$\mu(\mathbf{X}) = \sum_{1}^{n_{node}} N_i(\mathbf{X}) \bar{\mu}_i \quad \forall \mathbf{X} \in V_n \qquad (2)$$

where N_i and $\bar{\mu}_i$ stand for the shape/interpolation function and the nodal value of the property, respectively, at node i.

Then, using standard Galerkin's procedures, and a Newmark's time integration scheme (Hughes, 2000), the discrete counterpart of the continuum boundary value problem in equations (1) at the space-time slice $V_n \times [t_n, t_{n+1}] \subset \mathsf{R}^2 \times [0, T]$ reads, in matrix form:

$$\begin{aligned} &\mathbf{M}_{n+1} \cdot \bar{\mathbf{a}}_{n+1} + \mathbf{F}^{int}_{n+1}(\bar{u}_{n+1}) - \mathbf{F}^{ext}_{n+1}(\bar{\mathbf{u}}_{n+1}) = 0 \\ &\bar{\mathbf{u}}_{n+1} = \bar{\mathbf{u}}_n + \Delta t \bar{\mathbf{v}}_n + \tfrac{1}{2} \Delta t^2 [(1 - 2\beta)\bar{\mathbf{a}}_n + 2\beta \bar{\mathbf{a}}_{n+1}] \\ &\bar{\mathbf{v}}_{n+1} = \bar{\mathbf{v}}_n + \Delta t [(1 - \gamma)\bar{\mathbf{a}}_n + \gamma \bar{\mathbf{a}}_{n+1}] \end{aligned} \tag{3}$$

where $\mathbf{M}_{n+1}$, $\mathbf{F}^{int}_{n+1}$ and $\mathbf{F}^{ext}_{n+1}$ are, respectively, the mass-matrix, the internal forces vector and the external forces vector, $\bar{\mathbf{u}}_{n+1}$, $\bar{\mathbf{v}}_{n+1}$ and $\bar{\mathbf{a}}_{n+1}$ are, respectively, the *nodal* displacement, velocities and accelerations and β and γ are the classical Newmark's integration parameters. Equations (3) are a set of non-linear equations that can be solved for $\bar{\mathbf{u}}_{n+1}$, $\bar{\mathbf{v}}_{n+1}$ and $\bar{\mathbf{a}}_{n+1}$.

Combining interpolations, according to equation (2), and substitution in equations (1) allows determining the stresses and the internal variables $\boldsymbol{\sigma}_{n+1}(\mathbf{X}), \mathbf{q}_{n+1}(\mathbf{X})$ at points $\mathbf{X} \in V_n$ as required in next step.

Step II: Spatial information transfer

All the information necessary in subsequent time steps has now to be transferred to the nodes/particles of Ω_n. This is achieved by standard extrapolation (smoothing) procedures (Zienkiewicz et al., 2000) from the element Gauss points to the nodes. For instance, the nodal values, $\bar{\boldsymbol{\sigma}}_{n+1}$ and $\bar{\mathbf{q}}_{n+1}$, to be considered as initial values for the next time step, $n + 1$, are computed as:

$$\begin{aligned} \bar{\boldsymbol{\sigma}}_{n+1} &= \bar{\boldsymbol{\sigma}}_n + [\mathbf{M}_{\sigma_n}]^{-1} \cdot \int_{V_n} \mathbf{N}_{\sigma_n} \cdot \Delta\boldsymbol{\sigma}_{n+1} \, dV \\ \bar{\mathbf{q}}_{n+1} &= \bar{\mathbf{q}}_n + [\mathbf{M}_{q_n}]^{-1} \cdot \int_{V_n} \mathbf{N}_{q_n} \cdot \Delta\mathbf{q}_{n+1} \, dV \end{aligned} \tag{4}$$

where $\mathbf{M}_{\sigma_n}$ and $\mathbf{M}_{q_n}$ are standard "mass-like matrices", and $\mathbf{N}_{\sigma_n}$ and $\mathbf{N}_{q_n}$ are "transfer" matrices, with dimensions appropriated to the set of variables, $\boldsymbol{\sigma}$ and $\mathbf{q}$, computed in terms of the interpolation functions N_i of the finite element mesh on the domain V_n, and $\Delta\boldsymbol{\sigma}_{n+1}$ and $\Delta\mathbf{q}_{n+1}(\mathbf{X})$ are the point-wise corresponding increments computed at the present time step $[t_n, t_{n+1}]$.

Step III: Update

Finally the set of particles of Ω_n are updated to the new positions according to:

$$(\bar{\mathbf{X}}_{\mathsf{P}_i})_{n+1} = (\bar{\mathbf{X}}_{\mathsf{P}_i})_n + (\bar{\mathbf{u}}_{\mathsf{P}_i})_{n+1} \quad \forall \mathsf{P}_i \in \mathsf{P} \tag{5}$$

and the new particle configuration is determined as:

$$\Omega_{n+1} := \{(\bar{\mathbf{X}}_{\mathsf{P}_i})_{n+1}, (\bar{\mathbf{X}}_{\mathsf{P}_2})_{n+1}, \ldots, (\bar{\mathbf{X}}_{\mathsf{P}_{n_{part}}})_{n+1}\} \tag{6}$$

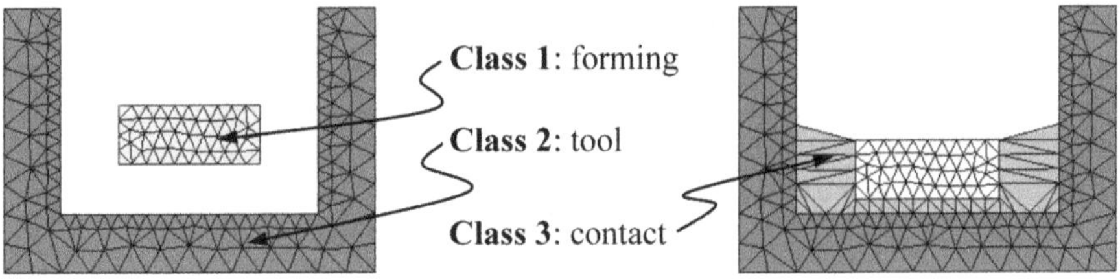

Fig. 3. Anticipating contact interface strategy

Then, the boundary of the new cloud of positions of the set of particles P is recognized via de corresponding alpha-shape strategy, and a new triangularization, determining the new spatial domain V_{n+1}, is performed. The algorithm proceeds to Step I of the next time step.

3 Contact Strategy: Anticipating Interface Mesh

One of the main difficulties found in standard contact algorithms in solid mechanics, is the identification of the interacting parts of two contacting bodies (master-slave based algorithms). The previously described PFEM setting provides a very interesting feature to be exploited in this sense: the possibility of anticipating the contact boundaries and of imposing the corresponding contact constraints in a diffuse manner, without the necessity of a precise identification of the contact topologies.

Let us consider, for illustration purposes, a forming process characterized by a forming material and some (elastic) tooling, amenable to experience mutual contact (see Figure 3). We can consider each of them as a specific class, constituted by its own cloud of particles. At every time step, a Delaunay triangularization is performed for every class and its boundary (in terms of the boundary particles) is recognized as indicated in Section 2 by the alpha-shape procedure.

Then, an additional triangularization is performed: the particles of the identified boundaries are defined as a new class (*the contact interface class*) and sent to the mesher that returns an interface mesh, which connects anticipating contacting particles of different classes. The value of the supplied alpha-shape value determines the maximum size of the resulting interface elements and, therefore, rules the degree of anticipation of the contacts.

The contact interface mesh constructed in this way, enjoys some specific properties:

- It is an interface mesh in the sense that there is no interior node (all the nodes are placed at the boundary). In consequence, all the computations done in that finite element are *naturally* condensed out to the boundary nodes.

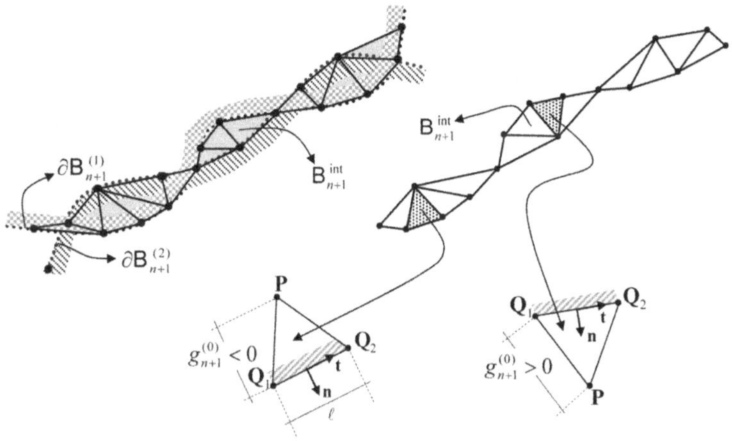

Fig. 4. Penalty strategy at the contact interface

- Typical properties of the Delaunay tessellation procedure ensure that the contact interface connects nearest particles in the contacting bodies.
- The identification of the contact topologies is trivially done via the mesh topology: potentially contacting nodes are those belonging to the same element of the mesh. This overcomes many of the difficulties and computational challenges found in classical contact recognition procedures.
- The contact condition, imposing that nodal gaps should be positive (no penetration), can be fulfilled in weak form in terms of strain measures defined at the elements of the interface mesh (see Section 3.1).

Generalization of this procedure to multiple contacting bodies is trivial. As for the imposition of the contact conditions there are several options. In next sections, a penalty strategy for this purpose is presented.

3.1 A Penalty Strategy at the Contact Interface

Let us consider the time step $[t_n, t_{n+1}]$ and two contacting boundaries, $\partial \mathsf{B}^{(1)}_{n+1}$ and $\partial \mathsf{B}^{(2)}_{n+1}$, and the resulting interface mesh (see Figure 4) defining the interface B^{int}_{n+1}.

We realize that every element of B^{int}_{n+1} has one node (node $\mathbf{P}$) placed on one of the contacting bodies, and two nodes, $\mathbf{Q}_1$ and $\mathbf{Q}_2$, on the other (see Figure 4b), which can be trivially identified. The *signed* distance of $\mathbf{P}$ to the segment $\mathbf{Q}_1 - \mathbf{Q}_2$, according to the normal[2] $\mathbf{n}$ of the boundary corresponding to this segment, is defined as the initial elemental normal gap:

[2]Information about the normals corresponding to the boundary segments is generated, during the boundary recognition procedure, for every body and transmitted to the boundary interface class.

$$g_{n+1}^{(e)^{(0)}} = (\mathbf{X_P} - \mathbf{X_{Q_1}}) \cdot \mathbf{n} = (\mathbf{X_P} - \mathbf{X_{Q_2}}) \cdot \mathbf{n} \quad \forall e \in \mathsf{B}_{n+1}^{int} \tag{7}$$

Notice that $g_{n+1}^{(e)^{(0)}} \geq 0$ corresponds to no initial penetration of node $\mathbf{P}$ on the opposite boundary. Therefore, negative values of the gap correspond to boundary penetrations.

Let us now define the *current* normal gap for the element e, at the iteration i of the linearization procedure, as:

$$\begin{aligned} g_{n+1}^{(e)^{(i)}}(\mathbf{u}_{n+1}^{(i)} &= g_{n+1}^{(e)^{(0)}} + \Delta g_{n+1}^{(e)^{(i)}} \\ &= g_{n+1}^{(e)^{(0)}} + \underbrace{\left|g_{n+1}^{(e)^{(0)}}\right|(\mathbf{n} \cdot \nabla^S \mathbf{u}_{n+1}^{(e)^{(i)}} \cdot \mathbf{n})}_{\Delta g_{n+1}^{(i)}} \quad \forall e \in \mathsf{B}_{n+1}^{int} \end{aligned} \tag{8}$$

which involves the iterative displacements[3] $\mathbf{u}_{n+1}^{(e)^{(i)}}$ at the element e. Now, let us define a strain measure associated to that gap as:

$$\tilde{\varepsilon}_{n+1}^{(e)} = \frac{g_{n+1}^{(e)}}{\ell} \tag{9}$$

where ℓ stands for the typical local particle distance (see Figure 4). The contact condition for boundaries, $\partial \mathsf{B}_{n+1}^{(1)}$ and $\partial \mathsf{B}_{n+1}^{(2)}$, can be immediately written as the positive character of the strain measure $\tilde{\varepsilon}(\mathbf{u}_{n+1})$:

$$g_{n+1}^{(e)} \geq 0 \Leftrightarrow \tilde{\varepsilon}_{n+1}^{(e)} \geq 0 \quad \forall e \in \mathsf{B}_{n+1}^{int} \tag{10}$$

In a variational context, condition (10) can be imposed via a Lagrange multiplier procedure:

$$\begin{aligned} &\Pi(\mathbf{u}_{n+1}, \lambda_{n+1}) = \Pi_{\mathsf{B}_{n+1}^{(1)}}(\mathbf{u}_{n+1}) + \Pi_{\mathsf{B}_{n+1}^{(2)}}(\mathbf{u}_{n+1}) + \int_{\mathsf{B}_{n+1}^{int}} \lambda_{n+1} \tilde{\varepsilon}(\mathbf{u}_{n+1})\, d\Omega \\ &\delta\Pi(\mathbf{u}_{n+1}, \lambda_{n+1}) = 0; \quad g(\mathbf{u}_{n+1}) \geq 0; \quad \lambda_{n+1} \geq 0; \quad \lambda_{n+1} g(\mathbf{u}_{n+1}) = 0 \end{aligned} \tag{11}$$

where $\Pi_{\mathsf{B}_{n+1}^{(1)}}(\mathbf{u}_{n+1})$ and $\Pi_{\mathsf{B}_{n+1}^{(2)}}(\mathbf{u}_{n+1})$ stand for the original minimizing functionals for the contacting bodies $\mathsf{B}_{n+1}^{(1)}$ and $\mathsf{B}_{n+1}^{(2)}$, and λ_{n+1} are the corresponding Lagrange multipliers. Now, all the typical strategies for solving the problem in equations (11) can be used. In particular, penalty-based strategies will be recovered by imposing a specific format to the Lagrange multipliers as:

$$\begin{aligned} &\lambda_{n+1}^{(e)} = \tfrac{1}{2} E^{int} \tilde{\varepsilon}_{n+1}^{(e)} \\ &E^{int} = \begin{cases} 0 \; for \; \tilde{\varepsilon}_{n+1}^{(e)} \geq 0 \\ K \; for \; \tilde{\varepsilon}_{n+1}^{(e)} < 0 \end{cases} \quad \forall e \in \mathsf{B}_{n+1}^{int} \end{aligned} \tag{12}$$

[3]Superindex $(\cdot)^S$ stands for the *symmetric part* of $(\cdot)$.

where $K > 0$ is the penalty parameter. The resulting problem is formally equivalent to the introduction of an elastic-type constitutive law, $\sigma - \tilde{\varepsilon}$, in the normal direction, defined in terms of the parameter, E^{int}, as a *pseudo*-elastic modulus:

$$\boldsymbol{\sigma}_{n+1}^{(e)} = \underbrace{\tilde{\sigma}_{n+1}^{(e)} \mathbf{n} \otimes \mathbf{n}}_{contact\ stress}$$
$$\tilde{\sigma}_{n+1}^{(e)} = E^{int} \tilde{\varepsilon}_{n+1}^{(e)} \quad \forall e \in \mathsf{B}_{n+1}^{int} \tag{13}$$

and *extending the boundary value problem in equations* (1) *to the interface domain* B_{n+1}^{int}.

Inclusion of Frictional Effects

Frictional effects can be now included in a very simple manner. For instance, on the basis of a simple Coulomb model, additional tangential stresses can be considered in the constitutive model for every element of the interface (see Figure 5):

$$\boldsymbol{\sigma}_{n+1}^{(e)} = \underbrace{\tilde{\sigma}_{n+1}^{(e)} \mathbf{n} \otimes \mathbf{n}}_{contact\ stress} + \underbrace{2\tilde{\tau}_{n+1}^{(e)} (\mathbf{t} \otimes \mathbf{n})^S}_{frictional\ stress}$$
$$\tilde{\sigma}_{n+1}^{(e)} = E^{int} \tilde{\varepsilon}_{n+1}^{(e)} \quad \forall e \in \mathsf{B}_{n+1}^{int} \tag{14}$$
$$\tilde{\tau}_{n+1}^{(e)} = \mu |\tilde{\sigma}_{n+1}^{(e)}| \, sign(\tilde{\gamma}_{n+1}^{(e)}) \quad (\tilde{\gamma}_{n+1}^{(e)} = 2\mathbf{t} \cdot \nabla^S \mathbf{u}_{n+1}^{(e)} \cdot \mathbf{n})$$

where $\mathbf{t}$ stands for the tangent vector associated to the normal $\mathbf{n}$ in anticlockwise sense.

Numerical Treatment

Resolution of the variational problem (11), in the context of the finite element method, leads to the following expression for the equivalent nodal forces, $\mathbf{F}_i^{cont}$, at the nodes i of the interface mesh, due to contact/friction actions:

$$\delta \Pi = 0 \Rightarrow$$
$$\mathbf{F}_i^{cont} = - \int_{\mathsf{B}_{n+1}^{int}} \nabla N_i \cdot \boldsymbol{\sigma}_{n+1} \, d\Omega \quad \forall i \in \partial \mathsf{B}_{n+1}^{(1)} \cup \partial \mathsf{B}_{n+1}^{(2)} \tag{15}$$

Besides, it can be readily proven that the elemental counterparts, $\mathbf{F}_i^{cont^{(e)}}$, of those contact/friction forces in a given element e, constitute a set of self-equilibrated forces:

$$\sum_{i=1}^{i=3} \mathbf{F}_i^{cont^{(e)}} = - \int_{\mathsf{B}_{n+1}^{int^{(e)}}} \nabla N_i^{(e)} \cdot \boldsymbol{\sigma}_{n+1}^{(e)} \, d\Omega = \mathbf{0} \tag{16}$$

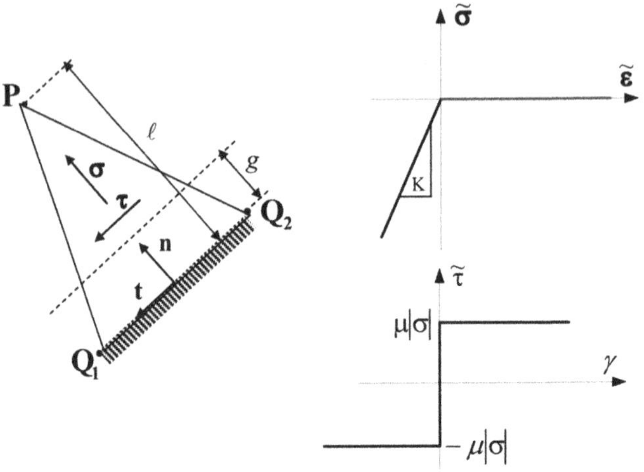

Fig. 5. Contact/friction effects: interface constitutive model

which proves the self-equilibrated character of those forces as action-reactions at the contacting boundaries $\partial \mathsf{B}_{n+1}^{(1)}$ and $\partial \mathsf{B}_{n+1}^{(2)}$, i.e.:

$$\sum_{j \in \partial \mathsf{B}_{n+1}^{(1)}} \mathbf{F}_j^{cont} + \sum_{j \in \partial \mathsf{B}_{n+1}^{(2)}} \mathbf{F}_j^{cont} = \sum_{e \in \mathsf{B}_{n+1}^{int}} \sum_{i=1}^{i=3} \mathbf{F}_i^{cont^{(e)}} = \mathbf{0} \tag{17}$$

Finally, in view of the structure of the problem in equations (11), the contact/friction nodal forces, $\mathbf{F}_i^{cont}$, once computed at every node of the interface mesh B_{n+1}^{int} according to equation (15), can be formally added to the original problem (1), for every contacting body, as external prescribed point forces acting on the boundary $\partial_\sigma V_{n+1}$.

In this way, the contact/friction problem gets reduced to computing nodal/friction forces at all the generated interface meshes, and incorporating them, as external point forces, at the boundaries of the contacting classes. *No additional degrees of freedom are involved.*

4 Representative Examples

In order to illustrate the modeling capacities of the methodology described so far, in next sections some applications to solid mechanics problems are presented.

4.1 Flexible Spring with Multiple Self-contacts

In Figure 6 an elastic flexible spring (Figure 6a) is pushed down vertically as to impose multiple contacts, as displayed in Figures 6b and 6c. In the figures,

(a)

(b)

(c)

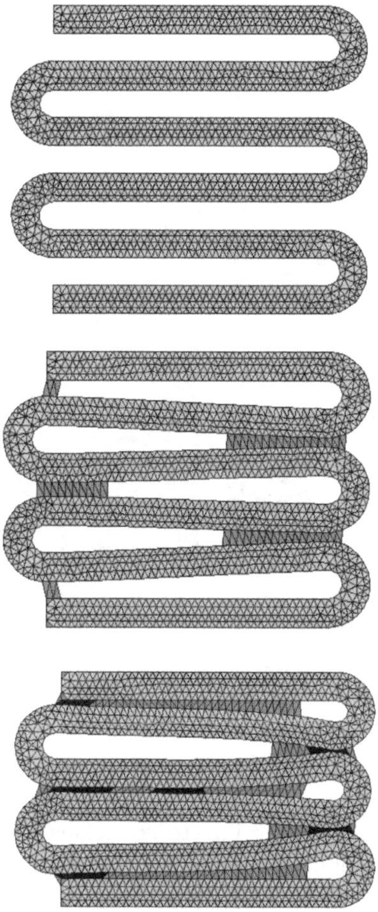

Fig. 6. Flexible spring: (a) initial configuration, (b) and (c) shrunk deformed configurations displaying the contact interfaces

the contact interface mesh is artificially amplified, for visualization purposes, and the contacting bodies are shrunk in correspondence.[4] The actual contacts, keep very little gaps that would not allow that visualization. Two facts should be noticed there:

- The contact interface mesh only affects a limited part of the boundary. This is ruled by the value of the alpha shape parameter precluding interface elements larger than the specified size.

[4]This criterion will be kept from now on, to allow visualizing the contact interface mesh.

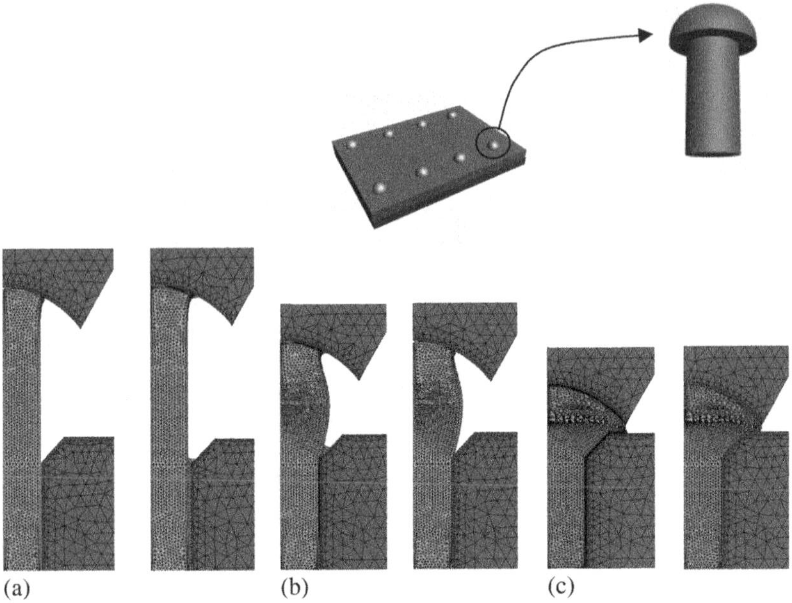

Fig. 7. Rivet forming. Top: Riveting parts. Bottom: (a), (b) and (c) deformed finite element mesh at different stages of the forming process. At every stage: actual mesh (left) and shrunk mesh (right), showing the contact interface is displayed

- Only few elements of the contact interface mesh are *active* in the sense that $g_{n+1}^{(e)} = 0 \Leftrightarrow \tilde{\varepsilon}_{n+1}^{(e)} = 0$ and, therefore, there is a positive reaction at the contact points according to equations (12) and (13). They are displayed in grey in Figure 6.

This example displays that multiple self-contacts, as the ones occurring in this problem, are trivially solved by the proposed procedure.

4.2 Riveting Process

In Figure 7, the geometrical description of a metallic rivet, and the deformation of the finite element mesh (axisymmetric cross-section) for different stages of the forming simulation process, are presented.

This is a very representative example of application of the proposed methodology to a variety of *metal forming* processes.[5] It is remarkable the capacity of the method to handle large strains in the deformed body, minimizing the angular distortions of the resulting finite element mesh. This is an advantageous property of the Delaunay triangulation based meshing procedure.

[5]Certainly, the reliability of the simulation requires the use of appropriate constitutive models, typically J_2 plasticity models and large strain kinematics.

(a)
(b)
(c)
(d)
(e)

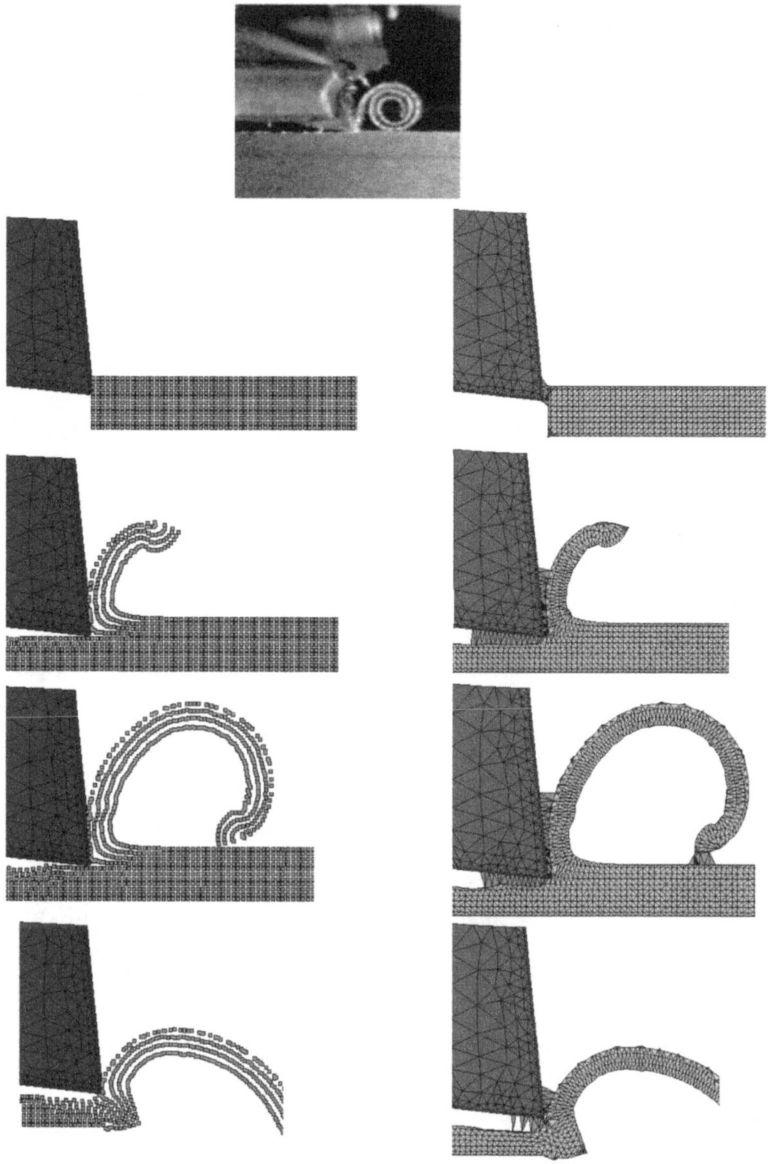

Fig. 8. Machining process: (a) chip forming process, (b)–(e) particle configurations (left: actual particle configurations, right: shrunk finite element meshes displaying the anticipating contact interfaces)

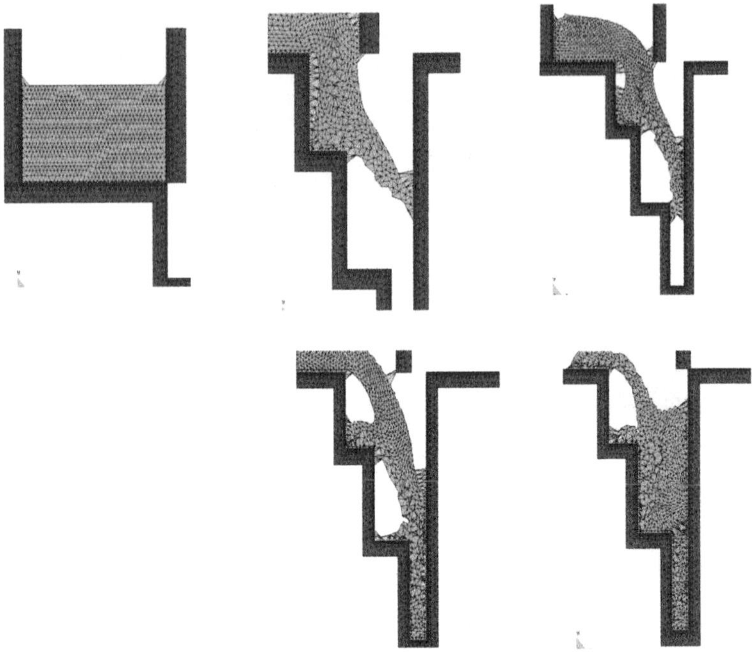

Fig. 9. Filling of a two-stepped cavity with powder material: finite element meshes displaying the contact interfaces at different stages

4.3 Machining Process

In Figure 8, simulation of an orthogonal cutting process on a metallic part is considered.

This is, again, a very illustrative example of the proposed methodology potential to handle challenging solid mechanics problems. In this case, the formation of the chip, very relevant for the quality of the mechanical process, generates a new boundary at the cutting tool edge. The alpha shape method automatically generates the new boundary as the deformed size at the elements, close to the cutting edge, exceeds the alpha-shape tolerance. The contact/friction model allows sliding the formed chip on the upper surface of the machined part. Both the boundary recognition and the imposition of the dynamic contacts are automatically done by the mesher.

4.4 Powder Filling Process

There are many interesting solid mechanics processes, involving flows of granular materials, where PFEM methods can be advantageously applied. Powder metallurgy manufacturing processes or casting processes are typical examples.

In Figure 9 the numerical simulation of a powder filling process is presented using a specific solid-type constitutive model (Oliver et al., 1996; Cante et al., 2005). The feeder moves horizontally, from left to right, dropping the powder into a two steps cavity. The goal of the analysis is to obtain, through a physically meaningful simulation of the process, the density distribution of the powder at the end of the process. This type of powder filling processes has been often modelled by the so called discrete *element methods*, on the basis of models made of a finite number of interacting balls or cylinders representing the powder grains (Wu et al., 2004; Coube et al., 2005). However, the large number of particles involved in the physical process translates into unaffordable computations when a realistic modelling is intended.

Here, the powder is treated as a continuum medium, ruled by an elastoplastic constitutive model, and modelled via the PFEM by tracking a limited number of the constituent particles. In the figure it can be observed as the powder/tool contact conditions are correctly imposed by the contact interface (also displayed in the figure). In addition, the alpha shape method allows reproducing the complex dynamics of the generated free surface of the powder during the filling process.

5 Concluding Remarks

In the previous sections, PFEM techniques applied to solid mechanics problems have been explored. In addition to their appealing properties, already explored for fluid mechanics problems, some additional advantages can be obtained when they are applied to solids i.e.:

- Delaunay tessellation techniques, for constructing finite element meshes on the basis of the selected cloud of points, minimize the angular distortion effects of the resulting elements, thus providing additional robustness to the modelling in problems dominated by large strains. The relatively low computational cost (Calvo et al., 2003) of those techniques allows frequent remeshing without leading to unaffordable computational costs.
- Alpha-shape techniques for boundary recognition purposes, constitute a powerful tool in those solid problems where new physical boundaries are generated (for instance, cutting processes) or when there is a continuous creation and disappearance of boundaries, as a natural part of the physics of the problem (i.e. powder filling processes).
- Contacting boundaries problems can be solved by constructing *anticipating contact interfaces* as it is presented in Section 3. Construction of those contact interfaces becomes an easy and automatic task by using, Delaunay tessellation techniques on the basis of particles of the contacting boundaries. Typical highly computationally demanding techniques, for recognizing point/segment contacts, can then be overcome by imposing the contact conditions on the generated contact interfaces by using, for instance, penalty techniques (Section 3.1). Then, friction effects can be trivially implemented (Section 3.1.1). This confers larger computational efficiency and robustness to simulation of problems involving a large number of contacts.

Although the formulation, and the corresponding examples, has been here restricted to 2D cases, extension to 3D problems seems straightforward and in essence the benefits of PFEM should be similar to the ones mentioned above.

Acknowledgements

Financial support from the Spanish Ministry of Science and Technology trough grants DPI203-00629 and DPI2004-07666-C02-02 and the Catalan Department of Universities, Research and Society of Information (DURSI) through grant 2005SGR0084 are gratefully acknowledged.

References

1. Belytschko T, Krongauz Y, Organ D, Fleming M, Krysl P (1996) Meshless methods: An overview and recent developments. Comput. Meth. in Appl. Mech. Engng. 139:3-47
2. Belytschko T, Lu YY, Gu L (1994) Element-Free Galerkin Methods. International Journal for Numerical Methods in Engineering 37:229-256
3. Calvo N, Idelsohn SR, Onate E (2003) The extended Delaunany tessellation. Engineering Computations 20:583-600
4. Cante JC, Oliver J, Gonzalez C, Calero JA, Benitez F (2005) On numerical simulation of powder compaction processes: powder transfer modelling and characterisation. Powder Metallurgy 48:85-92
5. Coube O, Cocks ACF, Wu CY (2005) Experimental and numerical study of die filling, powder transfer and die compaction. Powder Metallurgy 48:68-76
6. George PL (1991) Automatic Mesh Generation, Applications to Finite Methods, New York, Wiley
7. Hughes TJR (2000) The Finite Element Method: Linear Static and Dynamic Finite Element Analysis, Mineola, NY, Dover Publications Inc
8. Idelsohn SR, Onate E, Calvo N, Del Pin F (2003a) The meshless finite element method. Int. J. Num. Meth. Engng. 58:893-912
9. Idelsohn SR, Onate E, Del Pin F (2003b) A Lagrangian meshless finite element method applied to fluid-structure interaction problems. Computers & Structures 81:655-671
10. Idelsohn SR, Onate E, Del Pin F (2004) The particle finite element method: a powerful tool to solve incompressible flows with free-surfaces and breaking waves. Int. J. Num. Meth. Engng. 61:964-989
11. Lohner R, Sacco C, Onate E, Idelsohn SR (2002) A finite point method for compressible flow. Int. J. Num. Meth. Engng. 53:1765-1779
12. Malvern L (1969) Introduction to the Mechanics of a Continuous Medium. Englewood Cliffs, New Jersey, USA, Prentice-Hall
13. Oliver J, Oller S, Cante JC (1996) A plasticity model for simulation of industrial powder compaction processes. Int. J. Solids and Structures 33:3161-3178
14. Onate E, Idelsohn S (1998) A mesh-free finite point method for advective-diffusive transport and fluid flow problems. Comput. Mechanics 21: 283-292.

15. Onate E, Idelsohn SR, Zienkiewicz OC, Taylor RL (1996) A finite point method in computational mechanics. Applications to convective transport and fluid flow. Int. J. Num. Meth. Engng 39:3839-3866
16. Wu CY, Cocks ACF (2004) Flow behaviour of powders during die filling. Powder Metallurgy 47:127-136
17. Xu XL, Harada K (2003) Automatic surface reconstruction with alpha-shape method. Visual Computer.19:431-443
18. Zienkiewicz OC, Taylor RL (2000) The Finite Element Method. Oxford UK, Butterworth-Heinemann

Micro-Meso-Macro Modelling of Composite Materials

P. Wriggers and M. Hain

Institute of Mechanics and Computational Mechanics
University of Hannover,
Appelstr. 9a, 30167 Hannover, Germany
wriggers@ibnm.uni-hannover.de

Summary. Multi-scale models can be helpful in the understanding of complex materials used in engineering practice. Applications related to this class of problems covers different length scales in the range from μm to m. Using the concept of representative volume elements (RVE), the theoretical background is discussed in this contribution as well as the numerical treatment of the resulting three-dimensional RVEs.

The developed methodology is applied to a specific engineering material which is concrete. This construction material has to be investigated on three different scales: the hardened cement paste (hcp), the mortar and finally the concrete. Here, a successive two-stage approach is followed in which first the multi-scale model of hcp and mortar is applied. The resulting homogenization can then be used in the next step for a multi-scale mortar-concrete model.

At the micro-scale of hcp, a finite element mesh based on a three-dimensional computer-tomography with different constitutive equations for the three parts unhydrated residual clinker, pores and hydrated products is introduced. With respect to the finite element solution, homogenization techniques are used in order to calculate effective elastic material properties.

The constitutive equations at the micro-scale contain inelastic parameters, which cannot be obtained through experimental testings. Therefore, one has to solve an inverse problem which yields the identification of these properties. For computational efficiency and robustness, a combination of the stochastic genetic algorithm and the deterministic LEVENBERG-MARQUARDT method is used. In order to speed-up the computation time significantly, all calculations are distributed automatically within a network environment.

Inelastic behavior occurs when the micro-structure hcp is filled partly with water and a freezing process takes place. A constitutive model for ice is applied to the water filled parts of the micro-structure. The expansion of the ice leads to damage in the hcp which is associated with inelastic material behavior. If such a calculation is performed for different moisture and temperatures, a correlation between moisture, temperature and the inelastic material behavior can be obtained. The effective constitutive equation of hcp will serve as a basis for a multi-scale mortar-concrete model.

Eugenio Oñate and Roger Owen (eds.), Computational Plasticity, 105–122.

1 Introduction

A deeper understanding of the constitutive behavior of complex materials such as concrete can be achieved on one hand by experimental investigations and on the other hand by numerical simulations. While the experimental techniques are applied successfully since several decades, the use of numerical simulation techniques is quite young. This is due to the fact that the underlying three-dimensional models are very complex and need considerable computer power which only in the last decade has reached a sufficient state. The advantage of a numerical simulation is that one can look inside a specimen during the loading process and also is able to resolve fast processes on a different time scale. Figure 1 depicts the multi-scale modeling process which can be used for a material like concrete. The underlying nano-scale is not considered in this contribution.

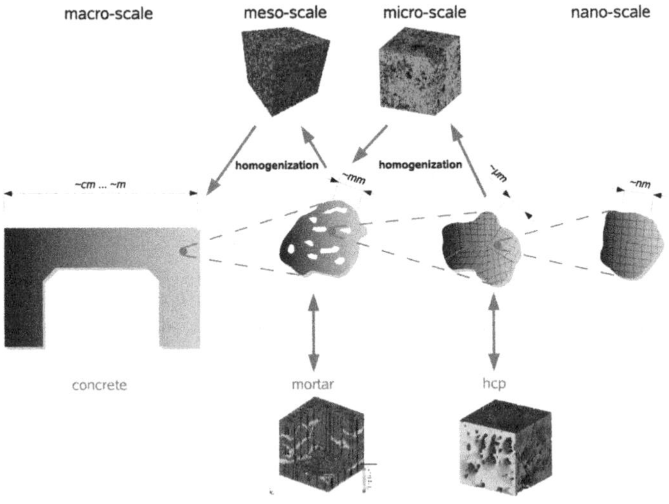

Fig. 1. Multi-scale model of concrete

Within the above mentioned multi-scale approach, different three-dimensional mechanical models are applied on each length scale in order to describe the constitutive behavior on that scale. These are called representative volume elements (RVEs). Normally, the material on a specific length scale consists of different phases which have to be taken into account in order to characterize the material with sufficient accuracy.

The RVE is then subjected to different loading conditions which lead to a material response. Based on these results a homogenization process can be initiated to describe the material behavior averaged over the whole RVE. The resulting homogenized constitutive equation is then applied within the the next scale to model the constituents of the RVE belonging to that scale.

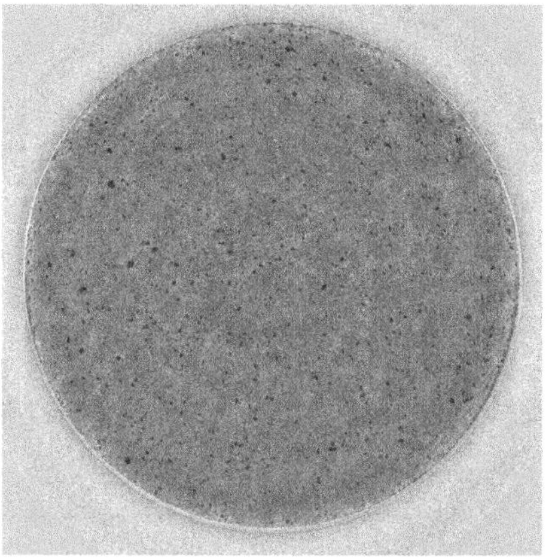

Fig. 2. Cut through of a CT-scan ($1750 \times 1750\mu m$)

In this contribution, a multi-scale model for hardened cement paste (hcp) is introduced. The scale of the superior system is denoted meso-scale, the underlying microscopic system of hcp micro-scale [17], [5]. The homogenized properties of hcp serves as a basis for the constitutive model of mortar at the meso-scale.

2 Micro-structure of Hardened Cement Paste (hcp)

The FE-analysis of hcp requires the discretization of the micro-structure. Here, the geometry on the micro-scale is obtained from a three-dimensional computer tomography (CT) which provides a spatial resolved distribution of the density. In this contribution, CT-scans of several specimens with an edge length of $1750\mu m$ and a resolution of $1\mu m$ have been performed by the *Bundesanstalt für Materialforschung und -prüfung* in Berlin, Germany.

The CT-scans have been applied for a hcp with the water-cement ratio of $w_c = 0.45$ and a degree of hydration of $h = 0.945$. In Fig. 2 a cut through of such a CT-scan is shown. The unhydrated residual clinker is black and the hydrated part is middle gray colored. One can see the accumulation of pores, which are marked in light gray. Each CT-file contains 1750^3 data points (each corresponds to $1\mu m^3$) and allocates totally 5GB of memory.

For the mechanical analysis, the surrounding synthetic tube is not relevant. Hence, only the interior part of the CT-scan is used for further mechanical simulations which contains 1008^3 data points (see Fig. 3).

Due to noise, a median filter has to be applied to the raw data. Based on the theory of Powers and coworkers [13], [12] the fractional volume of the pores c_p and the unhydrated residual clinker c_u are calculated

$$c_p = \frac{w_c - 0.36h}{w_c + 0.32}\,, \quad c_u = \frac{0.32(1 - c_p)}{w_c + 0.32}\,. \tag{1}$$

Fig. 3. Cut through of the raw data ($1008 \times 1008\mu m$)

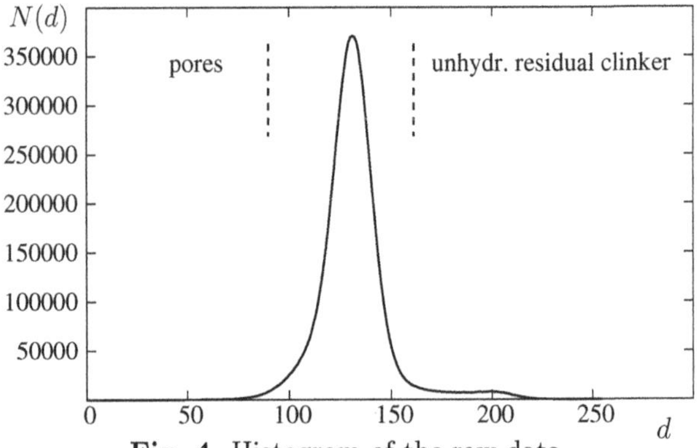

Fig. 4. Histogram of the raw data

In order to distinguish unhydrated residual clinker, hydration products and porosity a histogram is generated, where $N(d)$ describes the number of voxels at a specific density d (see Fig. 4).

Threshold values for the pores and the unhydrated residual clinker are calculated with respect to (1). Hence, the raw data can be segmented in the hydrated part, unhydrated residual clinker and porosity. In Fig. 5 a cross section of the filtered data is shown which is obtained from the raw data (see Fig. 3). The hydrated part has a volume fracture of 84Vol.%, the pores of approximately 14Vol.% and the unhydrated residual clinker of roughly 2Vol.%.

The micro-structure of hcp can also be generated by simulating the chemical reactions of the hydration. Result of such a simulation is an artificial micro-structure with different phases. The chemical reaction of the hydration can be simulated with respect to a cellular automata, as the programme from NIST does [6]. Herein, the whole micro-structure is divided into topological, geometrical and physical uniform cells. Each cell has an individual state, corresponding to its current property. A set of rules calculate the new state for each cell at time $n+1$. The result of this calculation depends on the state at time n, the states of the neighbors and a random part. The rules are applied at the same time to all cells.

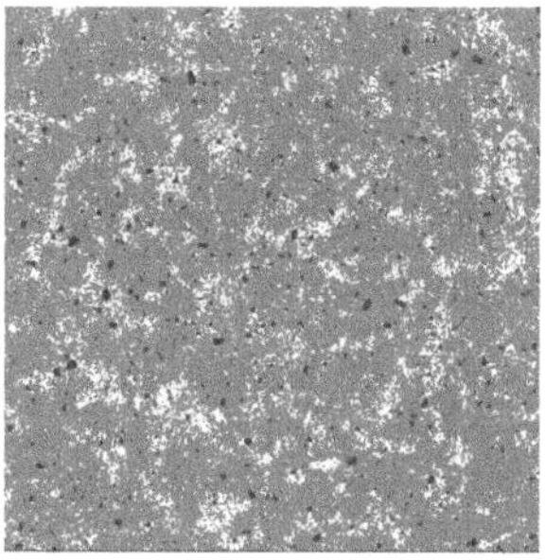

Fig. 5. Cross section of the filtered data ($1008 \times 1008\mu m$)

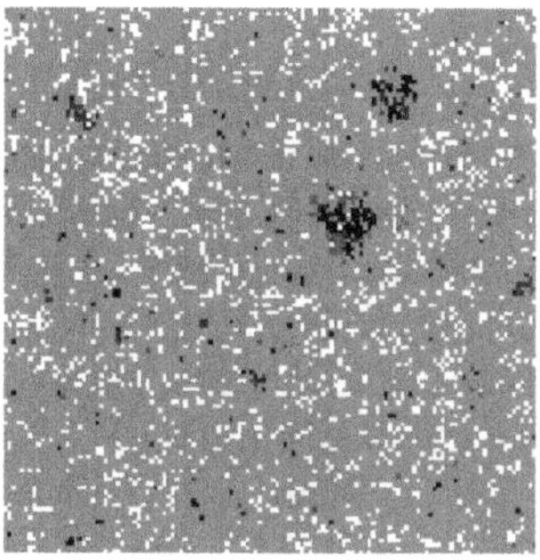

Fig. 6. Simulated micro-structure ($128 \times 128\mu m$)

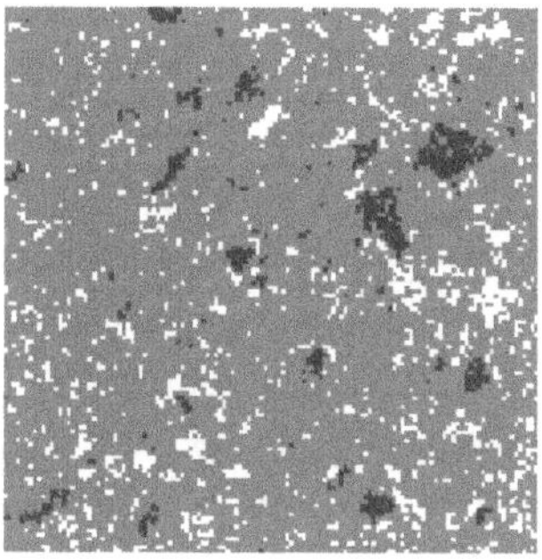

Fig. 7. CT-based micro-structure ($128 \times 128\mu m$)

If the hydration is simulated with the above mentioned cellular automata, one obtains a different geometry as those obtained from CT-scans, compare Figs. 6 and 7. Due to the random based rules of the programme, the porosity is evenly distributed. There are no accumulations of pores as the corresponding CT-scan depicts, but for a reliable simulation of frost heave a proper geometry of the pores is necessary. Therefore, only CT-based micro-structures are applied within the further calculations.

3 Constitutive Equations

Numerical results on the micro-scale are obtained by use of the finite element method which is applied to discretize the representative volume elements (RVEs). The RVEs are generated directly from the filtered CT-data and have different material properties for each phase. In this contribution RVEs with an edge length of $64\mu m$ are used. Since each voxel is represented by one finite element this leads to a 64^3 finite element mesh. Hence each RVE has about 800 000 degrees of freedoms (DOFs).

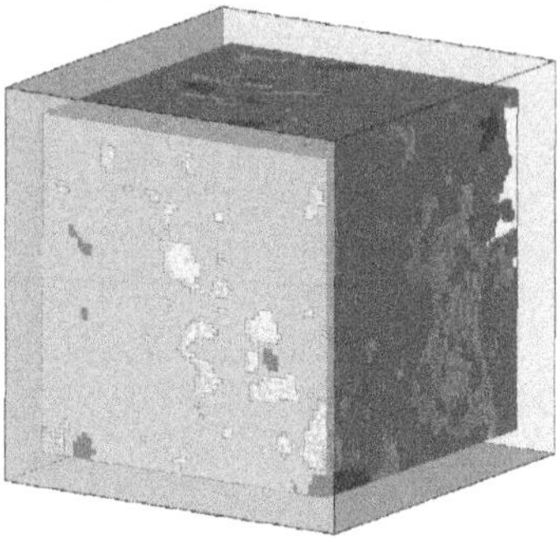

Fig. 8. Embedded RVE

Due to the different properties classical constant strain boundary conditions – leading to Voigt bounds – will cause problems at the boundary. Constant stress boundary conditions – leading to Reuss bounds – are also not be used. Instead the RVEs are embedded in a matrix of average stiffness, see Fig. 8. This corresponds to the self-consistency method from analytical micro-mechanics and is sometimes referred to as window-technique.

Material properties for the elastic behavior of the different phases from hcp can directly be taken from the literature [1], [2], [3]. To avoid numerical problems and further mechanical influences the pores are described using an elastic material of negligible stiffness. With respect to the minor volume fraction, averaged properties can be chosen for the unhydrated phase.

Table 1. Elastic properties of the phases

phase	E $\frac{N}{mm^2}$	ν –
unhydrated phase	$\approx$ 132 700	0.30
hydrated phase	20 000	0.21
pores	1	0.00

Experimental tests of hcp show that the inelastic effects of hcp are related to damage behavior combined with visco-plastic effects. Due to that a model for the hydrated phase is selected which consists of a visco-plastic constitutive relation of the classical PERZYNA-type combined with a damage model.

Due to mechanical loading, cracks occur at the nano-scale. These nanocracks are statistically distributed and oriented, thus isotropic damage can be chosen for the micro-structure. Stress-strain relations for isotropic damage can be found in [9], here we select

$$\boldsymbol{\sigma} = (1 - D)\mathbb{C} : \boldsymbol{\epsilon} \,. \tag{2}$$

The scalar variable D describes the damage and $\mathbb{C}$ refers to the material tensor of the undamaged elastic material. An evolution law for the damage variable

$$\dot{D} = \Delta\xi \frac{\partial S(\epsilon^{\mathrm{eq}})}{\partial \epsilon^{\mathrm{eq}}} \tag{3}$$

and a damage surface $S(\epsilon^{\mathrm{eq}})$ are introduced. The damage surface determines, whether the damage increases or not. It is defined by an exponential relation and depends on the equivalent strain ϵ^{eq}. The parameter a, b and c constitute the material properties of this model

$$S(\epsilon^{\mathrm{eq}}) := 1 - \exp\left[-\left(\frac{\epsilon^{\mathrm{eq}} - a}{b}\right)^{c}\right] - D \leqslant 0\,. \tag{4}$$

For the visco-plastic part, a penalty formulation $\mathcal{P}$ is introduced, see e.g. [14], based on the second fundamental theorem of thermodynamics. Here, the scalar variable η describes the viscosity, $\phi(f)$ denotes a penalty-function of power type

$$\begin{aligned} \mathcal{P} &= \boldsymbol{\sigma} : \dot{\boldsymbol{\epsilon}}^{\mathrm{pl}} + \tfrac{1}{\eta}\phi(f)\,, \\ \phi(f) &= \begin{cases} 0 & ; f \leq 0 \\ \frac{1}{m+1} f^{m+1} & ; f > 0 \end{cases}\,. \end{aligned} \tag{5}$$

In order to keep the constitutive model at the micro-scale as simple as possible, a DRUCKER-PRAGER model with the yield surface f is chosen. Therefore, different behavior of hcp in compression and tension can be described

$$f := \alpha \mathrm{tr}\, \boldsymbol{\sigma} + \|\mathrm{dev}\, \boldsymbol{\sigma}\| - \sqrt{\tfrac{2}{3}} k_f \leq 0\,. \tag{6}$$

Within the associated theory of visco-plasticity, the evolution equation for the inelastic strains $\dot{\boldsymbol{\epsilon}}^{\mathrm{pl}} = \Delta\dot{\lambda}\, \dfrac{\partial f}{\partial \boldsymbol{\sigma}}$ has to be integrated. Here we apply an EULER backward integration leading to

$$\boldsymbol{\epsilon}^{\mathrm{pl}}_{n+1} = \boldsymbol{\epsilon}^{\mathrm{pl}}_{n} + \Delta\lambda \left.\frac{\partial f}{\partial \boldsymbol{\sigma}}\right|_{n+1}\,. \tag{7}$$

The increment $\Delta\lambda = \frac{\Delta t}{\eta}\phi^{+}$ is locally, at every GAUSS point, calculated by a NEWTON-scheme

$$\Delta\lambda^{k+1} = \Delta\lambda^{k} - \left[\frac{\partial G^{k}}{\partial \Delta\lambda^{k}}\right]^{-1} G^{k}\,, \tag{8}$$

where G^k describes the nonlinear equation resulting from the evaluation of (5) and (6)

$$G^{k} = \frac{\eta \Delta\lambda^{k}}{\Delta t} - \left(f^{\mathrm{trial}} - 9\alpha^{2}\kappa\Delta\lambda^{k} - 2\mu\Delta\lambda^{k}\right)^{m} \stackrel{!}{=} 0\,. \tag{9}$$

For computational efficiency, the above mentioned set of constitutive equations is linearized consistently.

4 Homogenization

In order to determine effective elastic material properties for each RVE, homogenization is needed and the effective material tensor has to be calculated. The effective material tensor $\mathbb{C}_{\text{eff}}$ maps the average of the strains $\langle \boldsymbol{\epsilon} \rangle$ (effective strains) onto the average of the stress $\langle \boldsymbol{\sigma} \rangle$ (effective stresses) by

$$\langle \boldsymbol{\sigma} \rangle = \mathbb{C}_{\text{eff}} : \langle \boldsymbol{\epsilon} \rangle \,, \tag{10}$$

where $\langle \bullet \rangle$ denotes the volume average within the RVE, Ω being the volume of the RVE.

$$\langle \bullet \rangle = \frac{1}{|\Omega|} \int_{\Omega} \bullet \, \mathrm{d}\Omega \,. \tag{11}$$

In order to calculate the effective stresses and strains, one has to define boundary conditions and solve the boundary value problem on the micro-scale. The latter will be performed by using finite elements. Since the application of a pure stress field to the RVE leads to some difficulties regarding the suppression of rigid body modes, we apply here only pure displacement boundary conditions (constant stress state). Subsequently, the volume average of the strain can be evaluated from the finite element solution of the RVE by

$$\langle \boldsymbol{\epsilon} \rangle = \sum_{e=1}^{n_e} \frac{1}{\Omega_e} \int_{\Omega_e} \boldsymbol{\epsilon}_h \, d\Omega \tag{12}$$

where $\boldsymbol{\epsilon}_h$ is the strain computed from the finite element solution, Ω_e the volume of a finite element and n_e the total number of finite elements of the RVE. The same relation holds for the computation average of the stresses.

Once the effective strains and stresses are known the effective material tensor $\mathbb{C}_{\text{eff}}$ can be computed within the homogenization step.
In this this step a linear elastic effective material of

$$\boldsymbol{\sigma}(\langle \boldsymbol{\epsilon} \rangle) = \lambda_{\text{eff}} \operatorname{tr} \langle \boldsymbol{\epsilon} \rangle \, \mathbf{1} + 2\mu_{\text{eff}} \langle \boldsymbol{\epsilon} \rangle$$

is chosen. A least square functional is introduced to fit the averaged stresses with those obtained from the effective constitutive equation $\boldsymbol{\sigma}(\langle \boldsymbol{\epsilon} \rangle)$

$$\Pi := \| \, \langle \boldsymbol{\sigma} \rangle - \boldsymbol{\sigma}(\langle \boldsymbol{\epsilon} \rangle) \, \| \rightarrow \min \,. \tag{13}$$

The differentiation of this relation with respect to the effective material properties yields a symmetric linear equation for the two LAME constants λ^{eff} and μ^{eff}

$$\begin{pmatrix} 2\operatorname{tr} \langle \boldsymbol{\epsilon} \rangle \operatorname{tr} \langle \boldsymbol{\sigma} \rangle \\ 4 \langle \boldsymbol{\sigma} \rangle : \langle \boldsymbol{\epsilon} \rangle \end{pmatrix} = \begin{bmatrix} 6\operatorname{tr}^2 \langle \boldsymbol{\epsilon} \rangle & 4\operatorname{tr}^2 \langle \boldsymbol{\epsilon} \rangle \\ 4\operatorname{tr}^2 \langle \boldsymbol{\epsilon} \rangle & 8 \langle \boldsymbol{\epsilon} \rangle : \langle \boldsymbol{\epsilon} \rangle \end{bmatrix} \cdot \begin{pmatrix} \lambda_{\text{eff}} \\ \mu_{\text{eff}} \end{pmatrix} \,. \tag{14}$$

Due to the randomly distributed phases within the micro-structure of hcp, a Monte Carlo method was selected to compute the average values for the constitutive parameters. Hence a sufficiently large number of different RVEs has

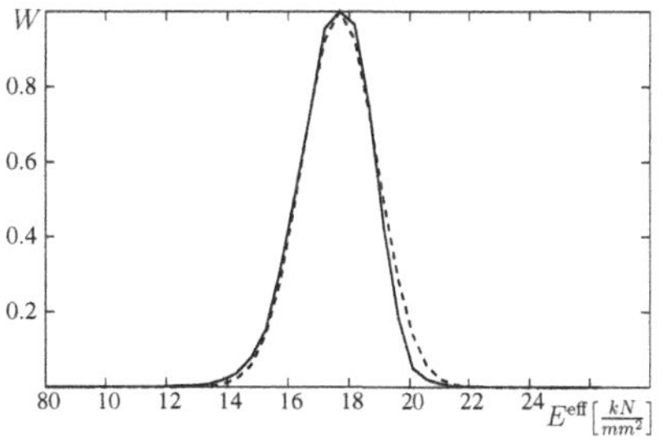

Fig. 9. Probability density of E_{eff}

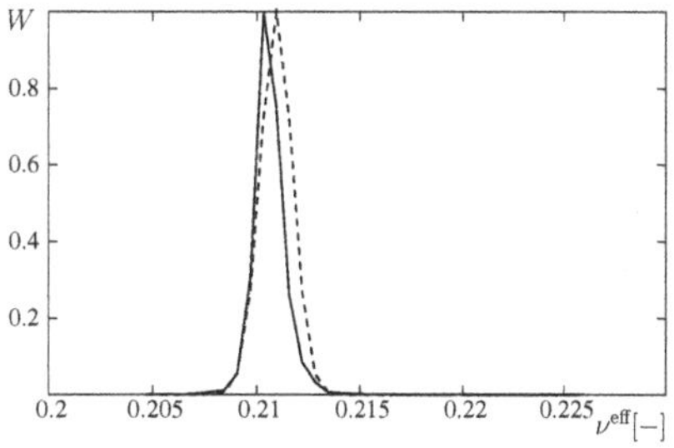

Fig. 10. Probability density of μ_{eff}

to be investigated. Each homogenization yields the effective material properties λ_{eff} and μ_{eff} of the specific RVE. Therefore, one obtains statistical distributed effective parameters, e.g. a probability density.

The probability density W of the effective YOUNG'S modulus E_{eff} and effective POISSON'S ratio μ_{eff} are shown in Figs. 9 and 10 for a set of 9200 RVEs, respectively. Both are close to a GAUSSian distribution (dashed lines). Furthermore, the above mentioned numerical results are in excellent agreement with results obtained from experimental tests of hcp, see Table 2.

Table 2. Experimental versus numerical results

test	$E_{\text{eff}}^{\text{med}}$	$\nu_{\text{eff}}^{\text{med}}$
	$\frac{N}{mm^2}$	–
numerical	17 665	0.211
experimental series 1	17 455	0.210
experimental series 2	17 727	0.210

By applying loading in different directions one can test whether the material response of a RVE is isotropic or anisotropic. For loadings only in the x-direction (constant strain ϵ_{xx}), the above mentioned homogenization procedure yields nearly the same results as for simultaneously loading in x, y and z-direction (full strain tensor). The same holds for loadings only in the y or z-direction (see Table 3). This depicts clearly that the response of the micro-structure (RVE) is isotropic. This result corresponds to the irregular geometry of the phases and the large number of finite elements for each RVE.

The above mentioned homogenization is a result from a real multi-scale approach. The geometry has been obtained from CT-scans of hcp. The material properties of the different phases of the RVE were taken directly from the literature, without any modification. The accuracy of this procedure is very good since the effective properties, obtained numerically, agree excellently with results from experimental tests.

Table 3. Results for different loadings

loading direction	E_{eff}	ν_{eff}
	$\frac{N}{mm^2}$	–
every direction	17 665	0.2107
only x-direction	17 952	0.2108
only y-direction	17 779	0.2110
only z-direction	17 740	0.2100

5 Parameter Identification

The inelastic material model on the micro-scale described in Section 3 contains constitutive parameters $\boldsymbol{\kappa}$, which neither can be found in the literature, nor obtained through experimental testing. Therefore, one has to compute these by parameter identification.

The displacement of the micro-structure $\langle \boldsymbol{u}(\boldsymbol{\kappa})\rangle$ depends, under a given loading, upon the unknown material properties. For an assumed constitutive model, see Section 3, the material parameters can be identified. The identification of these parameters is carried out, such that the displacements $\langle \boldsymbol{u}(\boldsymbol{\kappa})\rangle$ from the numerical simulation fit with the displacements $\boldsymbol{d}$ from an experimental test. This yields an ill-posed inverse problem which cannot be solved directly. Instead, an optimization problem is carried out.

The material parameters of the constitutive model presented in Section 3 have to be calculated, such that an objective function is minimized. The objective function A is defined by a least-square sum between each numerical and experimental value

$$A(\boldsymbol{\kappa}) = \sum_{i=1}^{n} \left(\langle u(\boldsymbol{\kappa})\rangle_i - d_i\right)^2 \rightarrow \min \, . \tag{15}$$

The identification is obtained by solving the above mentioned optimization problem. For computational efficiency and robustness, a combination of the stochastic genetic algorithm [16], [15] and the deterministic LEVENBERG-MARQUARDT method is used [4]. Here, the gradient information of the objective function is obtained numerically from a central difference quotient

$$\nabla_{\boldsymbol{\kappa}} A(\boldsymbol{\kappa})_i \approx \frac{A(\boldsymbol{\kappa} + \epsilon \boldsymbol{e}_i) - A(\boldsymbol{\kappa} - \epsilon \boldsymbol{e}_i)}{2\epsilon} \, . \tag{16}$$

In a first step a pre-optimization with a genetic algorithm is performed in order to get close to the global minimum. Once the value of the objective function falls below a certain threshold value, the optimization procedure switches to the more efficient LEVENBERG-MARQUARDT method.

Due to the complex micro-structure, the inelastic material behavior and the large number of loading steps, each examination of the objective function $A(\boldsymbol{\kappa})$ needs approximately 12 hours on a standard PC computer system.

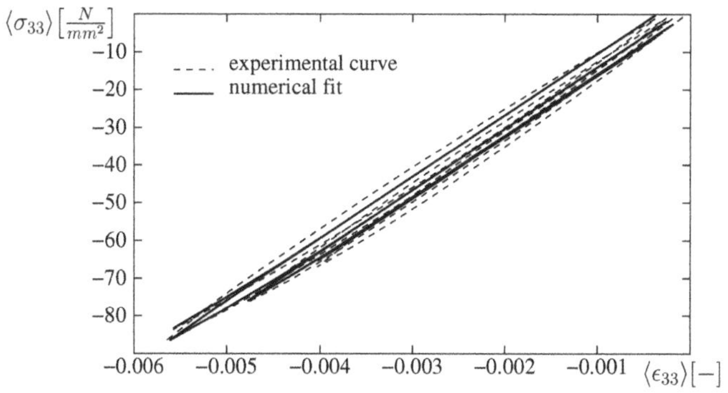

Fig. 11. Numerical fit of the experimental data

Therefore, a typical parameter identification with 120 evaluations requires roughly two month CPU time on a modern standard computer.
In order to keep the overall computing time within reasonable bounds, the calculations, which can be performed stand alone, are distributed within a network environment. A client-server based system has been implemented, a detailed introduction of those systems can be found in [11].

For the optimization procedure, a client is started. This client does not evaluate the objective function itself, but distributes the computation to the compute-servers via a network connection. The evaluation of the objective function $A(\boldsymbol{\kappa})$ is performed on those servers. Once the necessary output data is produced on the servers, it is transported back to the client automatically. The results are then used on the client for the next iteration step of the optimization procedure.

Within the network environment using 11 standard computers, the identification can be completed within six days. This yields a speed-up factor of approximately 10. In Fig. 11 a result of the parameter identification is shown. One can see a satisfying accuracy between the experimental and the numerical results. The experimental load-deflection curve varies with respect to different specimens, therefore it is not reasonable to fit the experimental data more exactly.

The identified parameters are presented in the subsequent table. The properties marked with $\bullet^\star$ have not been identified, but were chosen in advance. They describe the exponent of the equivalent strain and the penalty function, respectively. For their definition see Section 3.

A real multi-scale approach cannot be set up for the inelastic behavior since there do not exist experimental tests which yield values for the constitutive behaviour of the different phases within the micro-structure. Only effective load-deflection curves can be obtained through experiments. In order to get the desired inelastic properties of the micro-structure, one has to solve the inverse problem via parameter identification. Hence, the experimental effective behavior is used to calculate the inelastic properties at the micro-scale which

Table 4. Identified parameters

parameter	value
a	0.488
b	0.1
$c^{\star}$	3.0
α	0.538
k_f	48.27
$m^{\star}$	3
η	84 650

is not a multi-scale approach. However, the obtained inelastic constitutive data for the micro-structure are needed within the next step which then is related to a multi-scale approach.

6 Thermo-mechanical Coupling

Hardened cement past is a part of mortar which in itself is part of concrete. Concrete structures are often subjected to fluids such as rain. When this water permeates into the concrete, it actually fills the pores of the hcp to a certain degree. In freezing condition this water changes its phase to ice and with it expands. This expansion process may lead to damage in some parts of hcp. This process will be investigated in this section. For that we have to know the water contents of the hcp, its constitutive behaviour, see last section, and the constitutive behaviour of the ice which depends on temperature. A numerical simulation of this process has to take into account thermal effects besides the mechanical ones.

Considerable work related to the simulation of water permeability and water vapor diffusion in hcp has been performed in [8]. These results are used in the further calculations to obtain a realistic distribution of water on the micro-structural level. In Figs. 12-15 cuts through the RVE with different moisture contents w_h are shown. The water filled parts of the micro-structure are colored in blue. As expected the water prefers the boundaries of a pore.

For the coupled thermo-mechanical simulation, transient thermal conduction is introduced. The mechanical and the thermal behavior is coupled in a weak sense. We assume that cooling down of hcp leads to strains, but strains due to mechanical loading do not change the temperature.

The strains are separated into a mechanical part $\boldsymbol{\epsilon}^{\boldsymbol{u}}$ and a thermal part in an additive way

$$\boldsymbol{\epsilon} = \boldsymbol{\epsilon}^{\boldsymbol{u}} + \alpha_t(\Theta - \Theta_0)\mathbf{1} \, . \tag{17}$$

The parameter α_t denotes the coefficient of thermal expansion. With respect to the above mentioned split, the weak form of the internal part of the

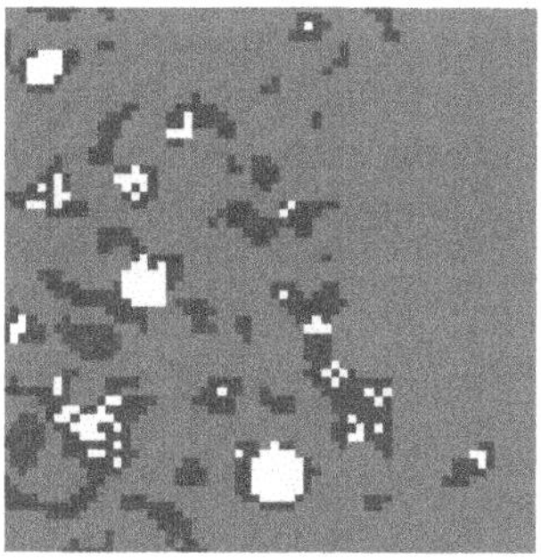

Fig. 12. Cut through of a RVE, $w_h = 80\%$ $(64 \times 64\mu m)$

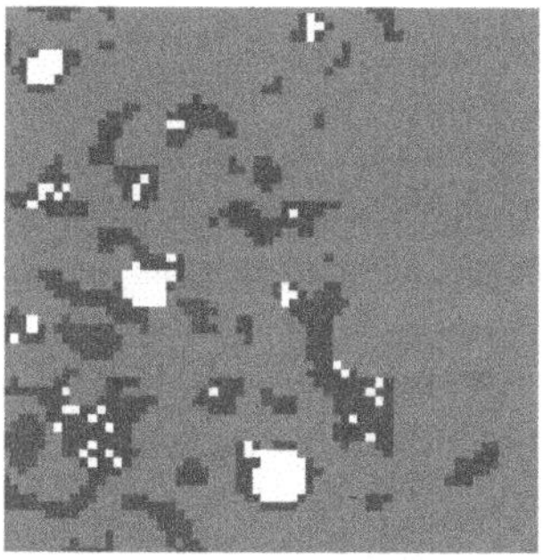

Fig. 13. Cut through of a RVE, $w_h = 85\%$ $(64 \times 64\mu m)$

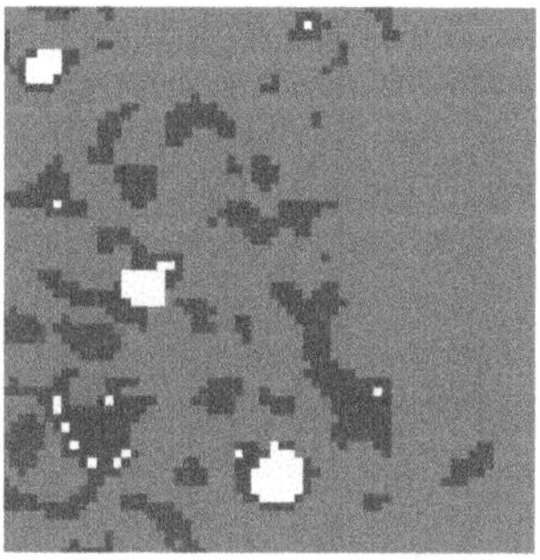

Fig. 14. Cut through of a RVE, $w_h = 90\%$ $(64 \times 64\mu m)$

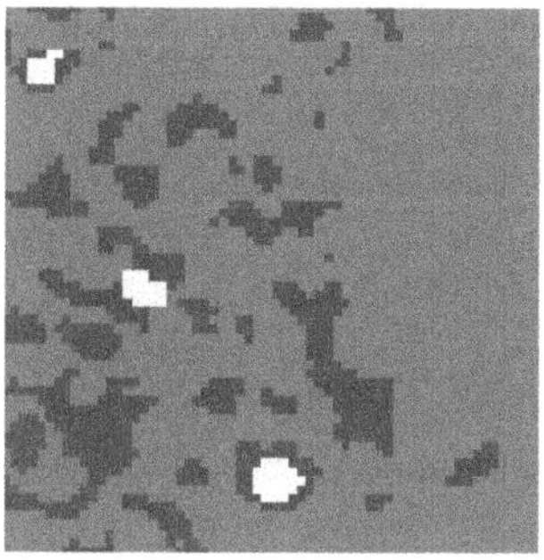

Fig. 15. Cut through of a RVE, $w_h = 95\%$ $(64 \times 64\mu m)$

mechanical equilibrium is given by

$$G^{\boldsymbol{u}}_{int} = \int_{\Omega} \text{grad}\, \boldsymbol{\eta}^{\boldsymbol{u}} : \mathbb{C} : \big(\boldsymbol{\epsilon} - \alpha_t(\Theta - \Theta_0)\mathbf{1}\big)\, \mathrm{d}\Omega\,. \tag{18}$$

Hence, the mechanical equilibrium depends both on the temperature Θ and the displacement $\boldsymbol{u}$

$$G_{\boldsymbol{u}} = G^{\boldsymbol{u}}(\Theta, \boldsymbol{u}) \tag{19}$$

while the thermal weak form only depends on the temperature

$$G_{\Theta} = G^{\Theta}(\Theta)\,. \tag{20}$$

The linearization and subsequent discretization of the thermal and mechanical weak yields the tangential stiffness matrix for the finite element method. It contains of the sub-matrices of the thermal and of the mechanical problem

$$\begin{bmatrix} \boldsymbol{K}_{uu} & \boldsymbol{K}_{u\Theta} \\ \mathbf{0} & \boldsymbol{K}_{\Theta\Theta} \end{bmatrix} \begin{Bmatrix} \Delta \boldsymbol{u} \\ \Delta \boldsymbol{\Theta} \end{Bmatrix} = - \begin{Bmatrix} \boldsymbol{G}_u \\ \boldsymbol{G}_{\Theta} \end{Bmatrix} \tag{21}$$

The thermo-mechanical coupling is described by the couple-matrix $\boldsymbol{K}_{u\Theta}$.

In general it is possible to subject the RVE to thermal and mechanical loading conditions which leads to a specific material response. For this analysis a constitutive model for ice has to be introduced and applied to the water filled parts of the micro-structure. In this contribution, the micro-structural ice is assumed to have the same properties as macroscopic ice. The material model for ice is chosen as visco-plastic model of PERZYNA type (see section 3) and is coupled thermo-mechanically. The previously mentioned expansion of ice is described through a negative coefficient of thermal expansion α_t.

In this contribution YOUNG'S modulus of ice is assumed to depend on the current temperature. The other constitutive parameters μ, k_f, α and ν are assumed to be constant.

It is well known that water filled pores with a hydraulic diameter of $r < 0.5\mu m$ freeze at temperatures of $\Theta \approx -20°C$. The pores in the micro-structure of hcp have this size. Hence YOUNG'S modulus of ice reaches a certain limiting value E_∞ which is kept constant below this temperature

$$E = \begin{cases} E_\infty & ; \Theta \leq -20°C \\ \approx 0 & ; \Theta \geq 0°C \end{cases} . \tag{22}$$

In order to describe this experimental evidence an equation is introduced in which YOUNG'S modulus depends upon the temperature and the four constitutive parameters E_0, E_∞, e and f.

$$E(\Theta) := E_0 + \frac{E_\infty - E_0}{\mathrm{e}} \exp\left[1 - \exp\left(\frac{\Theta - f}{e}\right)\right] . \tag{23}$$

The parameter E_0 describes the stiffness of unfrozen ice (respectively water) and has a negligible value. E_∞ denotes the stiffness of ice. The parameters e and f effect the development of YOUNG'S modulus and are selected such that the stiffness E_∞ is reached at a temperature of $\Theta \approx -20°C$ (see Fig. 16).

Within the analysis of the damage due to frost within the hcp the constitutive equation for ice is applied to the water filled pores of hcp. The material properties of ice are given in the Table 5. They are taken directly from the literature, see e.g. [7]. The properties indicated with $\bullet^\star$ are chosen in order to describe the temperature dependency of YOUNG'S modulus using (23).

Within the numerical simulations thermal loading is increased linearly such that the temperature inside the micro-structure reaches $\langle\Theta\rangle \approx -20°C$. Mechanical loadings are not applied within this investigation. However they are present e.g. due to dead loads and have to be considered for a determination of damage within hcp in a real structure. Here we want to compare with experiments of hcp in which we can neglect these loads. For the mechanical properties of hcp the identified parameters from Section 5 are used.

The principal stress due to thermal expansion is depicted in Fig. 17. Compression stress is marked in dark gray and tension stress in light gray. Obviously, the water filled pores are in compression state due to the expansion of ice.

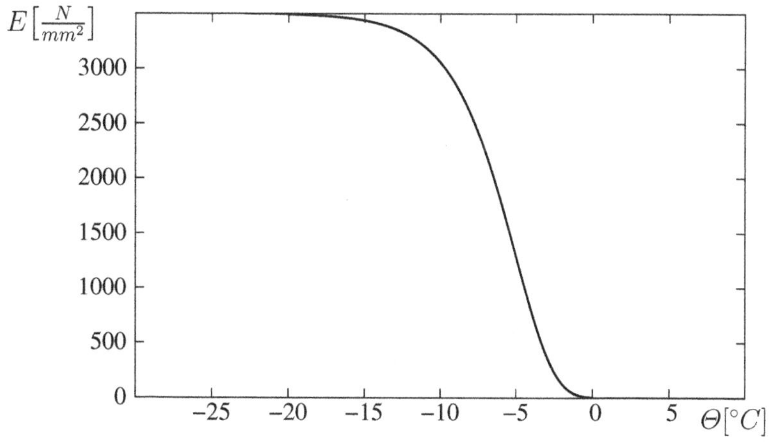

Fig. 16. Temperature dependency of Young's modulus ($e = 2.5$, $f = -5.0$)

Table 5. Material properties for ice

parameter	value
$E_0^\star$	1.0
E_∞	3 500
ν	0.35
k_f	1.633
α	0.333
η	1.0
m	0
$e^\star$	2.5
$f^\star$	-5.0

The thermal loading (freezing) leads to damage in the micro-structure and hence is related to an inelastic material response. For the given loading, one can observe damage zones in Fig. 6. The undamaged material is marked in light gray and the damaged material in dark gray. As one can expect, damage occurs between the water filled pores and will reduce the load carrying capacity of the hcp and thus the mortar. So far these simulations have only be performed for one sample related to the degree of water filling. Hence these results can be used to understand the mechanical behaviour of the freezing process, but not for quantitative evaluations leading to new constitutive relations for the effective material properties. For this purpose statistical computations have to be made with different samples as performed in section 4.

Based on such statistical evaluations we will compare the numerical results with those obtained from experimental tests of frozen hcp in future research.

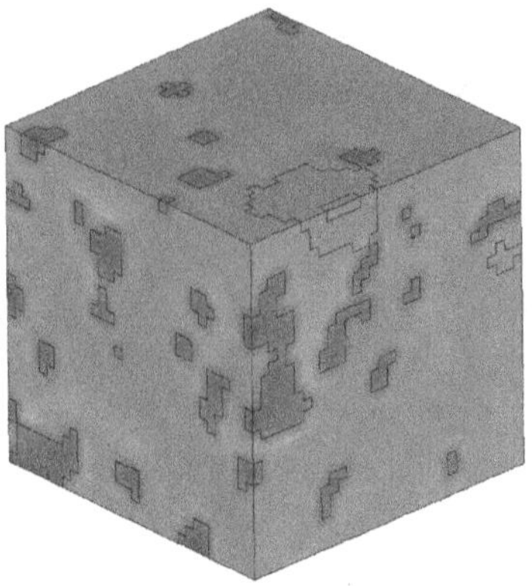

Fig. 17. Principal stress due to frost

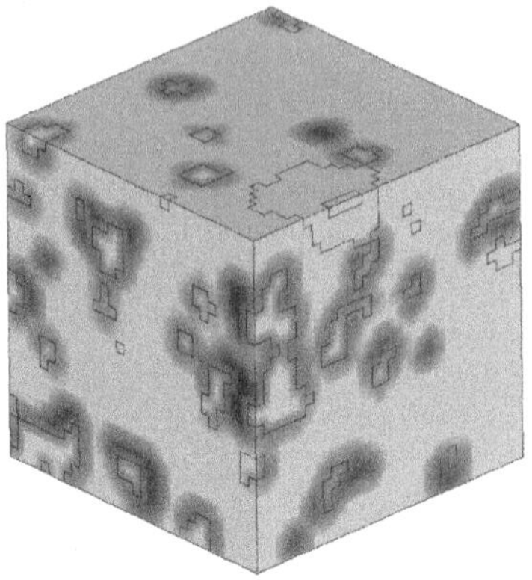

Fig. 18. Damage due to frost

When the above mentioned numerical simulations are performed for different moisture and temperatures states, a correlation between moisture, temperature and inelastic material behavior is obtained. In order to describe such inelastic mechanical phenomena at the meso-scale, an effective material model has to be constructed. In such meso-scale model the constitutive parameters have to depend now upon the mechanical strains, the moisture and the temperature.
Results for two different degrees of moisture are depicted in Fig. 19. From such numerical simulation one obtains the development of the effective damage variable $\langle D \rangle$ as a function of the effective temperature $\langle \Theta \rangle$. The effective values are computed within the RVE by volume averaging of the local damage and local temperature using the procedure described in equation (12).

Such curves have now to be computed for different degrees of moisture in order to get a function on meso-scale which relates the state of effective damage to temperature and moisture.

7 Outlook

On the next length scale, a model for mortar will be introduced. This model consists of an effective material for hcp, aggregates, pores and the cohesive zone around the aggregates. The geometry of mortar will be based on CT-scans. But additionally, a generated structure of mortar is also used. A particle generator has been implemented, which shuffles spherical particles in order to get close to a given sieve curve (see Fig. 20). With respect to this particle distribution, one obtains an artificial mortar. Further comparisons with CT-scans will prove, if such an artificial structure is realistically.

A corresponding FE-mesh is generated using the hanging nodes technique. The hanging node technique leads to a more accurate mesh without any dis-

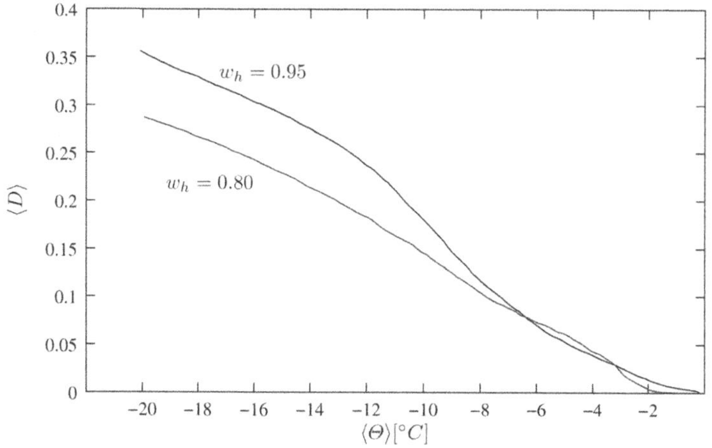

Fig. 19. Damage evolution due to frost

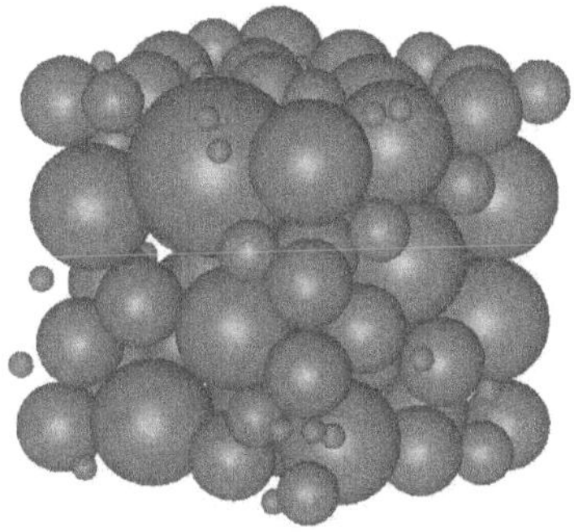

Fig. 20. Generated particle distribution

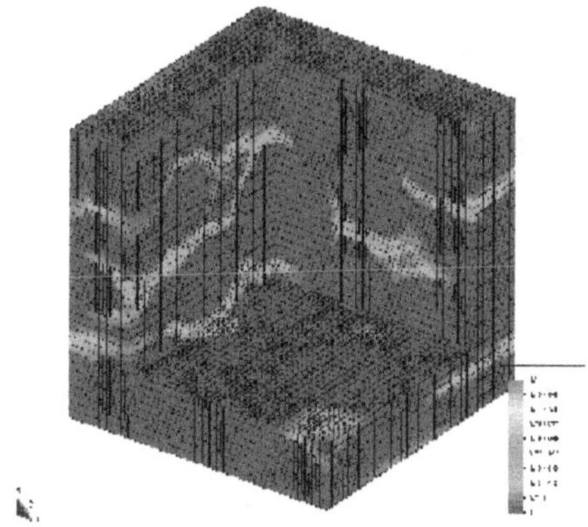

Fig. 21. Damage in artificial mortar

torted elements. In a first approach the aggregates are described with an elastic constitutive equation and the cohesive zone is neglected. Further research will prove, if these assumptions yield reliable results.

Due to mechanical loading, damage zones occur between the aggregates. A first result was calculated in [10] and is shown in Fig. 21. As expected, damage zones occur around the particles. Further research will compare those numerical results with accompanying experimental tests of mortar.

Acknowledgments

Financial support from the *Deutsche Forschungsgemeinschaft* (German Research Council) DFG is gratefully appreciated. All experimental results of hcp were performed by the *Institut für Bauforschung* (director Prof. Brameshuber) at Aachen, Germany. The *Bundesanstalt für Materialforschung und -prüfung* in Berlin, Germany provides the CT-scans of hcp.

References

1. Acker P (2001) Micromechanical analysis of creep and shrinkage mechanisms. Creep, Shrinkage and Durability Mechanics of Concrete and other Quasi-Brittle materials: proceedings of the sixth international conference 1:15–25
2. Bernard O, Ulm F-J, Lemarchand E (2003) A multiscale micromechanics-hydration model for the early-age elastic properties of cement-based materials. Cement and Concrete Research 33:1293–1309
3. Constantinides G, Ulm F-J (2004) The effect of two types of csh on the elasticity of cement-based materials: Results from nanoindentation and micromechanical modeling. Cement and Concrete Research 34:67–80
4. Geiger C, Kanzow C (1999) Numerische Verfahren zur Lösung unrestringierter Optimierungsaufgaben. Springer Verlag
5. Gross D, Seelig T (2001) Bruchmechanik. Springer Verlag
6. Internetlink. National institute of standards and technology; http://fire.nist.gov/bfrlpubs/build00/art060.html.
7. Kolari K, Kouhia R, Kärnä T (2002) Ice failure simulation–softening material model. Proceedings of the 16th IAHR International Symposium on Ice, Dunedin, New Zealand, pages 154–159
8. Koster M, Hannawald J, Brameshuber W (2006) Simulation of water permeability and water vapor diffusion through hardened cement paste. Computational Mechanics 37(2):163–172
9. Lemaitre J (1996) A course on damage mechanics. Springer Verlag
10. Löhnert S (2004) Computational homogenization of microheterogeneous materials at finite strains including damage. PhD thesis, Universität Hannover
11. Pollakowski M (2004) Grundkurs Socketprogrammierung mit C unter Linux. Vieweg Verlag
12. Powers TC (1962) Physical properties of cement paste. Proceedings of the Fourth International Symposium on the Chemistry of Cement; Paper V-1: Physical Properties of Cement Paste 2(43):577–613
13. Powers TC, Brownyard TL (1948) Studies of the physical properties of hardened portland cement paste. Research Laboratories of the Portland Cement Association Bulletin 22:101–992
14. Simo JC, Hughes TJR (1998) Computational Inelasticity. Springer, New York, Berlin
15. Thielecke F (1997) Parameteridentifizierung von Simulationsmodellen für das viskoplastische Verhalten von Metallen -Theorie, Numerik, Anwendung-. PhD thesis, Technische Universität Braunschweig
16. Wedemeier K (1990) Beiträge zur Theorie und Numerik von Materialien mit innerer Reibung am Beispiel des Werkstoffes Beton. PhD thesis, Universität Hannover
17. Zohdi TI, Wriggers P (2004) An Introduction to Computational Micromechanics. Springer Verlag, New York

Numerical Modeling of Transient Impact Processes with Large Deformations and Nonlinear Material Behavior

Ekkehard Ramm[1], Tobias Erhart[2] and Wolfgang A. Wall[3]

[1] Institute of Structural Mechanics, University of Stuttgart, Pfaffenwaldring 7, 70569 Stuttgart, Germany
eramm@statik.uni-stuttgart.de
[2] DYNAmore GmbH, Industriestr. 2, 70565 Stuttgart, Germany
tobias.erhart@dynamore.de
[3] Chair for Computational Mechanics, Technical University of Munich, Boltzmannstr. 15, 85748 Garching, Germany
wall@lnm.mw.tum.de

Summary. A robust computational approach for transient dynamic impact processes will be presented. Main focus will be on an adaptive remeshing strategy and constitutive models for metals and geomaterials under impact loading. Practical relevant numerical examples will complete this contribution.

Key words: Impact, Large Deformations, Adaptive Remeshing, Metals, Geomaterials

1 Introduction

The present study is concerned with a special case of solid mechanics, where structures are exposed to short-time, highly concentrated loading. Such transient impact processes appear in civil and military security technology, dynamic soil compaction, vehicle crash or fastening and demolition technology. They are characterized by varying non-linearities, as e.g. large deformations and strains, highly non-linear material behavior, frictional contact between multiple bodies and stress wave propagation. Development and combination of different methods in the fields of adaptivity, constitutive modeling, element technology, efficient time discretization and contact are essential for reliable computations and predictions in engineering practice. Accuracy, robustness and efficiency are mandatory requirements for the solution of those complex problems.

In this contribution, we will mainly focus on two issues, namely adaptive remeshing and constitutive modeling of metals and geomaterials especially

Eugenio Oñate and Roger Owen (eds.), Computational Plasticity, 123–144.

elaborated for impact loading. The proposed methods are tested for model problems and their performance in relevant industrial applications is verified.

2 Class of Problems

The specific class of problems focussed in this study excel themselves through a large complexity. Essentially these are fast transient plane strain or axisymmetric problems that are treated in an explicit way. Adopted material models range from simple elastic up to large strain thermo-elastic-viscoplastic models. The problems exhibit large deformations and may even include severe mesh distortion. Frictional contact is an essential feature of the related physical processes.

3 Adaptive Remeshing

Since large deformations occur in impact simulations and a Lagrangean description is used, repeated remeshing of individual domains is necessary [6] [21]. To achieve a quality controlled solution and an optimal deployment of used computational resources at the same time, an adaptive strategy is applied [12]. The essential ingredients of an adaptive remeshing strategy in this highly nonlinear regime are (i) a mesh quality check for triggering automatic remeshing, (ii) a reasonable assessment of discretization errors and derivation of a corresponding mesh density distribution, (iii) an automatic mesh generation tool for graded meshes and (iv) methods for the transfer of state variables from old to new discretization. The core of this strategy is the assessment of discretization errors by adequate indicators. Different indicators are presented and new methods are developed, which are especially suited for the simulation of transient impact processes. Theoretical definitions and numerical treatment of these aspects are dealt with.

3.1 Mesh Quality Check

In our approach, mesh quality is evaluated for each element by a combination of two geometrical criteria, the so-called 'corner angle criterion' and the ratio of the radius of the inscribed circle to the radius of the circumscribed circle [12]. If the combination of these two measures, e.g. the mean value falls below a prescribed threshold value for a prescribed number of elements, the remeshing procedure starts automatically. With this mesh quality check, it is guaranteed, that the FE solution is prevented from occurence of a negative Jacobian.

3.2 Error Assessment and Mesh Density Distribution

In our approach, three categories of error indicators, gradient-based, local quantity-based as well as geometric, are used for adaptive remeshing [12]. In the following, all indicators that have been studied and implemented are presented in more detail along with their associated refinement strategy, i.e. the determination of a corresponding mesh density distribution.

Gradient-Based Indicators

Gradient-based indicators use the fact that there is a connection between the size of local gradients and the quality of solution. Here, 'gradients' not only mean stresses, but local variations of arbitrary quantitites like strains, plastic work, etc.. The corresponding adaptive remeshing strategy is characterized by an equi-distribution - with respect to the element length - of those gradients in the domain of solution. This leads to a refinement in high gradient zones on the one hand and to a coarsening in low gradient areas on the other hand. Now the question is, which relevant quantities for the gradient calculation should be used to indicate a meaningful error and subsequently govern the mesh density? We have chosen two different elastic and one plastic variable, whose gradients are tightly connected to some of the specific physical phenomena or processes occurring in complex problems like in our applications, depending on the specific example at hand. The first quantity is the von Mises stress

$$\psi_1 := \mu \, \sqrt{2(\operatorname{dev}(\boldsymbol{\epsilon}^e) : \operatorname{dev}(\boldsymbol{\epsilon}^e))} \tag{1}$$

which especially tracks elastic shape changing behavior, whereas the second one, the hydrostatic stress

$$\psi_2 := \kappa \, \operatorname{tr}(\boldsymbol{\epsilon}^e) \tag{2}$$

mainly grasps elastic volumetric changes. Plastic regions in a structure, e.g. localization zones, can be detected with the third quantity 'plastic work'

$$\psi_3 := \int_{t_0}^{t} J^{-1} \boldsymbol{\tau} : \dot{\boldsymbol{\epsilon}}^p \mathrm{d}t \tag{3}$$

Gradients of these quantities need to be calculated at all nodes of the old mesh. Therefore, the superconvergent patch recovery technique in an element patch [31] is used to get information about the local gradient at the particular assembly node, i.e. the central node of the patch. According to this method, a least square fit between a polynomial approximation and discrete values (e.g. at quadrature points) leads to a minimization problem. After solution of the corresponding system of linear equations, the norm of the gradient can be calculated at the position of node N (central node of patch)

$$g_k := \|\nabla\psi_k(\mathbf{x}_N)\| = (\nabla\psi_k(\mathbf{x}_N) \cdot \nabla\psi_k(\mathbf{x}_N))^{1/2} \tag{4}$$

The local element size of the new mesh has to be inversely proportional to the size of the local gradient due to the above mentioned reasons. In addition, the magnitude of refinement needs to be controlled in order to adapt the element sizes to the different indicator and/or examples. This can be achieved by scaling the gradient with a prescribed value $\Delta\psi_k$, which describes the maximal allowed change of the considered quantity ψ_k per element. At the end the new element size $h_{e,k}^{new}$ results in

$$h_{e,k}^{new} = \frac{\Delta\psi_k}{g_k} \tag{5}$$

Local Quantity Indicators

On the basis of pure physical considerations, local refinement indicators can be developed that control the mesh density in the interior mesh. For this purpose, a physically relevant quantity is calculated for each element in the old mesh. The new mesh is then generated provided that this quantity becomes equi-distributed so that every element experiences roughly the same 'physical action'. This obviously results in refinement for high values and in coarsening for small values of this quantity per old element. For these indicators, no gradients need to be calculated. In the following we sketch some of these local quantity indicators that are given in the literature and that we adopted for our study. For problems of strain localization, Ortiz and Quigley [19] proposed an adaptive strategy, which is based on the equi-distribution of variation of the velocity field $\mathbf{v}$ over the elements of the mesh. Such localization problems can be characterized by large differences between the mechanical behavior of the global structure and a local zone, e. g. shear bands, bringing along a loss of ellipticity of the governing equations. For a two-dimensional element, this indicator can be expressed as the maximum deviation for the velocity v_i of two nodes a and b

$$\phi_1 := \max_i\{\max_{a,b} |v_{ia}^e - v_{ib}^e|\} \tag{6}$$

This is a suitable indicator for the analysis of transient processes with wave propagation as well as localization. It is however restricted to dynamic calculations. Singularities in quasi-static zones of transient problems or poor approximations of quasi-static quantities are not treated appropriately.

Another physical indicator was used by Batra and Ko [2] for shear bands in plane strain compression and by Camacho and Ortiz [6] for different impact and penetration calculations. The intention of this indicator is that the integral of the second invariant of the deviatoric strain-rate tensor over an element

$$\phi_2 := \int_{\varphi(\Omega_e)} \sqrt{\frac{1}{2}\mathrm{dev}\dot{\boldsymbol{\epsilon}} : \mathrm{dev}\dot{\boldsymbol{\epsilon}}}\;\; \mathrm{d}v \tag{7}$$

is the same for all elements. This indicator is capable of resolving strain localizations in dynamic calculations.

A third indicator, based on the equi-distribution of plastic power, was developed by Marusich and Ortiz [17] for simulation of high-speed machining:

$$\phi_3 := \int_{\varphi(\Omega_e)} \dot{W}^p \;\mathrm{d}v = \int_{\varphi(\Omega_e)} J^{-1}\boldsymbol{\tau} : \dot{\boldsymbol{\epsilon}}^p \mathrm{d}v \tag{8}$$

This indicator is especially suited for nonlinear, highly dynamic problems with large plastic deformations, since it leads to refinement in high strain rate regions. In order to be able to combine different indicators later, we adopt here a similar strategy as before. At first, the elementwise calculated quantities ϕ_l are related to the respective old element size h_e^{old}

$$\tilde{g}_l = \frac{\phi_l}{h_e^{old}} \tag{9}$$

Now we use the same refinement strategy as we did for the gradient-based error indicators, i.e. the new element size is inversely proportional to the 'variation per element' $\tilde{g}_l$ and it is scaled by a prescribed value $\Delta\phi_l$

$$h_{e,l}^{new} = \frac{\Delta\phi_l}{\tilde{g}_l} \tag{10}$$

Geometric Indicator

Strongly curved surfaces have to be discretized with sufficiently small elements to avoid significant 'geometrical errors'. This is especially important with respect to mass conservation and in contact situations. The curvature of the boundary is used as a geometric indicator to control this problem. However, this indicator is only activated if it suggests refinement of the 'mechanical' indicator driven mesh density. In case of a discrete boundary description, e.g. the polygonal boundary of a finite element mesh, the discrete curvature κ^h at the position of a finite element node is given as

$$\kappa^h := \frac{2\vartheta}{s_1 + s_2} \tag{11}$$

where ϑ is the smaller angle between two boundary segments with lengths s_1 and s_2, respectively. If this measure is calculated for boundary nodes of the old mesh, the element size in the new mesh results in

$$h_{e,g}^{new} = \frac{\vartheta_{adm}}{\kappa^h} \tag{12}$$

where ϑ_{adm} is a prescribed value, which can be interpreted as admissible angle between two adjacent boundary segments. A similar indicator was used by Camacho and Ortiz [9]. In general, through application of this indicator, there is no connection between internal element size and boundary element size. In such cases, a mesh smoothing technique is used to avoid large differences between element sizes in the interior and at the boundary.

Selection of Indicators and Final Mesh Density Distribution

The objective of the different indicators and possible application areas have been given above. For real and complex applications the question of the final selection of indicator remains. Which specific indicator or combination of

different indicators should be used? The answer is both trivial and difficult at the same time. It should be rather clear from the above discussion which indicators are not suited in which cases. The final selection however can only be based on engineering intuition and on a deep insight into the physics of the problem or the class of problems. It is crucial to know what the governing phenomena and the dominating processes are. One should be aware that the answers not only depend on the problem itself but also on the respective quantities of interest of a specific simulation.
To obtain the final mesh density, two further steps are needed. As already pointed out it is in some cases useful to combine different error indicators. For this purpose, the minimum of the different element sizes $h_{e,k}^{new}$ (which belong to the five gradient-based indicators), $h_{e,l}^{new}$ (which belong to the three local quantity indicators) and $h_{e,g}^{new}$ (the geometric indicator for the boundary) has to be chosen as final element size

$$h_e^{new} = \min\{h_{e,k}^{new}, h_{e,l}^{new}, h_{e,g}^{new}\} \tag{13}$$

On the other hand, the element size has to be restricted by some prescribed threshold values, because undesirably small elements (time step size in explicit scheme!) or large elements (approximation quality!) should be avoided:

$$h_e^{\min} \leq h_e^{new} \leq h_e^{\max} \tag{14}$$

3.3 Mesh Generation

Generating a graded mesh needs two sources of information: the boundary of the domain and the mesh density inside the domain mostly given on a background mesh. In our case of large deformations and changing topologies the background mesh is preferably the old mesh of the preceding time step, for which the error assessment is done and the mesh density distribution is calculated. Secondly, a polygonal description of the boundary has to be provided. Boundary nodes must be generated adaptively, i.e. subject to the calculated mesh density. Special care has to be taken of boundaries in contact situations. The re-ordering of nodes on convexly curved boundaries during remeshing can result in sudden gaps between contacting bodies, which results in so-called 'remeshing shocks': oscillations induced by unphysically high contact forces. This is remedied by a smooth contact formulation combined with a procedure for precise surface recovery [9]. After that, a new 'almost-all-quad' mesh is generated inside the domain by an 'advancing front method' starting from these previously generated boundary nodes and based on the given mesh density found at the nodes of the old mesh [16] [20]. In exceptional cases, e.g. for sharp corners, triangular elements (e.g. CST - constant strain triangles) are also allowed. This has the advantage, that no 'bad quads' have to be generated and therefore the time span until next remeshing is increased.

3.4 Transfer of State Variables

Since results are determined in an iterative or in an incremental process for nonlinear or transient problems, a transfer of state variables between old and new spatial discretization is essential [21] [6]. Examples of variables, which have to be 'mapped' are strains, stresses, velocities, boundary conditions, etc., i.e. a mapping strategy for both nodal and integration point data is needed. The location and number of nodes and integration points changes totally in case of complete remeshing of a domain. Therefore, the location of new points in the old mesh has to be determined by an intelligent search algorithm, e.g. directional search [20], and thereafter, state variables have to be transfered by a suitable mapping method like inter/extrapolation [21], inverse distance weighting [8] or moving least square techniques [31] [26]. Especially in the context of explicit time integration the quality of the mapping scheme is crucial. Since mapping cannot be supplemented by equilibrium iterations in these cases mapping errors may propagate and pollute the whole simulation. Other aspects like minimization of diffusion should be met as in implicit situations.

4 Constitutive Modeling

Based on the theory of finite plasticity, constitutive models for thermoviscoplastic metals and cohesive as well as non-cohesive frictional materials are presented and developed. One focus will be on a formulation for loose, granular media under high pressure loadings. For this a Drucker-Prager-Cap model is modified and enhanced. The properties and effects of the developing powder will be examined.

Our implementation of finite strain plasticity is based on the multiplicative split of the deformation gradient into elastic and plastic parts $\mathbf{F} = \mathbf{F}^{\mathrm{e}}\mathbf{F}^{\mathrm{p}}$ and an isotropic Eulerian formulation in eigenvalues [24] [23] [18]. Using a spectral decomposition of the Finger-tensor $\mathbf{b} = \mathbf{F}\mathbf{F}^{\mathbf{T}}$ and the Kirchhoff stress tensor $\boldsymbol{\tau} = J\boldsymbol{\sigma}$, a general return mapping scheme with elastic predictor and closest point projection algorithm formulated in principal logarithmic strains $\boldsymbol{\epsilon} = 1/2\ln(\mathbf{b}) = [\epsilon_1, \epsilon_2, \epsilon_3]^T$ and in principal Kirchhoff stresses $\boldsymbol{\tau} = [\tau_1, \tau_2, \tau_3]^T$ is applied. With this approach, which is valid for isotropic materials, the functional form of the return mapping is identical to the algorithm of the infinitesimal theory [23]. Therefore, the whole scheme with elastic predictor, plastic corrector etc. is not shown here, but only the most important parts of our models are presented briefly.

4.1 Metals

The mechanical behavior of metals under transient impact loadings is mainly affected through strain rate and temperature. Therefore, a Johnson-Cook plasticity model [15] is used, where the yield limit is a function of effective plastic

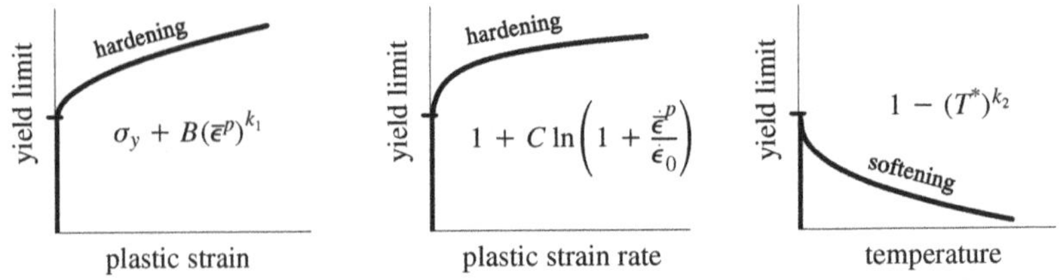

Fig. 1. Qualitative influence of strain, strain rate and temperature on yield limit in Johnson-Cook plasticity model

strain $\bar{\epsilon}^p$, effective plastic strain rate $\dot{\bar{\epsilon}}^p$ and absolute temperature T:

$$f(\boldsymbol{\tau}, q) = \|\mathbf{s}\| - \sqrt{\frac{2}{3}} q(\bar{\epsilon}^p, \dot{\bar{\epsilon}}^p, T) \leq 0 \tag{15}$$

with norm of stress deviator $\|\mathbf{s}\| = \|\text{dev}(\boldsymbol{\tau})\|$ and yield limit

$$q(\bar{\epsilon}^p, \dot{\bar{\epsilon}}^p, T) = \left(\sigma_y + B(\bar{\epsilon}^p)^{k_1}\right) \left(1 + C \ln\left(1 + \frac{\dot{\bar{\epsilon}}^p}{\dot{\bar{\epsilon}}_0}\right)\right) \left(1 - (T^*)^{k_2}\right) \tag{16}$$

σ_y indicates the static yield limit and material parameters B and k_1 describe the strain hardening behavior (see Fig. 1). The increase of strength due to high strain rates, i.e. the viscosity, is considered through the second term of eq. (16). Factor C quantifies the strain rate dependency, whereas the rate of plastic strain $\dot{\bar{\epsilon}}^p$ is normalized by $\dot{\bar{\epsilon}} = 1.0$ [1/s]. With the last part in (16), the effect of decreasing strength for high temperatures is reproduced: The homologous temperature $T^* = (T - T_0)/(T_m - T_0)$ is a measure for the mobility of crystal grid components and is modeled as a function of current temperature T, room temperature T_0 and melting temperature T_m. This non-dimensional variable ranges from 0 (current = room temp.) to 1 (current = melt. temp.). Thermal softening is controlled by exponent k_2. Since transient impact and penetration processes are of very short duration, adiabatic temperature changes are assumed in our study. In addition, for the local temperature evolution, the empirical assumption is used, that most (90-100 percent) of the plastic work W^p is transformed into heat

$$\dot{T} = \frac{\eta_d}{\varrho c_v} \dot{W}^p = \frac{\eta_d}{\varrho c_v} J^{-1} \boldsymbol{\tau}^T \dot{\boldsymbol{\epsilon}}^p, \quad 0.9 \leq \eta_d \leq 1.0 \tag{17}$$

with density ϱ, grade of dissipation η_d and specific heat capacity c_v. The return mapping algorithm ends up in a local Newton iteration, where plastic strain and plastic strain rate are the implicitely considered unknowns, whereas the temperature is accounted for in an explicit manner, i.e. it is computed from previous time step values.

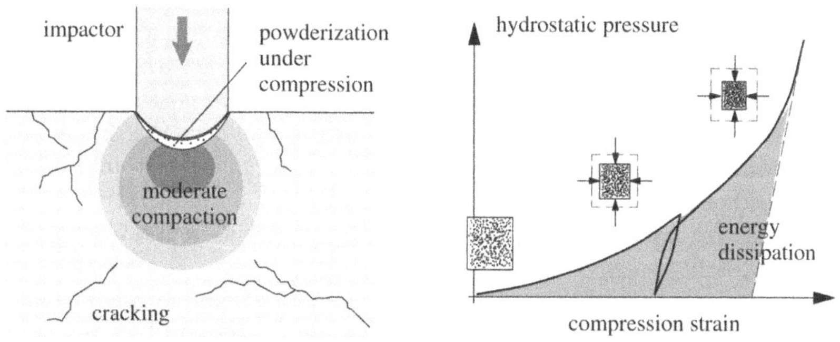

Fig. 2. Failure modes of geomaterial under impact loading (left) and typical compaction behavior of powder (right)

4.2 Geomaterials

A constitutive model for cohesive and non-cohesive frictional materials will be presented, which is especially suited for dynamic impact processes. During a high velocity impact acting on cohesive frictional material (e.g. sandstone), a crushed zone appears directly under the impactor nose. In this region of highly confined loading, the intact material disintegrates and loose, granular or powder-like media develops. Realistic description of this powderization and the material modeling of powder itself are important topics of our study.

Characterization of Geomaterials Under Impact Loading

Considering a solid structure made of intact, cohesive frictional material (e.g. concrete, sandstone, rock) under transient impact loading, different states of stress and therefore different states of failure can be observed, which we divided into three zones (Fig. 2): (i) high-pressure driven crushing (or powderization) directly under the impactor, (ii) moderate compaction due to mixed compression/shear loading and (iii) tensile fracture with low shearing. As already stated above, our main focus will be on the first part, where compact cohesive frictional material is going to be totally disintegrated to loose, granular powder. The material will obviously change its mechanical properties during this crushing process. Due to comminution, a higher density can be achieved than with the original material. This densification reserve leads to a temporary decrease of hardening behavior and an increase of maximal density. Additionally, crushing causes total loss of internal cohesion due to particle and matrix breaking, which results in full decrease of cohesion and tensile strength on macroscopic level. The effect becomes obvious, when the evolved powder gets unloaded and loaded in the opposite direction (i.e. tension): Without tensile strength and cohesive capability, unlimited motion of powder particles is possible and dilatant loosening of the material will be observed. Since powder-like materials are distinctly compressible, depending on the initial density, they can have severe damping and energy absorbing effects in dynamic impact processes that are of interest in a number of different applications. Herein, the kinetic energy of a system is transformed into internal energy by a relative volume change of the powder (Fig. 2). A better understanding of this phenomenon is needed to extenuate or to intensify the damping and energy absorbing influence of powder in a quantitative and predictable way. In our investigations, the powder material of interest is of dry fine sand type.

In this study, a continuum cap model for the numerical simulation of geomaterials under quasi-static as well as dynamic loading is presented. A nonsmooth multisurface plasticity model with tension cutoff, Drucker-Prager failure envelope and strain hardening cap provides the reproduction of the relevant phenomena: tensile failure, material flow under shearing and compaction under pressure. For the comminution process, i.e. the transition from intact, cohesive frictional material to loose, granular media, this model will be enhanced. It includes a criterion for the powder development, different evolution laws for material parameters like tensile strength, cohesion and compressibility and a modified hardening law.

Modified Drucker-Prager-Cap Model

The original multi-surface plasticity model consists of a nonlinear Drucker-Prager cone for shear failure, an elliptical cap for compressive strain hardening and a limit in the tension region for tensile failure [22] [7] [13]. Therefore, the elastic region is bounded by following three yield criteria:

$$\begin{aligned} f_1(\boldsymbol{\tau}) &:= \|\mathbf{s}\| - F_e(I_1) \leq 0 \\ f_2(\boldsymbol{\tau}, q) &:= F_c(\|\mathbf{s}\|, I_1, q) - F_e(q) \leq 0 \\ f_3(\boldsymbol{\tau}) &:= I_1 - T \leq 0 \end{aligned} \tag{18}$$

These yield surfaces $f_1 = 0$, $f_2 = 0$ and $f_3 = 0$ (see Fig. 3) are functions of stress invariants $I_1 = \mathrm{tr}(\boldsymbol{\tau})$ and $\|\mathbf{s}\| = \|\mathrm{dev}(\boldsymbol{\tau})\|$ as well as functions

$$F_e(I_1) := \alpha - \lambda \exp(\beta I_1) - \theta I_1 \tag{19}$$

and

$$F_c(\|\mathbf{s}\|, I_1, q) := \sqrt{\|\mathbf{s}\|^2 + \frac{1}{R^2}(q - I_1)^2} \tag{20}$$

F_e defines the nonlinear Drucker-Prager cone and F_c describes the elliptical cap. The transitions between yield surfaces is discontinuous with two so-called corner regions. At the position, where cap and shear failure envelope intersect, we have

$$F_e(I_1 = q) = \frac{q - X(q)}{R} \tag{21}$$

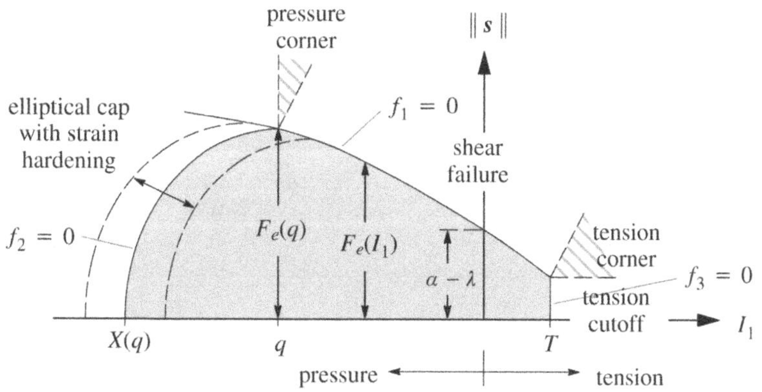

Fig. 3. Yield surfaces of Drucker-Prager Cap plasticity in meridian plane

with hydrostatic compressive strength

$$X(q) = q - R(\alpha - \lambda \exp(\beta q) - \theta q) \tag{22}$$

Material parameters of this formulation are the Drucker-Prager parameters α, λ, β and θ, the cap ellipticity R, the initial hydrostatic compressive strength X_0 and the hydrostatic tensile limit T. Internal variable $q < 0$ is a function of volumetric plastic strain and it is responsible for the position and movement of the cap (hardening).

The original Drucker-Prager-Cap (DPC) model of Hofstetter et al. [13] comes with associated flow rules for all three failure surfaces. However, we will use non-associated flow rules for the Drucker-Prager cone section and the tensile region to get rid of discontinuous flow directions in both corners and to receive less dilatant behavior for shear failure, which seems to be more realistic from an experimental point of view [25]. Finally, our three flow directions $\mathbf{m_1}$, $\mathbf{m_2}$ and $\mathbf{m_3}$ are given by:

$$\begin{aligned}
\mathbf{m_1} &:= \frac{\mathbf{s}}{\|\mathbf{s}\|} - \left(\frac{I_1 - q}{T - q}\right) \frac{\partial F_e(I_1)}{\partial I_1}\, \mathbf{1} = \frac{\mathbf{s}}{\|\mathbf{s}\|} + (\lambda\beta \exp(\beta I_1) + \theta) \left(\frac{I_1 - q}{T - q}\right) \mathbf{1} \\
\mathbf{m_2} &:= \frac{\|\mathbf{s}\|}{F_c(\|\mathbf{s}\|, I_1, q)} \frac{\mathbf{s}}{\|\mathbf{s}\|} - \frac{q - I_1}{R^2 F_c(\|\mathbf{s}\|, I_1, q)}\, \mathbf{1} \\
\mathbf{m_3} &:= \frac{\|\mathbf{s}\|}{F_e(T)} \frac{\mathbf{s}}{\|\mathbf{s}\|} + \left[1 - \left(1 + \frac{\partial F_e(T)}{\partial I_1}\right) \frac{\|\mathbf{s}\|}{F_e(T)}\right] \mathbf{1}
\end{aligned} \tag{23}$$

A comparison between associated and non-associated flow directions for the DPC model is shown in Fig. 4.

Assuming that the relation between pressure and volume change is known from triaxial experiments, a strain hardening law can be defined, which describes the dependence of internal variable q from volumetric plastic strain $\epsilon^p_{vol} = \mathrm{tr}(\boldsymbol{\epsilon}^p)$. Here, two similar functions for that relation are provided. The

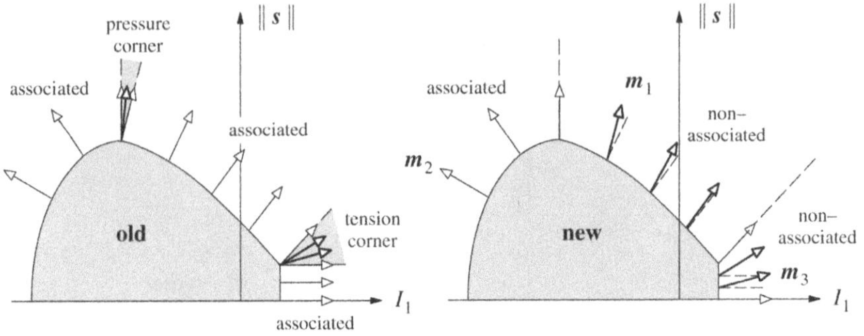

Fig. 4. Flow directions of DPC model: associated and non-associated

first one was proposed by Sandler and Rubin [22], whereas the second one was developed for better agreement with our experimental results for dry fine sand type:

$$\epsilon^p_{vol} = \begin{cases} W\left(\exp\left(D(X(q) - X_0)\right) - 1\right) & \text{Sandler and Rubin [22]} \\ DW\frac{X(q)-X_0}{1-D(X(q)-X_0)} & \text{Erhart et al. [11]} \end{cases} \tag{24}$$

In Fig. 5 both hardening functions are shown. It can be observed, that the second curve (representing the second hardening law [11]) approaches the asymptote at $\epsilon^p_{vol} = -W$ in a more moderate way.
In triaxial experiments, cohesive as well as non-cohesive geomaterials show a dependence from third invariant of stress deviator $J_3 = \det(\mathbf{s})$ when it comes to failure. To reproduce this behavior in simulations, we scale all three yield surfaces with the Willam-Warnke function [28]

$$r(\upsilon, e) = \frac{2(1-e^2)\cos\upsilon + (2e-1)\sqrt{4(1-e^2)\cos^2\upsilon + 5e^2 - 4e}}{4(1-e^2)\cos^2\upsilon + (2e-1)^2} \tag{25}$$

with Lode angle

$$\upsilon = \frac{1}{3}\arccos\left(\frac{3\sqrt{3}}{2}\frac{J_3}{(J_2)^{1.5}}\right) \tag{26}$$

In our algorithm, the third stress invariant is considered in an explicit sense, i.e. Lode angle υ is calculated at t_n and then yield surfaces are scaled with the Willam-Warnke function $r(\upsilon, e)$ at t_{n+1}. Therefore, the stress update algorithm is not affected here.

Powderization Under High Pressure

With the plasticity model presented above, it is possible to reproduce the mechanical behavior of compact, cohesive geomaterials (sandstone, concrete,

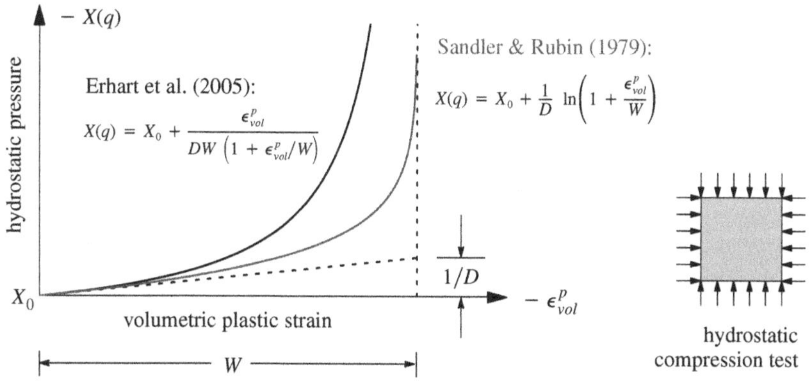

Fig. 5. Strain hardening laws for modified DPC model

mortar, etc.) as well as loose, granular media (sand, powder, etc.). The distinction between them is achieved by appropriate choice of material parameters. Now, this model will be enhanced in order to describe the transition from compact to loose material, i.e. the powderization under high pressure. At first, a criterion is needed, which indicates the beginning of powder formation. On the other hand, a consistent transition from 'compact' to 'loose' is necessary. Therefore, a strategy is developed, where a new internal variable is used to modify the material parameters tensile strength (T) and cohesion (α, λ). Additionally, hardening response is modified in loading and unloading to grasp the different behavior of cohesive and non-cohesive material. In our approach, a stress dependent criterion is used as 'powderization indicator'

$$\frac{1-2\nu}{1+\nu}\delta < \delta_{crush} \quad \text{and} \quad I_1 < I_{1,crush} \tag{27}$$

where δ_{crush} and $I_{1,crush}$ are material parameters and factor δ ('triaxiality') is a function of stress invariants [14]:

$$\delta = -\frac{I_1}{2\sqrt{3}\sqrt{J_2}cos\upsilon} \tag{28}$$

With this criterion, beginning of crushing is detected for very high pressures and highly confined material (e.g. under impactor nose). If the criterion is fulfilled, tensile strength and cohesion will decrease, which is reproduced by softening in tension and shear regime in our constitutive model. Therefore, material parameters α, λ and T are now functions of new internal crushing variable q_c. Following the hardening law (section 4.3), this history variable is chosen in such a way that it is identical with the $1/D$-scaled volumetric plastic strain, which develops after the beginning of powderization:

$$\dot{q}_c = \begin{cases} \frac{1}{D}\dot{\epsilon}^p_{vol} & \text{if } \dot{\epsilon}^p_{vol} > \dot{\epsilon}^p_{vol,crush} \\ 0 & \text{else} \end{cases} \tag{29}$$

Shear strength $\alpha - \lambda$ at $I_1 = 0$ is reduced by modification of the nonlinear Drucker-Prager yield surface with following exponential softening (given in incremental form)

$$\begin{aligned} \lambda_{n+1} &= \alpha_n - (\alpha_n - \lambda_0) \exp\left(-\frac{q_{c,n}}{q_{cu}}\right) \\ \alpha_{n+1} &= \alpha_n + (\lambda_{n+1} - \lambda_n) \exp(\beta q_n) \end{aligned} \tag{30}$$

with material parameter q_{cu} controlling the reduction intensity. With these two equations it is guaranteed, that the position of the cap is not changed in the current time step. To receive the same amount of softening for the tensile limit, following nonlinear equation has to be solved for unknown T_n

$$\frac{F_e(T_n)}{T_n} = \frac{F_e(T_0)}{T_0} \tag{31}$$

The increase of hydrostatic compressive strength $X(q)$ with increasing plastic compression (i.e. the expansion of the cap) is controlled by hardening law (24). On the other hand, the contraction of the cap due to dilatant plastic flow in shear or tension depends on the material at hand. Loose, granular media like sand or powder is able to break up quite easily due to missing cohesion, whereas for intact cohesive geomaterials like concrete this phenomenon is not very distinct. We incorporated this observation in our model by allowing cap contraction only for powder-like materials (see Fig. 6).

As already mentioned above, comminution of initially compact material has the consequence that a higher compaction is possible, clearly depending on the initial density. This phenomenon was observed in experiments [30] [29] and is incorporated in our model by modification of the original hardening law. In Fig. 7, this modified hardening/loosening behavior is shown schematically. At point **A**, the intact material begins to deform plastically under hydrostatic pressure. Further increase of loading leads to an increase of density (contraction). At point **B**, powderization begins and hardening slope is less than before. Now, the material is crushed and condensed and therefore, higher compaction is possible (**C**) compared to a load path without powderization (**C'**). The transition from **B** to **C** is defined analog to the usual hardening law

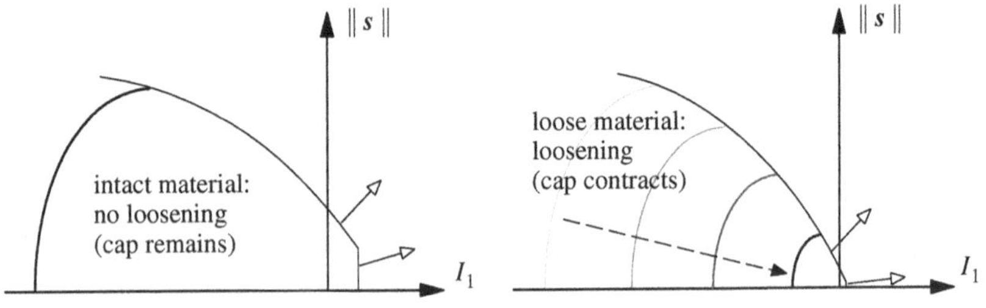

Fig. 6. Different cap treatment for dilatant plastic flow

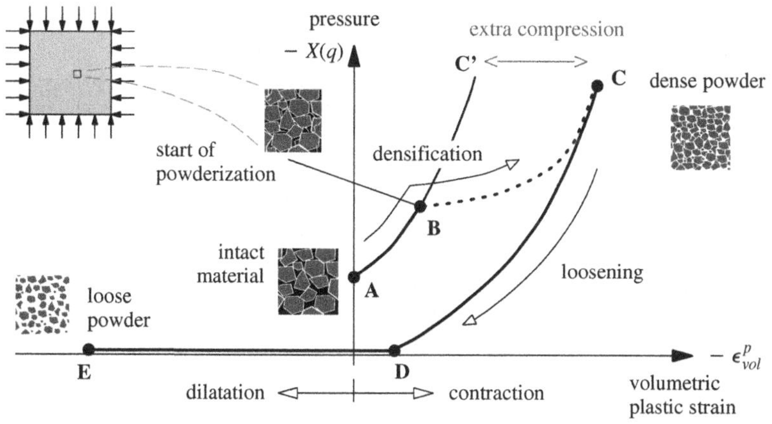

Fig. 7. Modified hardening for geomaterials with powderization

with two new material parameters W_{crush} and D_{crush}. If the highly condensed powder is now loaded in shear or tensile direction (dilatant plastic flow), loosening of the material and a decrease of hydrostatic compressive strength can be observed (**C** → **D**). Further reduction of powder density is achieved with nearly no effort, i.e. unlimited flow of the material is possible.

5 Numerical Examples

Practical relevant applications will demonstrate the performance of our overall approach. In Sec. 5.1, adaptive simulations with high-speed deformations of metals will be presented, followed by several geomaterial computations with our modified DPC model in Sec. 5.2.

5.1 Adaptive Computations

To demonstrate the effectiveness of our adaptive strategy and to examine the suitability of different error indicators for large deformation transient problems, three numerical examples will be presented: Taylor bar impact, high-strain rate compression of WHA block and penetration of a steel cylinder by WHA long rod. For a detailed description of model settings, material parameters, etc. we refer to [12]. Here, only short descriptions of most interesting results are given.

Taylor Bar Impact

A cylindrical rod made of soft metal hits a rigid wall with high velocity [27] [6]. During the impact, elastic and slower plastic waves are initiated, propagating

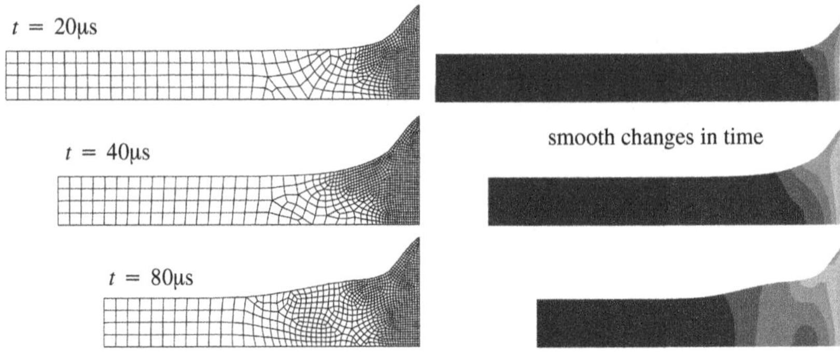

Fig. 8. Taylor bar impact with indicator 'gradient of plastic work': Adaptive meshes and plastic energy (0–1500 Nmm)

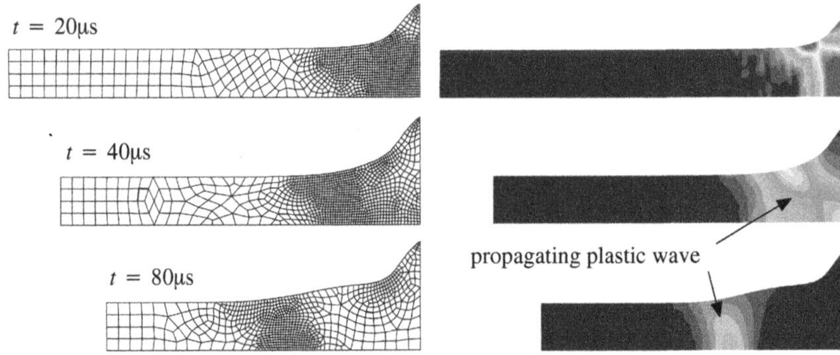

Fig. 9. Taylor bar impact with indicator 'plastic power': Adaptive meshes and plastic strain rates (0-75000 1/s)

in axial direction and reflecting at the bar tails. Therefore, the initial kinetic energy is transformed into internal energy and large plastic deformations can be observed.

In our analysis, we compared different error indicators. In Figs. 8 and Fig.9 the interrelation between adaptive meshes and choice of indicators is illustrated. The indicator 'gradient of plastic work' causes a smooth and continuous growth of fine mesh regions being consistent with the evolving plastic zone (Fig. 8). The indicator 'rate of plastic work' also depicts the connection between adaptive discretizations and the corresponding physical/mechanical process: Refinement occurs where rate of plastic strain is very high, i.e. where current material flow takes place. Therefore, the propagating plastic wave is accompanied by a shift of fine mesh region (Fig. 9).

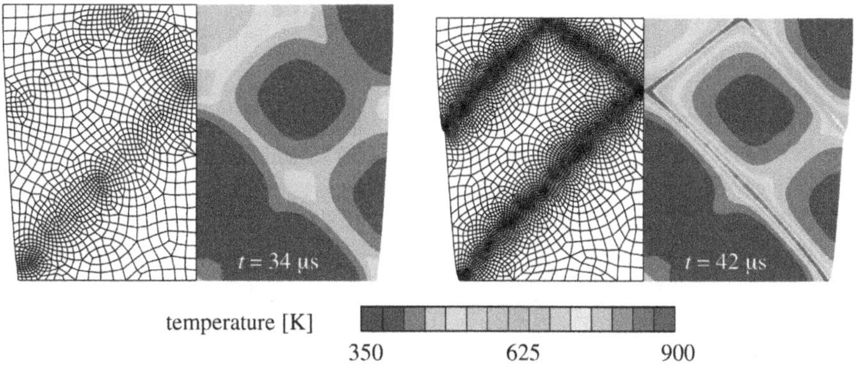

Fig. 10. WHA block: Adaptive meshes and temperature distribution

High-Strain Rate Compression of WHA Block

In the second example, a prismatic metal block undergoes highly dynamic compression, which causes large plastic strains, which again effects significant temperature rise [3]. The involved thermal softening finally results in adiabatic shear bands, i.e. strain localization in consequence of heat. Figure 10 shows the evolution of shear bands represented by temperature distributions and appropriate adaptive meshes. For error assessment, the local quantity indicator 'variation of velocity' was used as refinement indicator. This indicator is well suited here, since dynamically developing strain localizations are detected automatically. The development of localizations results in expedient mesh refinements, which again provide high resolutions in corresponding regions.

Penetration of a Steel Cylinder by WHA Long Rod

A cylindrical rod made of tungsten heavy alloy (WHA) hits an externally clamped cylindrical block made of high-strength steel at high velocity [6] [1]. Figure 11 shows deformed meshes and appropriate equivalent plastic strain distributions at time instances 20 μs and 40 μs with a logarithmic spectrum. The physical process gets obvious: The impactor penetrates the steel cylinder and produces a deep crater. At the same time, the melted tungsten is pressed outwards and the bar almost completely erodes. Plastic deformations are very high in both parts. These phenomena can be observed in experiments [1] and in corresponding simulations [6] in the same way. In our calculation with repeated remeshings, the spatial discretization adaptively follows the transient process. At the beginning, strong refinement in the process zone under the impactor is ascertained. Beyond this area of high plastic deformation, the material behaves elastically and therefore the mesh density is relatively low. The fine mesh region increases with proceeding penetration depth. The error indicator

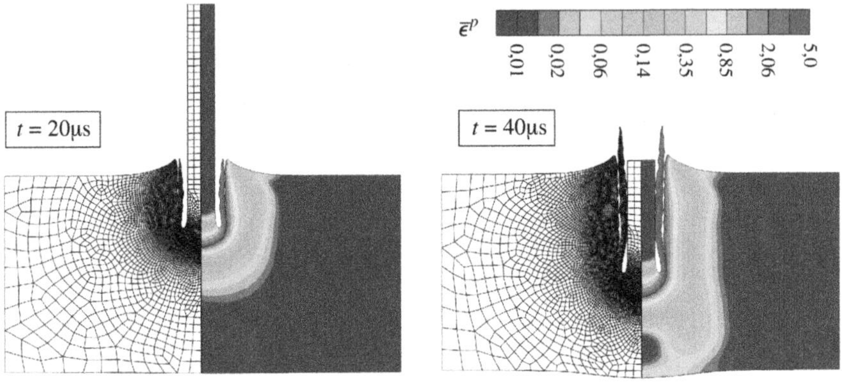

Fig. 11. Adapive meshes and equivalent plastic strains of WHA long rod penetration

'gradient of plastic work' is apparently well suited to follow the physics of this problem in an intelligent way.

5.2 Geomechanical computations

In this section, we present some results from numerical simulations with our enhanced DPC model for geomaterials. More detailed descriptions of these examples can be found in [11] and [10].

Quasi-Static Tests

First, verification and evaluation was done on different quasi-static tests like confined or hydrostatic compression for several geomaterials [4] [5] [29]. After identification of appropriate material parameters good correlation with experimental results can be achieved (see Fig. 12). Here, especially the third diagram of a hydrostatic compression test on sandstone [29] has to be pointed out, since the distinct change of hardening behavior due to powderization is well captured by our modified DPC model.

Triple Impact on Microconcrete

A steel rod with spherical tail hits the surface of a fine grade concrete [10]. Due to the impacts, the material powderizes in a small zone directly under the impactor, which is kept in our model through the criterion described in Sect. 4.2: In Fig. 13, 'powderized elements' are represented by dark grey coloring. Additionally, the evolution of the hydrostatic compressive stress is shown for point A: The densification and loosening of the comminuted material leads to increasing strength followed by decreasing strength for every single impact. On the other hand, this behavior is not discovered for the intact microconcrete

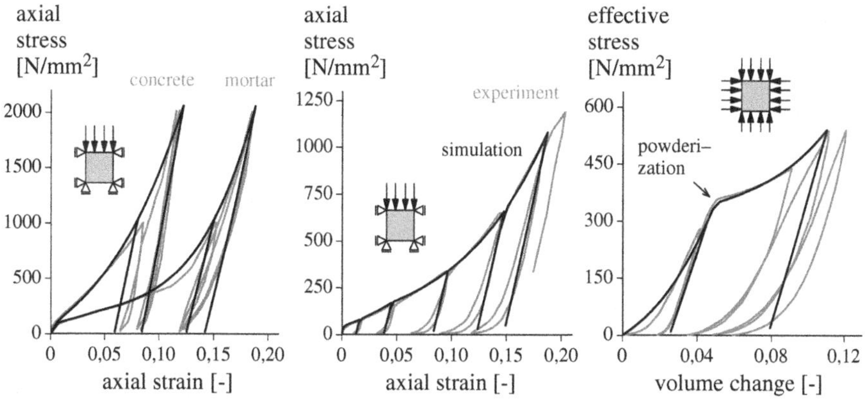

Fig. 12. Confined compression and hydrostatic tests [4] [5] [29]

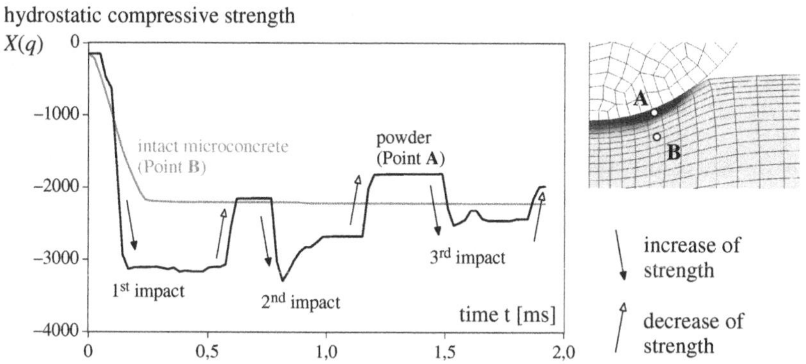

Fig. 13. Results from triple impact on fine grade concrete

(point B). Hence it follows that part of the kinetic energy is dissipated by the arised powder layer. Further inspection of this phenomenon will be done in our last example.

Dynamic Compaction of Powder

This example could be deemed to be a 'dynamic oedometer test', i.e. we carried out dynamic compression (with several impacts) of confined dry fine sand type powder measuring penetration depth and contact force [11]. In the penetration-time diagram (Fig. 14), the periodical indentation for three different layer heights can be observed as well as the decrease of penetration depth after a maximum compression was reached. As gets obvious from Fig. 14 the experimental and numerical results are in good agreement for all three thicknesses. The time dependent behavior of the contact force shows the damping effect of powder (Fig. 14). The periodical impact causes peak loads at the

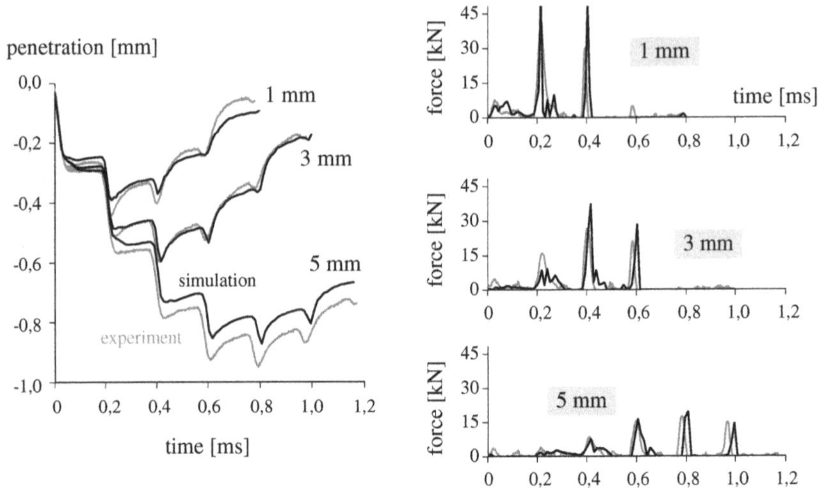

Fig. 14. Evolution of penetration and contact force in dynamic compression of powder

same instances but they are different in size depending on the layer thickness. Also the time of maximum contact force is delayed with increasing height. The amount of absorbed energy, i.e. the grade of damping by powder depends on layer thickness and initial density.

6 Conclusion

High-velocity impact problems are physically extremely complex and therefore numerically very demanding. The specific challenge is due to the interaction of several areas like large deformations exhibiting large strains needing efficient adaptive schemes, dynamic frictional contact algorithms avoiding non-physically oscillations of contact forces and in particular distinct non-linear material models. It was found that the applied explicit formulation was very successful. It came along with time step control and continuous updates of the FE meshes based on physically motivated error indicators. Impact automatically drives the constitutive response into the highly non-linear range. Despite this fact macroscopic continuum based material models turned out to be sufficient for a realistic simulation of the structural response, both for metals as well as for geomaterials. It is remarkable that one constitutive model, namely the Drucker-Prager-Cap plasticity model, is rich enough to represent the material disintegration of geomaterials from the intact phase via a powderization stage with a subsequent hardening to a potential loosening after unloading. It opens a large variety of applications from soil compaction to fasting problems, demolition techniques, crash simulations and other impact problems in civil and military engineering.

References

1. Anderson Jr CE, Hohler V, Walker JD, Stilp AJ (1995) Time-resolved penetration of long rods into steel targets. Int J Imp Engrg 16:1–18
2. Batra RC, Ko KI (1992) An adaptive mesh refinement technique for the analysis of shear bands in plane strain compression of thermoviscoplastic solid. Comput Mech 10:369–379
3. Batra RC, Peng Z (1995) Development of shear bands in dynamic plane strain compression of depleted uranium and tungsten blocks. Int J Imp Engrg 16:375–395
4. Bazant ZP, Bishop FC, Chang TP (1986) Confined compression tests of cement paste and concrete up to 300 ksi. J Amer Conc Inst 83:553–560
5. Burlion N, Pijaudier-Cabot G, Dahan N (2001) Experimental analysis of compaction of concrete and mortar. Int J Num Anal Meth Geomech 25:291–308
6. Camacho GT, Ortiz M (1997) Adaptive Lagrangian modelling of ballistic penetration of metallic targets. Comput Meth Appl Mech Engrg 142:269–301
7. Chen WF, Baladi GY (1985) Soil plasticity – Theory and implementation. McGraw Hill, New York
8. Dyduch M, Habraken AM, Cescotto S (1992) Automatic adaptive remeshing for numerical simulation of metalforming. Comput Meth Appl Mech Engrg 101:283–298
9. Erhart T, Taenzer L, Diekmann R, Wall WA (2001) Adaptive remeshing issues for fast transient, highly nonlinear processes. In Waszczyszyn Z, Pamin J (eds) Proceedings of ECCM 2001. Cracow, Poland
10. Erhart T (2004) Strategien zur numerischen Modellierung transienter Impaktvorgaenge bei nichtlinearem Materialverhalten. PhD Thesis, Institute of Structural Mechanics, University of Stuttgart, Germany
11. Erhart T, Wall WA, Ramm E (2005) A robust computational approach for dry powders under quasi-static and transient impact loadings. Comput Meth Appl Mech Engrg 194:4115–4134
12. Erhart T, Wall WA, Ramm E (2006) Robust adaptive remeshing strategy for large deformation, transient impact simulations. Int J Num Meth Engrg 65:2139–2166
13. Hofstetter G, Simo JC, Taylor RL (1993) A modified cap model: Closest point solution algorithms. Comput Struct 46:203–214
14. Hofstetter G, Mang HA (1995) Computational mechanics of reinforced concrete structures. Vieweg & Sohn, Braunschweig/Wiesbaden
15. Johnson GR, Cook WH (1985) Fracture characteristics of three metals subjected to various strains, strain rates, temperatures and pressures. Engrg Fract Mech 21:31–48
16. Lo SH (1985) A new mesh generation scheme for arbitrary planar domains. Int J Num Meth Engrg 21:1403–1426
17. Marusich TD, Ortiz M (1995) Modelling and simulation of high-speed machining. Int J Num Meth Engrg 38:3675–3694
18. Miehe C (1998) A formulation of finite elastoplasticity based on dual co- and contravariant eigenvector triads normalized with respect to a plastic metric. Comput Meth Appl Mech Engrg 159:223–260
19. Ortiz M, Quigley JJ (1991) Adaptive mesh refinement in strain localization problems. Comput Meth Appl Mech Engrg 90:781–804

20. Peraire J, Vahdati M, Morgan K, Zienkiewicz OC (1987) Adaptive remeshing for compressible flow simulations. J Comp Phys 72:449–466
21. Peric D, Vaz Jr. M, Owen DRJ (1999) On adaptive strategies for large deformations of elasto-plastic solids at finite strains: Computational issues and industrial applications. Comp Meth Appl Mech Engrg 176:279–312
22. Sandler IS, Rubin D (1979) An algorithm and a modular subroutine for the cap model. Int J Num Anal Meth Geomech 3:173–186
23. Simo JC (1992) Algorithms for static and dynamic multiplicative plasticity that preserve the classical return mapping schemes of the infinitesimal theory. Comput Meth Appl Mech Engrg 99:61–112
24. Simo JC, Miehe C (1992) Associative coupled thermoplasticity at finite strains: Formulation, numerical analysis and implementation. Comput Meth Appl Mech Engrg 98:41–104
25. Smith SS, Willam KJ, Gerstle KH, Sture S (1989) Concrete over the top, or: Is there life after peak? ACI Mat J 86:491–497
26. Tabbara M, Blacker T, Belytschko T (1994) Finite element derivative recovery by moving least square interpolants. Comput Meth Appl Mech Engrg 117:211–223
27. Taylor G (1948) The use of flat-ended projectiles for determining dynamic yield stress. I. Theoretical considerations. Proc Roy Soc London, Series A, Math Phys Sci 194:289–299
28. Willam KJ, Warnke E (1975) Constitutive model for triaxial behavior of concrete. In: Proc. Concrete structures subjected to triaxial stresses, International Association for Bridge and Structural Engineering, Section III: 1–30
29. Wong TF, Szeto H, Zhang J (1992) Effect of loading path and porosity on the failure mode of porous rocks. Appl Mech Rev 45:281–293
30. Zhang J, Wong TF, Davis DM (1990) Micromechanics of pressure-induced grain crushing in porous rocks. J Geophys Res 95:341–352
31. Zienkiewicz OC, Zhu JZ (1992) The superconvergent patch recovery and a posteriori error estimates. Part 1: The recovery technique. Int J Num Meth Engrg 33:1331–1364

A Computational Model For Viscoplasticity Coupled with Damage Including Unilateral Effects

D.R.J. Owen[1], F.M. Andrade Pires[2] and E.A. de Souza Neto[1]

[1] Civil and Computational Engineering Centre, School of Engineering University of Wales Swansea, Singleton Park, Swansea SA2 8PP, UK D.R.J.Owen@swansea.ac.uk, E.deSouzaNeto@swansea.ac.uk

[2] DEMEGI – Department of Mechanical Engineering and Industrial Management, Faculty of Engineering, Oporto University, Rua Dr. Roberto Frias, s/n, 4200-465 Porto, Portugal fpires@fe.up.pt

Summary. This contribution is concerned with the numerical modelling of non-linear solid material behaviour in the presence of ductile damage. The description of the complex inelastic material behaviour is accomplished by coupling the elasto-viscoplastic constitutive model, discussed by Perić (1993)[1], with a ductile damage evolution law, introduced by Ladevèze & Lemaitre (1984) [2]. The evolution of the damage internal variable includes the important effect of micro-crack closure, which may dramatically decrease the rate of damage growth under compression [3]. The theoretical basis of the material model and the computational treatment, within the framework of a finite element solution procedure, are presented. The resulting integration algorithm reduces to the solution of only *one scalar non-linear equation* and generalizes the standard *return mapping* procedures of the infinitesimal theory. Numerical tests of the integration algorithm, which rely in the analysis of iso-error maps, are provided.

1 Introduction

The numerical treatment of different material phenomena, in the context of finite element simulations, has been addressed in several publications (see [4, 5, 6, 7, 8, 9, 10, 11] and references therein) during the last three decades or so. As a result, a wide range of material models, incorporating elastic, viscoelastic and elasto-plastic material behaviour is currently available in standard commercial finite-element codes. The computational algorithms that model the inelastic material behaviour have achieved a high degree of maturity. This is particularly true for the isotropic material response and situations

Eugenio Oñate and Roger Owen (eds.), Computational Plasticity, 145–164.

in which different rheological phenomena (elasticity, viscoelasticity, plasticity) can be considered independently of each other.

Despite such developments, it is often necessary to enhance the constitutive description to describe noticeable features of the material behaviour and also to formulate models with greater predictive capability. Here, we are particularly interested in the inelastic constitutive description of materials subjected to forming operations. These processes are usually characterised by the presence of extreme deformations and strains, often resulting in localised material deterioration with possible fracture nucleation and growth. Rate sensitivity and strain rate effects are also known to have a significant role in the constitutive description. In many relevant practical problems, even when the material is initially isotropic, plastic flow is usually responsible for inducing anisotropy. In this case, the experimental identification of material parameters becomes a very difficult and complicated task, with very few examples in the published literature. Bearing in mind that a model intended to represent such phenomena should be simple enough to allow efficient numerical treatment and easy experimental verification of material parameters, this work is restricted to situations in which the overall behaviour can be regarded as *isotropic*. Therefore, a scalar damage variable is chosen to represent the material internal degradation. The assumption of isotropic damage in many cases is not too far from reality, as a result of the random shapes and distribution of the included particles that trigger damage initiation and growth.

The purpose of this contribution is the formulation and numerical implementation of a phenomenological constitutive model for elasto-viscoplastic solids, capable of handling regions of high rate-sensitivity to rate independent conditions in the presence of ductile damage. The description of the complex inelastic material behaviour is accomplished by coupling a power-law elasto-viscoplastic constitutive model [1, 12], which is widely accepted for the description of rate-dependent deformations of solids, with a ductile damage evolution law [2, 13]. The damage growth is influenced by the hydrostatic stress state and includes the important effect of micro-crack closure. The introduction of unilateral damage effects allows for a clear distinction between states of identical triaxiality but stresses of opposite sign (tension and compression) in the damage evolution. This effect may dramatically decrease the rate of damage growth under compression, which was highlighted by numerical tests carried out by the authors [3].

The chapter is organized as follows: Section 2 discusses the essential assumptions of the model and outlines the set of constitutive equations that govern the coupled elasto-viscoplastic damage behaviour. The algorithm for numerical integration of the model is described in detail in Section 3 and the closed form of the consistent tangent operator is presented. An assessment of the accuracy and stability of the elastic predictor-viscoplastic corrector algorithm is carried out relying on the analysis of iso-error maps in Section 4. The chapter ends with the concluding remarks presented in Section 5.

2 Elasto-Viscoplastic Damage Constitutive Model

In this section, the constitutive relations, represented by a set of equations in time, which govern the elasto-viscoplastic damage model with crack closure effects are presented. The undamaged phenomenological behaviour of the material is modelled by a von Mises type power-law elasto-viscoplastic model described in Section 2.1. The important concept of effective stress [14] is recalled in Section 2.2. In Section 2.3 the principle of *strain equivalence* is used to derive effective constitutive equations for the damaged material. The damage evolution law, which includes the important effect of crack closure, is presented in Section 2.4.

2.1 Viscoplastic Model

It is well known that the phenomenological behaviour of real materials is generally time-dependent in the sense that the stress response always depends on the rate of loading and/or the time scale considered. The effects of time dependent mechanisms are particularly visible at higher temperatures. Several different visco-plasticity models have been proposed in the past and, in practice, a particular choice should be dictated by its ability to model the dependency of the plastic strain rate on the state of stress for the material under consideration. This section provides a brief review of the equations governing the undamaged material. The elasto-viscoplastic model described is based on a von Mises yield criterion and a power-law isotropic hardening [1].

The model is defined by an elastic constitutive equation, i.e., a linear elastic relation between the stress tensor, $\boldsymbol{\sigma}$, and the elastic strain, $\boldsymbol{\varepsilon}^e$:

$$\boldsymbol{\sigma} = \mathsf{D}^e : \boldsymbol{\varepsilon}^e \tag{1}$$

where the symbol : denotes double contraction and D^e is the standard isotropic elasticity fourth order tensor given by

$$\mathsf{D}^e = 2G\left[\mathsf{I} - \tfrac{1}{3}\boldsymbol{I}\otimes\boldsymbol{I}\right] + K\,\boldsymbol{I}\otimes\boldsymbol{I} \tag{2}$$

where I, is the fourth order identity tensor. The material constants G and K are, respectively, the shear and bulk moduli. The conventional additive decomposition of the total strain rate, $\dot{\boldsymbol{\varepsilon}}$, into an elastic contribution, $\dot{\boldsymbol{\varepsilon}}^e$, and an inelastic contribution, $\dot{\boldsymbol{\varepsilon}}^{vp}$:

$$\dot{\boldsymbol{\varepsilon}} = \dot{\boldsymbol{\varepsilon}}^e + \dot{\boldsymbol{\varepsilon}}^{vp} \tag{3}$$

is assumed. Furthermore, an associative plastic flow rule is adopted:

$$\dot{\boldsymbol{\varepsilon}}^{vp} = \dot{\gamma}\,\frac{\partial\Phi(\boldsymbol{\sigma},\,\sigma_y)}{\partial\boldsymbol{\sigma}}\,, \tag{4}$$

where $\dot{\gamma}$ is the plastic multiplier whose expression is defined later. In the above, Φ is the von Mises yield function

$$\Phi(\boldsymbol{\sigma}, \sigma_y) \equiv q\,(\boldsymbol{s}\,(\boldsymbol{\sigma})) - \sigma_y = \sqrt{3\,J_2(\boldsymbol{s})} - \sigma_y\,, \tag{5}$$

where $\boldsymbol{s} \equiv \boldsymbol{\sigma} - \frac{1}{3}\,(\mathrm{tr}\boldsymbol{\sigma})\,\boldsymbol{I}$, with $\boldsymbol{I}$ the identity tensor, is the stress deviator, and

$$\sigma_y = \sigma_y(\bar{\varepsilon}^{\,vp}) \tag{6}$$

is the stress-like variable associated with isotropic hardening. In the present case (isotropic *strain* hardening), σ_y is an experimentally determined function of the equivalent plastic strain, $\bar{\varepsilon}^{\,vp}$, whose evolution is defined by the rate equation:

$$\dot{\bar{\varepsilon}}^{\,vp} = \sqrt{\tfrac{2}{3}}\,\|\dot{\boldsymbol{\varepsilon}}^{vp}\|\,. \tag{7}$$

The yield function Φ defines an elastic domain such that the material behaviour is purely elastic (no viscoplastic flow) whenever

$$q < \sigma_y.$$

Among the various possibilities for the definition of $\dot{\gamma}$, here, the following form of a power-type law is adopted [1]:

$$\dot{\gamma} = \frac{1}{\mu}\left\langle \left(\frac{q}{\sigma_y}\right)^{1/\epsilon} - 1\right\rangle, \tag{8}$$

where μ and ϵ are the *viscosity* and *rate-sensitivity*, respectively. These material parameters are, generally, temperature-dependent and can only assume positive values. The symbol $\langle\cdot\rangle$ represents the ramp function defined as

$$\langle x\rangle = (x + |x|)/2. \tag{9}$$

The evolution problem described by the set of constitutive equations (1)–(8), has a firm experimental basis and is widely accepted as a description of rate-dependent deformations of solids.

Remark 1. The elasto-viscoplastic model contains, as special limiting cases, two important models[1]:

(i) When $\mu \to 0$ (no viscosity) and/or $\epsilon \to 0$ (no rate-sensitivity), the standard rate-independent von Mises elasto-plastic model is recovered.
(ii) When $\mu \to \infty$ a form of viscoelastic model is recovered.

2.2 Concept of Effective Stress

An important step in the formulation of damage models is the introduction of damage effects without loosing the properties of well established models of elasto-plasticity and elasto-viscoplaticity. Therefore, several different concepts and postulates have been introduced in the literature in order to account for the material progressive internal deterioration. The most frequently used concept, which is crucial to the definition of the theory, is the concept of *effective stress* [15, 16].

Due to the diversity of forms in which internal damage manifests itself at the microscopic level, variables of different mathematical nature (scalars, vectors, tensors) possessing different physical meaning (reduction of load bearing area, loss of stiffness, distribution of voids) have been employed in the description of damage under various circumstances. Here, only one single scalar variable, D, will be used, representing the simplest possible isotropic formulation. According with the concept of effective stress an *effective stress tensor* is introduced as

$$\tilde{\boldsymbol{\sigma}} \equiv \frac{1}{1-D}\,\boldsymbol{\sigma}\,. \tag{10}$$

The damage variable assumes values between 0 (for the undamaged material) and 1 (for the completely damaged material). In practice, a critical value $D_c < 1$ usually defines the onset of a macro-crack (i.e., complete loss of load carrying capacity at a point).

Continuum damage mechanics relies on the postulate of *strain equivalence*, which states that "the strain behaviour of a damaged material is represented by constitutive equations of the virgin material (without damage) in the potential of which the stress is simply replaced by the effective stress" [14, 17]. This principle can be used to derive effective constitutive equations for the damaged material based on the equations which govern the undamaged material response, simply by replacing the stress tensor $\boldsymbol{\sigma}$ in these equations by the effective stress tensor $\tilde{\boldsymbol{\sigma}}$ according to (10).

2.3 Elasto-Viscoplasticity Coupled with Damage

A coupled elasto-viscoplastic model can be obtained by including the effect of damage in the power-law viscoplastic model described in Section 2.1. This can be accomplished by simply substituting Equation (10) in the definition of the von Mises yield function:

$$\Phi\left(\boldsymbol{\sigma}, \sigma_y, D\right) \equiv \frac{q}{1-D} - \sigma_y = \frac{\sqrt{3\,J_2(\boldsymbol{s})}}{1-D} - \sigma_y(\bar{\varepsilon}^{vp})\,. \tag{11}$$

It should be noted that (11) accounts for two competing effects: damaging, which shrinks (isotropically) the elastic domain (defined as the subset of stress space for which $\Phi \le 0$) as D grows; and hardening, which can expand the elastic domain (also isotropically) with the growth of σ_y. The von Mises yield function can be rewritten as

$$\Phi\left(\boldsymbol{\sigma}, \sigma_y, D\right) \equiv q - (1-D)\,\sigma_y = \sqrt{3\,J_2(\boldsymbol{s})} - (1-D)\sigma_y(\bar{\varepsilon}^{vp})\,. \tag{12}$$

If this particular form is used, the associative plastic flow rule (4) remains unchanged, i.e., is not directly affected by the introduction of damage. This equation is more convenient and will be used later for the computational implementation. In addition to Equation (12), damage effects will also be included in the definition of the viscoplastic multiplier:

$$\dot{\gamma} = \frac{1}{\mu} \left\langle \left[\frac{q}{(1-D)\sigma_y} \right]^{1/\epsilon} - 1 \right\rangle , \tag{13}$$

or equivalently,

$$\dot{\gamma} = \begin{cases} \dfrac{1}{\mu} \left[\left(\dfrac{q}{(1-D)\sigma_y} \right)^{1/\epsilon} - 1 \right] & \text{if} \quad \Phi(\boldsymbol{\sigma}, \sigma_y, D) > 0 \\ 0 & \text{if} \quad \Phi(\boldsymbol{\sigma}, \sigma_y, D) \leq 0 \, . \end{cases} \tag{14}$$

Note that the effect of internal damage on the elastic behaviour of the material is ignored in the present model. That is, the elasticity tensor is not a function of the damage variable or in other words, elasticity and damage are assumed to be decoupled. This simplification can be justified if the elastic strain remains truly infinitesimal in the type of problems addressed with this model.

Remark 2. The damage variable ranges between 0 and 1, with $D=0$ corresponding to the sound (undamaged) material and $D=1$ to the fully damaged state with complete loss of load carrying capacity. Note that damage growth induces softening, i.e., shrinkage of the yield surface defined by

$$\Phi = 0 \, .$$

For $D=0$ the yield surface reduces to that of the (pressure insensitive) von Mises type power-law elasto-viscoplastic model. In the presence of damage, i.e., for $D \neq 0$ the yield surface shrinks and its size reduces to zero for $D=1$.

2.4 Damage Evolution Law

The damage evolution law should reflect the nucleation and growth of voids and microcracks which accompany viscoplastic flow. Damage and viscoplasticity are undoubtedly coupled, as the presence of internal deterioration introduces local stress concentrations which may in turn drive viscoplastic deformation. The evolution of the damage internal variable is assumed to be governed by the relation:

$$\dot{D} = \begin{cases} 0 & \text{if } \bar{\varepsilon}^{vp} \leq \bar{\varepsilon}_D^{vp} \\ \dfrac{\dot{\gamma}}{1-D} \left(\dfrac{-Y}{r} \right)^s & \text{if } \bar{\varepsilon}^{vp} > \bar{\varepsilon}_D^{vp} \, , \end{cases} \tag{15}$$

where r, s and $\bar{\varepsilon}_D^{vp}$ are material constants. In the nucleation phase, experimental evidence reveals that there is no noticeable effect of damage on the mechanical properties, therefore the constant $\bar{\varepsilon}_D^{vp}$ is the so-called *damage threshold*, i.e., the value of accumulated plastic strain below which no damage evolution is observed. The quantity

$$
\begin{aligned}
Y = \frac{-1}{2E(1-D)^2}[(1+\nu)\boldsymbol{\sigma}_+ : \boldsymbol{\sigma}_+ - \nu \langle \mathrm{tr}\boldsymbol{\sigma}\rangle^2] \\
- \frac{h}{2E(1-hD)^2}[(1+\nu)\boldsymbol{\sigma}_- : \boldsymbol{\sigma}_- - \nu \langle -\mathrm{tr}\boldsymbol{\sigma}\rangle^2] ,
\end{aligned}
\tag{16}
$$

is the *damage energy release rate*, with E and ν denoting, respectively, the Young's modulus and the Poisson's ratio of the undamaged material. The tensors $\boldsymbol{\sigma}_+$ and $\boldsymbol{\sigma}_+$ are, respectively, the *tensile* and *compressive* components of $\boldsymbol{\sigma}$, defined as:

$$
\boldsymbol{\sigma}_+ = \sum_{i=1}^{3} \langle \sigma_i \rangle \, \boldsymbol{e}_i \otimes \boldsymbol{e}_i \tag{17}
$$

and

$$
\boldsymbol{\sigma}_- = \sum_{i=1}^{3} \langle -\sigma_i \rangle \, \boldsymbol{e}_i \otimes \boldsymbol{e}_i \, , \tag{18}
$$

with $\{\sigma_i\}$ and $\{\boldsymbol{e}_i\}$ denoting, respectively, the eigenvalues and an orthonormal basis of eigenvectors of $\boldsymbol{\sigma}$. The *crack closure parameter*, h, is an experimentally determined coefficient which satisfies:

$$
0 \leq h \leq 1 \, . \tag{19}
$$

This coefficient characterizes the closure of microcracks and micro-cavities and depends upon the density and the shape of the defects. It is a material dependent parameter and, for simplicity, h is considered as constant. A value $h \approx 0.2$ is typically observed in many experiments [18]. This definition of the energy release rate (16) was introduced by Ladevèze (1983)[13] and Ladevèze & Lemaitre (1984) [2]. Note that, for a state of purely tensile principal stresses, the damage energy release rate (16), can be simplified and rewritten as

$$
\begin{aligned}
Y &= \frac{-1}{2E(1-D)^2} \left[(1+\nu)\, \boldsymbol{\sigma} : \boldsymbol{\sigma} - \nu \, (\mathrm{tr}\, \boldsymbol{\sigma})^2\right] \\
&= \frac{-q^2}{2E(1-D)^2} \left[\frac{2}{3}(1+\nu) + 3(1-2\nu)\left(\frac{p}{q}\right)^2\right] .
\end{aligned}
\tag{20}
$$

For states with purely compressive principal stresses, (16) will give absolute values of Y smaller than those produced by (20), resulting in a decrease of damage growth rates. Also note that the limit $h = 1$ corresponds to no crack closure effect whereas the other extreme, $h = 0$, corresponds to a total crack closure, with no damage evolution under compression. Any other value of h describes a partial crack closure effect.

Remark 3. The particular form of the energy release rate (20), was initially proposed by Lemaitre (1983) [19] in order to describe the influence of stress triaxiality ratio, p/q, on the rate of damage growth. The inclusion of the hydrostatic component of $\boldsymbol{\sigma}$ in the definition of Y implies that $\dot{D}$ increases (decreases) with increasing (decreasing) triaxiality ratio.

One important feature of damage growth is the clear distinction between rates of damage growth observed for states of stress with identical triaxiality but stresses of opposite sign (tension and compression). Such a distinction stems from the fact that, under a compressive state, voids and micro-cracks that would grow under tension will partially close, reducing (possibly dramatically) the damage growth rate. This phenomenon can be crucially important in the simulation of forming operations, particularly under extreme strains. It is often the case that, in such operations, the solid (or parts of it) undergoes extreme compressive straining followed by extension or vice-versa [3].

3 Integration Algorithm

In this section the derivation of an integration algorithm for the elasto-visco-plastic damage constitutive model, described in the previous section is carried out in detail. Operator split algorithms are particularly suitable for numerical integration of constitutive equations and are widely used in the context of elasto-plasticity and also elasto-viscoplasticity [20, 21, 22, 1, 7].

Let us consider a typical time step over the time interval $[t_n, t_{n+1}]$, where the time and strain increments are defined in the usual way as

$$\Delta t = t_{n+1} + t_n, \qquad \Delta\boldsymbol{\varepsilon} \equiv \boldsymbol{\varepsilon}_{n+1} - \boldsymbol{\varepsilon}_n \,. \tag{21}$$

In addition, all variables of the problem, given by the set $\{\boldsymbol{\sigma}_n, \boldsymbol{\varepsilon}^e_n, \boldsymbol{\varepsilon}^{vp}_n, \bar{\varepsilon}^{vp}_n, D_n\}$, are assumed to be known at t_n. The operator split algorithm should obtain the updated set $\{\boldsymbol{\sigma}_{n+1}, \boldsymbol{\varepsilon}^e_{n+1}, \boldsymbol{\varepsilon}^{vp}_{n+1}, \bar{\varepsilon}^{vp}_{n+1}, D_{n+1}\}$ of variables at t_{n+1} consistently with the evolution equations of the model. The algorithm comprises the standard *elastic predictor* and the *visco-plastic return mapping* which, for the present model, has the following format.

Elastic Predictor

The first step in the algorithm is the evaluation of the *elastic trial state* where the increment is assumed purely elastic with no evolution of internal variables (internal variables *frozen* at t_n). The elastic trial strain and trial accumulated viscoplastic strain are given by:

$$\boldsymbol{\varepsilon}^{\mathrm{e\,trial}} = \boldsymbol{\varepsilon}^{\mathrm{e}}_n + \Delta\boldsymbol{\varepsilon}; \qquad \bar{\varepsilon}^{vp\,trial} = \bar{\varepsilon}^{vp}_n \,. \tag{22}$$

The corresponding elastic trial stress tensor is computed:

$$\boldsymbol{\sigma}^{\text{trial}} = \mathbf{D}^e : \boldsymbol{\varepsilon}^{e\,\text{trial}}, \tag{23}$$

where $\mathbf{D}^e$ is the standard isotropic elasticity tensor. Equivalently, in terms of stress deviator and hydrostatic pressure, we have:

$$\boldsymbol{s}^{\text{trial}} = 2G\,\boldsymbol{e}^{e\,\text{trial}}, \qquad p^{\text{trial}} = K\,v^{e\,\text{trial}}, \tag{24}$$

where

$$\boldsymbol{e}^{e\,\text{trial}} = \boldsymbol{e}^e_n + \Delta\boldsymbol{e}, \qquad v^{e\,\text{trial}} = v^e_n + \Delta v\,. \tag{25}$$

The material constants G and K are, respectively, the shear and bulk moduli, $\boldsymbol{s}$ and p stand for the deviatoric and hydrostatic stresses. The strain deviator and the volumetric strain are denoted, respectively, by $\boldsymbol{e}$ and v. The trial yield stress is simply

$$\sigma_y^{\text{trial}} = \sigma_y(\bar{\varepsilon}^{vp}). \tag{26}$$

The next step of the algorithm is to check whether $\boldsymbol{\sigma}^{\text{trial}}$ lies inside or outside of the trial yield surface. With variables $\bar{\varepsilon}^{vp}$ and D frozen at time t_n we compute:

$$\begin{aligned}\Phi^{\text{trial}} &:= q^{\text{trial}} - (1-D_n)\sigma_y(\bar{\varepsilon}^{vp}) \\ &= \sqrt{\tfrac{3}{2}}\|\boldsymbol{s}^{\text{trial}}\| - (1-D_n)\sigma_y(\bar{\varepsilon}^{vp})\,.\end{aligned} \tag{27}$$

If $\Phi^{\text{trial}} \leq 0$, the process is indeed elastic within the interval and the elastic trial state coincides with the updated state at t_{n+1}. In other words, there is no viscoplastic flow or damage evolution within the interval and

$$\begin{aligned}&\boldsymbol{\varepsilon}^e_{n+1} = \boldsymbol{\varepsilon}^{e\,\text{trial}}; \qquad \boldsymbol{\sigma}_{n+1} = \boldsymbol{\sigma}^{\text{trial}}; \qquad \bar{\varepsilon}^{vp}_{n+1} = \bar{\varepsilon}^{vp\,\text{trial}}; \\ &\sigma_{y\,n+1} = \sigma_y^{\text{trial}}; \qquad D_{n+1} = D^{\text{trial}}.\end{aligned} \tag{28}$$

Otherwise, we apply the viscoplastic corrector algorithm described in the following.

Visco-plastic corrector (or return mapping algorithm)

At this stage, we solve the evolution equations of the model with the elastic trial state as the initial condition. With the adoption of a backward Euler discretisation, the viscoplastic corrector is given by the following set of algebraic equations:

$$\begin{aligned}&\boldsymbol{\sigma}_{n+1} = \boldsymbol{\sigma}^{\text{trial}} - \Delta\gamma\,\mathbf{D} : \left.\frac{\partial\Phi}{\partial\boldsymbol{\sigma}}\right|_{n+1} \\ &\bar{\varepsilon}^{vp}_{n+1} = \bar{\varepsilon}^{vp}_n + \Delta\gamma \\ &D_{n+1} = \begin{cases} 0 & \text{if } \bar{\varepsilon}^{vp}_{n+1} \leq \bar{\varepsilon}^{vp}_D \\ D_n + \frac{\Delta\gamma}{1-D_{n+1}}\left(\frac{-Y_{n+1}}{r}\right)^s & \text{if } \bar{\varepsilon}^{vp}_{n+1} > \bar{\varepsilon}^{vp}_D\,, \end{cases}\end{aligned} \tag{29}$$

where the *incremental multiplier*, $\Delta\gamma$, is given by:

$$\Delta\gamma = \frac{\Delta t}{\mu} \left\{ \left[\frac{q(\boldsymbol{\sigma}_{n+1})}{(1 - D_{n+1})\, \sigma_y(\bar{\varepsilon}^{vp}_{n+1})} \right]^{1/\epsilon} - 1 \right\}, \tag{30}$$

with Δt denoting the time increment within the considered interval. After solving (29), we can update:

$$\boldsymbol{\varepsilon}^{vp}_{n+1} = \boldsymbol{\varepsilon}^{vp}_{n} + \Delta\gamma \left. \frac{\partial \Phi}{\partial \boldsymbol{\sigma}} \right|_{n+1} \qquad \boldsymbol{\varepsilon}^{e}_{n+1} = \boldsymbol{\varepsilon}^{e\ \text{trial}} - \Delta\gamma \left. \frac{\partial \Phi}{\partial \boldsymbol{\sigma}} \right|_{n+1} . \tag{31}$$

The visco-plastic corrector can be more efficiently implemented by reducing (29) to a *single* non-linear equation for the incremental multiplier $\Delta\gamma$.

3.1 Single-Equation Corrector

As we shall see in what follows, analogously to what happens to the classical von Mises model, the above system can be reduced by means of simple algebraic substitutions to a *single* non-linear equation having the incremental plastic multiplier, $\Delta\gamma$, as a variable. Firstly, we observe that the plastic flow vector:

$$\frac{\partial \Phi}{\partial \boldsymbol{\sigma}} = \sqrt{\tfrac{3}{2}}\, \frac{\boldsymbol{s}}{\|\boldsymbol{s}\|} \tag{32}$$

is deviatoric. The stress update equation $(29)_1$ can then be split as:

$$\begin{aligned} \boldsymbol{s}_{n+1} &= \boldsymbol{s}^{\text{trial}} - \Delta\gamma\, 2G \sqrt{\tfrac{3}{2}}\, \frac{\boldsymbol{s}_{n+1}}{\|\boldsymbol{s}_{n+1}\|} \\ p_{n+1} &= p^{\text{trial}} , \end{aligned} \tag{33}$$

where p denotes the hydrostatic pressure and G is the shear modulus. Further, simple inspection of $(33)_1$ shows that $\boldsymbol{s}_{n+1}$ is a scalar multiple of $\boldsymbol{s}^{\text{trial}}$ so that, trivially, we have the identity:

$$\frac{\boldsymbol{s}_{n+1}}{\|\boldsymbol{s}_{n+1}\|} = \frac{\boldsymbol{s}^{\text{trial}}}{\|\boldsymbol{s}^{\text{trial}}\|} , \tag{34}$$

which allows us to re-write $(33)_1$ as:

$$\boldsymbol{s}_{n+1} = \left(1 - \sqrt{\tfrac{3}{2}}\, \frac{\Delta\gamma\, 2G}{\|\boldsymbol{s}^{\text{trial}}\|} \right) \boldsymbol{s}^{\text{trial}} = \left(1 - \frac{\Delta\gamma\, 3G}{q^{\text{trial}}} \right) \boldsymbol{s}^{\text{trial}} \tag{35}$$

where q^{trial} is the elastic trial von Mises equivalent stress:

$$q^{\text{trial}} = q(\boldsymbol{s}^{\text{trial}}) = \sqrt{\tfrac{3}{2}}\, \|\boldsymbol{s}^{\text{trial}}\| . \tag{36}$$

Equation (35) results in the following update formula for q:

$$q_{n+1} = q^{\text{trial}} - 3G\,\Delta\gamma\,. \tag{37}$$

With the substitution of the above formula together with $(29)_2$ into (30) we obtain the following scalar algebraic equation for the incremental multiplier, $\Delta\gamma$:

$$\Delta\gamma - \frac{\Delta t}{\mu}\left\{\left[\frac{q^{\text{trial}} - 3G\,\Delta\gamma}{(1-D_{n+1})\,\sigma_y(\bar{\varepsilon}^{vp}_{n+1})}\right]^{1/\epsilon} - 1\right\} = 0\,, \tag{38}$$

or, equivalently, after a straightforward rearrangement,

$$D_{n+1} = D(\Delta\gamma) \equiv 1 - \frac{\sqrt{\frac{3}{2}}\,\|\boldsymbol{s}^{\text{trial}}\| - 3G\,\Delta\gamma}{\sigma_{y_0} + R(\bar{\varepsilon}^{vp}_n + \Delta\gamma)}\left(\frac{\Delta t}{\mu\,\Delta\gamma + \Delta t}\right)^{\epsilon}, \tag{39}$$

which expresses D_{n+1} as an explicit function of $\Delta\gamma$. Finally, by introducing the damage explicit function (39) into the discretised damage evolution equation $(29)_3$, the viscoplastic corrector is reduced to the solution of a *single* algebraic equation for the incremental multiplier, $\Delta\gamma$:

$$F(\Delta\gamma) \equiv \begin{cases} D(\Delta\gamma) = 0 & \text{if } \bar{\varepsilon}^{vp}_{n+1} \leq \bar{\varepsilon}^{\text{p}}_D \\ D(\Delta\gamma) - D_n - \frac{\Delta\gamma}{1-D(\Delta\gamma)}\left(\frac{-Y(\Delta\gamma)}{r}\right)^{s} = 0 & \text{if } \bar{\varepsilon}^{vp}_{n+1} > \bar{\varepsilon}^{vp}_D\,. \end{cases} \tag{40}$$

In $(40)_2$, the dependency of Y on $\Delta\gamma$ originates from its dependency on the updated values of D and $\boldsymbol{\sigma}$:

$$\begin{aligned} Y(\Delta\gamma) = &\frac{-1}{2E[1-D(\Delta\gamma)]^2}[(1+\nu)\boldsymbol{\sigma}_+(\Delta\gamma) : \boldsymbol{\sigma}_+(\Delta\gamma) - \nu\,\langle \text{tr}\boldsymbol{\sigma}(\Delta\gamma)\rangle^2] \\ &- \frac{h}{2E[1-hD(\Delta\gamma)]^2}[(1+\nu)\boldsymbol{\sigma}_-(\Delta\gamma) : \boldsymbol{\sigma}_-(\Delta\gamma) - \nu\,\langle -\text{tr}\boldsymbol{\sigma}(\Delta\gamma)\rangle^2]\,, \end{aligned} \tag{41}$$

The updated stress tensor, $\boldsymbol{\sigma}_{n+1}$, whose tensile and compressive components take part in the calculation of Y_{n+1}, is obtained as:

$$\boldsymbol{\sigma}_{n+1} = \boldsymbol{s}_{n+1} + p_{n+1}\,\boldsymbol{I}\,, \tag{42}$$

where $\boldsymbol{I}$ is the second order identity tensor and $\boldsymbol{s}_{n+1}$ is obtained from the standard implicit return mapping as a function of $\Delta\gamma$ according to update formula (35):

$$\boldsymbol{s}_{n+1} = \left(1 - \frac{\Delta\gamma\;3G}{q^{\text{trial}}}\right)\boldsymbol{s}^{\text{trial}}; \qquad p_{n+1} = p^{\text{trial}}\,. \tag{43}$$

The single-equation viscoplastic corrector comprises the solution of the above equation for $\Delta\gamma$, followed by the straightforward update of the relevant variables. The solution of the equation for $\Delta\gamma$ is, as usual, undertaken by the Newton-Raphson iterative scheme. The overall algorithm for the numerical integration of the elasto-viscoplastic damage model, which includes the effect of crack closure, is summarised in Box 1 in pseudo-code format.

(i) *Elastic predictor.* Given $\Delta\varepsilon$, Δt and the state variables at t_n, compute the *elastic trial state*:

$$\begin{aligned}
&\boldsymbol{\varepsilon}^{e\,trial} = \boldsymbol{\varepsilon}^{e}_{n} + \Delta\boldsymbol{\varepsilon}; \quad \boldsymbol{e}^{trial} = \text{dev}[\boldsymbol{\varepsilon}^{e\,trial}]; \; v^{trial} = \text{tr}[\boldsymbol{\varepsilon}^{e\,trial}] \\
&\bar{\varepsilon}^{vp\,trial} = \bar{\varepsilon}^{vp}_{n}; \qquad D^{trial} = D_n \\
&\boldsymbol{s}^{trial} = 2G\,\boldsymbol{e}^{trial}; \qquad p^{trial} = K\,v^{trial} \\
&q^{trial} = \sqrt{\tfrac{3}{2}}\,\|\boldsymbol{s}^{trial}\|,
\end{aligned}$$

(ii) Check for viscoplastic flow. First compute:

$$\Phi^{trial} = q^{trial} - (1 - D_n)\left[\sigma_{y_0} + R(\bar{\varepsilon}^{vp}_{n})\right],$$

IF $\Phi^{trial} \leq \varepsilon_{tol}$ THEN (elastic step)

Update $(\cdot)_{n+1} = (\cdot)^{trial}$ and EXIT

ELSE GOTO (iii)

(iii) *Visco-plastic corrector.* Solve the return mapping equation

$$F(\Delta\gamma) \equiv \begin{cases} D(\Delta\gamma) = 0 & \text{if } \bar{\varepsilon}^{vp}_{n+1} \leq \bar{\varepsilon}^{vp}_{D} \\ D(\Delta\gamma) - D_n - \frac{\Delta\gamma}{1 - D(\Delta\gamma)}\left(\frac{-Y(\Delta\gamma)}{r}\right)^{s} = 0 & \text{if } \bar{\varepsilon}^{vp}_{n+1} > \bar{\varepsilon}^{vp}_{D} \end{cases}$$

with $D(\Delta\gamma)$ defined by (39) and $Y(\Delta\gamma)$ defined through (16), (39) (42) and (43).

(iv) *Update the variables*:

$$\begin{aligned}
&\boldsymbol{s}_{n+1} = \left(1 - \frac{\Delta\gamma\,3G}{q^{trial}}\right)\boldsymbol{s}^{trial}; \qquad p_{n+1} = p^{trial}; \\
&\boldsymbol{\sigma}_{n+1} = \boldsymbol{s}_{n+1} + p_{n+1}\,\boldsymbol{I}; \qquad \bar{\varepsilon}^{vp}_{n+1} = \bar{\varepsilon}^{vp}_{n} + \Delta\gamma; \\
&\boldsymbol{\varepsilon}^{e}_{n+1} = \frac{1}{2G}\,\boldsymbol{s}_{n+1} + \frac{1}{3K}\,p_{n+1}\,\boldsymbol{I}; \quad D_{n+1} = D(\Delta\gamma).
\end{aligned}$$

(v) EXIT

Box 1: Elastic predictor/visco-plastic return mapping integration algorithm for the elasto-viscoplastic damage model with crack closure effect (over time interval $[t_n, t_{n+1}]$)

Remark 4. (computational implementation aspects) In the computer implementation of the model (as shown in Box 1), it is important to specify the damage function $D(\Delta\gamma)$, as expressed in equation (39). The reason for this lies in the fact that, for low rate-sensitivity, i.e., small values of ϵ, the Newton-Raphson scheme for solution of (38) becomes unstable as its convergence bowl is sharply reduced with decreasing ϵ. The reduction of the convergence bowl stems from the fact that large exponents $1/\epsilon$ can easily produce numbers which are computationally intractable. This fact has been recognised by Perić (1993) [1] in the context of a more general visco-plastic algorithm. In equation (39), on the other hand, the term to the power ϵ on the left hand side can only assume values within the interval $[0, 1]$ and causes no numerical problems within practical ranges of material constants.

3.2 Consistent Tangent Operator

To obtain the consistent tangent operator for the case of a strain-driven problem, all variables of the problem are considered as functions of the strain $\boldsymbol{\varepsilon}$. The exact linearization of the algorithm described in Box 1 is performed by a systematic application of the concept of directional derivative.

In the elastic case, the *elastic consistent tangent* at t_{n+1}, is simply the standard elasticity operator

$$\hat{\mathsf{D}} = \mathsf{D}^e = 2G \left[\mathsf{I} - \tfrac{1}{3} \boldsymbol{I} \otimes \boldsymbol{I} \right] + K \, \boldsymbol{I} \otimes \boldsymbol{I} \tag{44}$$

where I, is the fourth order identity tensor.

In the elasto-viscoplastic damage case, i.e., when it is assumed that viscoplastic flow occurs within the step, the tangent operator is called the *elasto-viscoplastic damage consistent tangent* and is denoted by $\hat{\mathsf{D}}^{vp}$. For the present model it is possible to obtain a closed form expression for the tangent operator. The details of derivation, which is rather lengthy, will be omitted here and we shall limit ourselves to show only its final expression which is given by:

$$\mathsf{D}^{vp} = a_1 \left[\mathsf{I} - \tfrac{1}{3} \boldsymbol{I} \otimes \boldsymbol{I} \right] + a_2 \, \bar{\boldsymbol{s}}_{n+1} \otimes \bar{\boldsymbol{s}}_{n+1} + a_3 \, \bar{\boldsymbol{s}}_{n+1} \otimes \boldsymbol{I} + K \, \boldsymbol{I} \otimes \boldsymbol{I} \,, \tag{45}$$

where $\bar{\boldsymbol{s}}_{n+1}$ is the normalised stress deviator:

$$\bar{\boldsymbol{s}}_{n+1} = \frac{\boldsymbol{s}_{n+1}}{||\boldsymbol{s}_{n+1}||} \,, \tag{46}$$

and the scalars a_1, a_2, a_3, are given by:

$$\begin{aligned} a_1 &= 2G \left(1 - \tfrac{\Delta\gamma \, 3G}{q^{\mathrm{trial}}} \right) \\ a_2 &= 6G^2 \left[\tfrac{\Delta\gamma}{q^{\mathrm{trial}}} + \tfrac{\partial F}{\partial \left(q^{\mathrm{trial}} \right)} \Big/ \tfrac{\partial F}{\partial \left(\Delta\gamma \right)} \right] \\ a_3 &= 2G \sqrt{\tfrac{2}{3}} \, K \left[\tfrac{\partial F}{\partial \left(p^{\mathrm{trial}} \right)} \Big/ \tfrac{\partial F}{\partial \left(\Delta\gamma \right)} \right] . \end{aligned} \tag{47}$$

In the definition of constants a_2 and a_3, the scalars $\partial F / \partial \left(\Delta\gamma \right)$, $\partial F / \partial \left(q^{\mathrm{trial}} \right)$ and $\partial F / \partial \left(p^{\mathrm{trial}} \right)$, correspond to the derivatives of the return mapping residual function defined by (40):

$$
\begin{aligned}
\frac{\partial F}{\partial(\Delta\gamma)} &= \frac{\partial D}{\partial(\Delta\gamma)} + \frac{1}{1-D_{n+1}}\left(-\frac{Y_{n+1}}{r}\right)^s \\
&\qquad \left\{\left[-\frac{\partial D}{\partial(\Delta\gamma)}/(1-D_{n+1}) - s\,\frac{\partial Y}{\partial(\Delta\gamma)}/Y_{n+1}\right]\Delta\gamma - 1\right\} \\
\frac{\partial F}{\partial(q^{\text{trial}})} &= \frac{\partial D}{\partial(q^{\text{trial}})} - \Delta\gamma\frac{\partial D}{\partial(q^{\text{trial}})}/(1-D_{n+1})^2\left(-\frac{Y_{n+1}}{r}\right)^s \\
&\qquad - \frac{s\,\Delta\gamma}{r\,(1-D_{n+1})}\left(-\frac{Y_{n+1}}{r}\right)^{s-1}\frac{\partial Y}{\partial(q^{\text{trial}})} \\
\frac{\partial F}{\partial(p^{\text{trial}})} &= \frac{s\,\Delta\gamma}{r\,(1-D_{n+1})}\left(-\frac{Y_{n+1}}{r}\right)^s\frac{\partial Y}{\partial(p^{\text{trial}})}.
\end{aligned}
\tag{48}
$$

where the scalars $\partial Y/\partial(\Delta\gamma)$, $\partial Y/\partial(q^{\text{trial}})$ and $\partial Y/\partial(p^{\text{trial}})$, represent the derivatives of the energy release rate function, defined by (16):

$$
\begin{aligned}
\frac{\partial Y}{\partial(\Delta\gamma)} &= -\frac{\frac{\partial D}{\partial(\Delta\gamma)}}{E\,(1-D_{n+1})^3}\,b_+ + \frac{2G\sqrt{\frac{3}{2}}}{E\,(1-D_{n+1})^2}\,\boldsymbol{C}_+ : \bar{\mathbf{s}}_{n+1} - \frac{h^2\,\frac{\partial D}{\partial(\Delta\gamma)}}{E\,(1-h\,D_{n+1})^3}\,b_- \\
&\quad + \frac{2G\sqrt{\frac{3}{2}}\,h}{E\,(1-h\,D_{n+1})^2}\,\boldsymbol{C}_- : \bar{\mathbf{s}}_{n+1} \\
\frac{\partial Y}{\partial(q^{\text{trial}})} &= -\frac{\frac{\partial D}{\partial(q^{\text{trial}})}}{E\,(1-D_{n+1})^3}\,b_+ - \frac{\sqrt{\frac{2}{3}}}{E\,(1-D_{n+1})^2}\,\boldsymbol{C}_+ : \bar{\mathbf{s}}_{n+1} - \frac{h^2\,\frac{\partial D}{\partial(q^{\text{trial}})}}{E\,(1-h\,D_{n+1})^3}\,b_- \\
&\quad + \frac{\sqrt{\frac{2}{3}}\,h}{E\,(1-h\,D_{n+1})^2}\,\boldsymbol{C}_- : \bar{\mathbf{s}}_{n+1} \\
\frac{\partial Y}{\partial(p^{\text{trial}})} &= -\frac{1}{E\,(1-D_{n+1})^2}\,\boldsymbol{C}_+ : \boldsymbol{I} - \frac{h}{E\,(1-h\,D_{n+1})^2}\,\boldsymbol{C}_- : \boldsymbol{I}
\end{aligned}
\tag{49}
$$

furthermore, the scalars b_+ and b_- and the second order tensors $\boldsymbol{C}_+$ and $\boldsymbol{C}_-$ introduced in (49), are given by

$$
\begin{aligned}
b_+ &= (1+\nu)\,\boldsymbol{\sigma}_+ : \boldsymbol{\sigma}_+ - \nu\,\langle \operatorname{tr}\boldsymbol{\sigma}\rangle^2 \\
b_- &= (1+\nu)\,\boldsymbol{\sigma}_- : \boldsymbol{\sigma}_- - \nu\,\langle -\operatorname{tr}\boldsymbol{\sigma}\rangle^2 \\
\boldsymbol{C}_+ &= (1+\nu)\,\frac{\partial\boldsymbol{\sigma}_+}{\partial\boldsymbol{\sigma}} : \boldsymbol{\sigma}_+ - \nu\,\langle \operatorname{tr}\boldsymbol{\sigma}\rangle\,\boldsymbol{I} \\
\boldsymbol{C}_- &= (1+\nu)\,\frac{\partial\boldsymbol{\sigma}_-}{\partial\boldsymbol{\sigma}} : \boldsymbol{\sigma}_- - \nu\,\langle -\operatorname{tr}\boldsymbol{\sigma}\rangle\,\boldsymbol{I}
\end{aligned}
\tag{50}
$$

Here, one should note that the tensors $\boldsymbol{C}_+$ and $\boldsymbol{C}_-$ defined in $(50)_3$ and $(50)_4$, respectively, contain the terms:

$$
\frac{\partial\boldsymbol{\sigma}_+}{\partial\boldsymbol{\sigma}}, \qquad \text{and} \qquad \frac{\partial\boldsymbol{\sigma}_-}{\partial\boldsymbol{\sigma}}.
$$

To compute such derivatives, we first note that (17) and (18) define $\boldsymbol{\sigma}_+$ and $\boldsymbol{\sigma}_-$ as *isotropic* tensor-valued functions of $\boldsymbol{\sigma}$. Such functions are particular cases of the families of functions discussed by Chadwick & Ogden (1971)[23] and Carlson & Hoger(1986) [24] and their derivatives are promptly available in closed form. Finally, the scalars $\partial D/\partial\,(\Delta\gamma)$ and $\partial D/\partial\,(q^{\text{trial}})$, which are the outcome of the derivation of the damage function (39) are defined by

$$\begin{aligned}
\frac{\partial D}{\partial\,(\Delta\gamma)} &= \left(\frac{\Delta t}{\mu\,\Delta\gamma+\Delta t}\right)^{\epsilon}\left[\frac{3G}{\sigma_y(\bar{\varepsilon}^{vp}_{n+1})}+\frac{H(q^{\text{trial}}-3G\Delta\gamma)}{\sigma_y(\bar{\varepsilon}^{vp}_{n+1})^2}+\frac{\mu\epsilon}{\mu\,\Delta\gamma+\Delta t}\left(\frac{q^{\text{trial}}-3G\Delta\gamma}{\sigma_y(\bar{\varepsilon}^{vp}_{n+1})}\right)\right]\\
\frac{\partial D}{\partial\,(q^{\text{trial}})} &= -\frac{1}{\sigma_y(\bar{\varepsilon}^{vp}_{n+1})}\left(\frac{\Delta t}{\mu\,\Delta\gamma+\Delta t}\right)^{\epsilon}
\end{aligned} \tag{51}$$

In the above, H denotes the derivative of the hardening function evaluated at t_{n+1}:

$$H = \left.\frac{\mathrm{d}\sigma_y}{\mathrm{d}\bar{\varepsilon}^{vp}}\right|_{\bar{\varepsilon}^{vp}_{n+1}}. \tag{52}$$

Remark 5. It is important to note that, the resulting elasto-viscoplastic tangent operator D^{ep} is generally unsymmetric so that, within the context of finite element computations, an unsymmetric solver is required in the global Newton-Raphson scheme.

4 Accuracy Analysis of the Integration Algorithm

To illustrate the accuracy of the integration algorithm in practical situations, this section presents some iso-error maps, produced with material constants covering a range of high rate-sensitivity to rate-independency. Iso-error maps have long been accepted as an effective and reliable (if not the only) tool for assessing the accuracy of constitutive integration algorithms under realistic finite time/strain steps [25, 26, 20].

The maps have been generated in the standard fashion. Using the three-dimensional implementation of the model, we start from a stress point at time t_n, $\boldsymbol{\sigma}_n$, lying on the yield surface (refer to Fig. 1) and apply a sequence of strain increments (within the interval $[t_n, t_{n+1}]$), corresponding to linear combinations of trial stress increments of the form

$$\Delta\boldsymbol{\sigma}^{\text{trial}} = \frac{\Delta\sigma_T}{q}\boldsymbol{T} + \frac{\Delta\sigma_N}{q}\boldsymbol{N}, \tag{53}$$

where $\boldsymbol{N}$ and $\boldsymbol{T}$ are, respectively, the unit normal and tangent vectors to the yield surface and q is the von Mises equivalent stress. For each increment of trial stress, we obtain a numerical solution, $\boldsymbol{\sigma}^{\text{num}}_{n+1}$, with the above described algorithm in one step. In addition, a solution assumed to be 'exact', $\boldsymbol{\sigma}^{\text{exact}}_{n+1}$, is obtained with the same algorithm by dividing the corresponding strain (and time) increment into 1000 sub-increments of equal size. For each point in

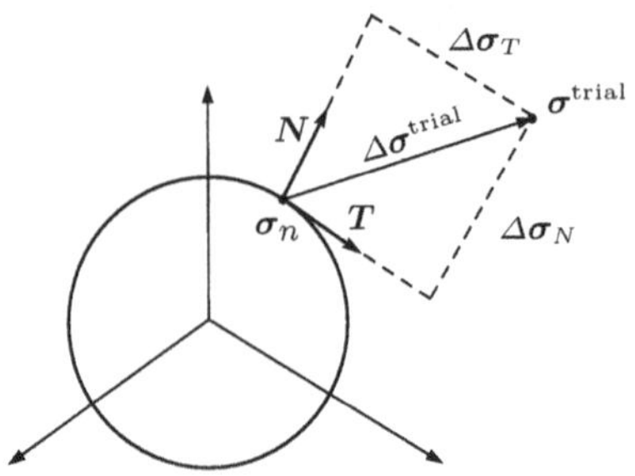

Fig. 1. Isoerror map. Trial stress increment directions

Table 1. Material data for aluminium alloy

Description	Symbol	Value
Elastic Modulus	E	69004 $[MN/m^2]$
Poisson's ratio	ν	0.3
Initial yield stress	σ_{y_0}	80.559 $[MPa]$
Hardening curve	$\sigma_y(\bar{\varepsilon}^{vp})$	$589 \cdot (10^{-4} + \bar{\varepsilon}^{vp})^{0.216}$
Damage data (exponent)	s	1.0
Damage data (denominator)	r	2.8 $[MPa]$

which a numerical and 'exact' solution is obtained, the error is computed as:

$$\text{ERROR} = \frac{\|\boldsymbol{\sigma}_{n+1}^{\text{exact}} - \boldsymbol{\sigma}_{n+1}^{\text{num}}\|}{\|\boldsymbol{\sigma}_{n+1}^{\text{exact}}\|} \times 100 \, .$$

The resulting iso-error map is the contour plot of the error field. The material properties adopted in the present analysis, are listed in Table 1. These parameters were taken from Reference [27] for an aluminium alloy, except for the value of the damage denominator, r. The value of this material constant has been calibrated by performing several numerical tests with a single axisymmetric finite element, such that critical value of damage ($D = 1$) is attained, for the same applied displacement. To preserve the constant rate of total strain $\Delta\varepsilon/\Delta t$ for the iso-error map under consideration, the time increment Δt is appropriately scaled. Figure 2 shows iso-error maps obtained at low and high strain rates with the non-dimensional rate

$$\mu \, \|\dot{\boldsymbol{\varepsilon}}\|$$

set respectively to 1 and 1000. For each non-dimensional rate, three values of rate-sensitivity parameter, ϵ, have been used: 10^0, 10^{-1} and 0. For $\epsilon = 0$ virtually identical maps are obtained for the two rates [see Figs. 2 (a)$_3$ and 2 (b)$_3$] and the algorithm recovers the rate-independent solution.

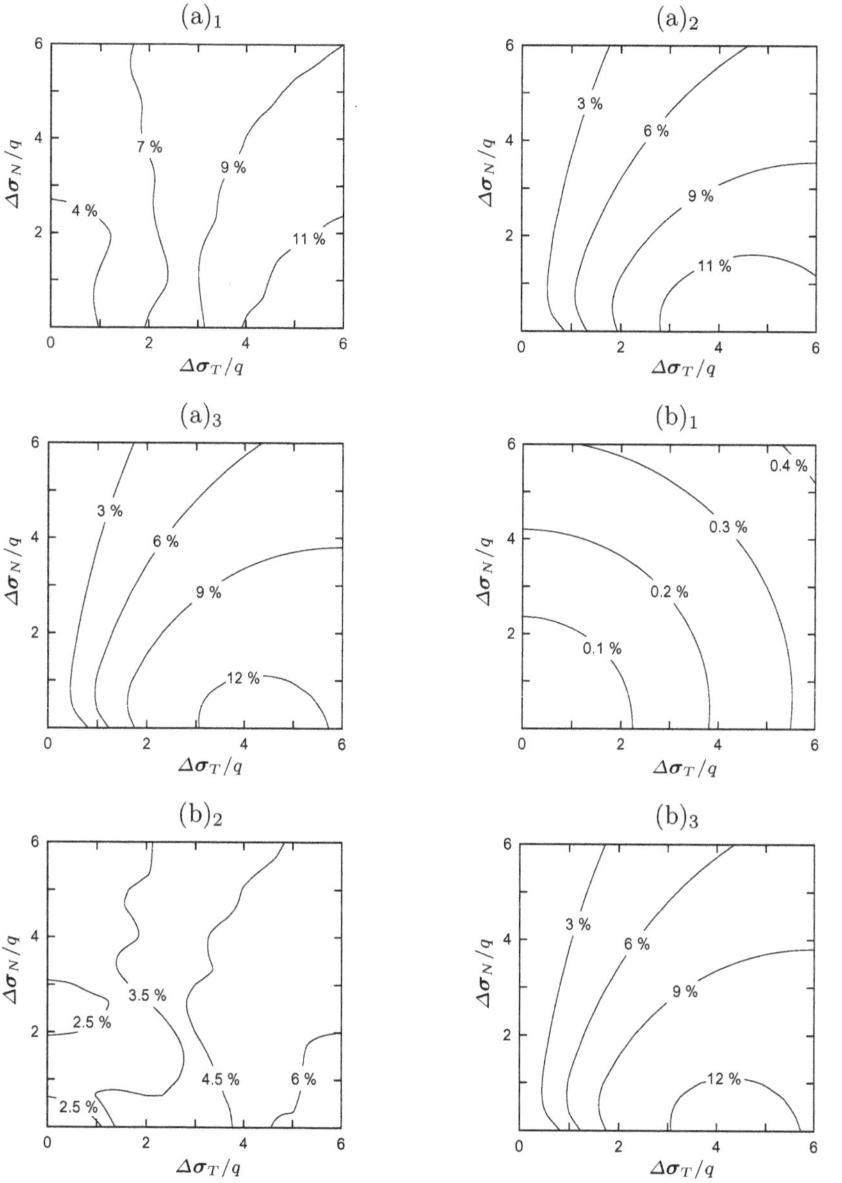

Fig. 2. Iso-error maps for $D = 0\%$ and $\bar{\varepsilon}^{\mathrm{p}} = 0$: $(\mathrm{a})_1$ $\mu\,\|\dot{\boldsymbol{\varepsilon}}\| = 1$, $\epsilon = 10^0$; $(\mathrm{a})_2$ $\mu\,\|\dot{\boldsymbol{\varepsilon}}\| = 1$, $\epsilon = 10^{-1}$; $(\mathrm{a})_3$ $\mu\,\|\dot{\boldsymbol{\varepsilon}}\| = 1$, $\epsilon = 0$; $(\mathrm{b})_1$ $\mu\,\|\dot{\boldsymbol{\varepsilon}}\| = 10^3$, $\epsilon = 10^0$; $(\mathrm{b})_2$ $\mu\,\|\dot{\boldsymbol{\varepsilon}}\| = 10^3$, $\epsilon = 10^{-1}$ and $(\mathrm{b})_3$ $\mu\,\|\dot{\boldsymbol{\varepsilon}}\| = 10^3$, $\epsilon = 0$

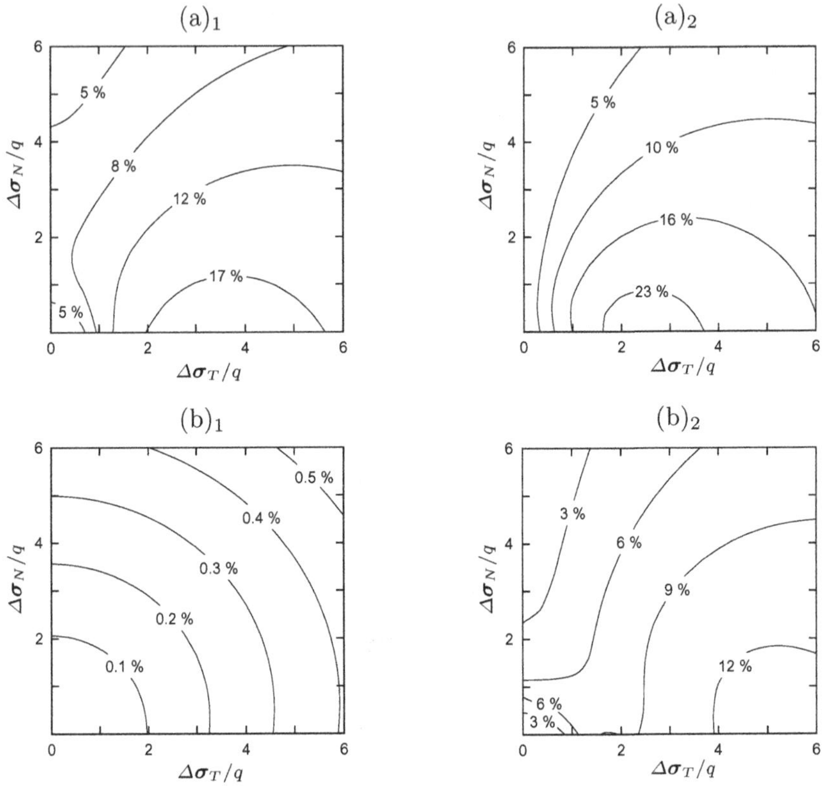

Fig. 3. Iso-error maps for $D = 15\%$ and $\bar{\varepsilon}^{\mathrm{p}} = 0.348$: (a)$_1$ $\mu\,\|\dot{\varepsilon}\| = 1$, $\epsilon = 10^0$; (a)$_2$ $\mu\,\|\dot{\varepsilon}\| = 1$, $\epsilon = 10^{-1}$; (b)$_1$ $\mu\,\|\dot{\varepsilon}\| = 10^3$, $\epsilon = 10^0$; and (b)$_2$ $\mu\,\|\dot{\varepsilon}\| = 10^3$, $\epsilon = 10^{-1}$

In Fig. 3 the error map of the algorithm is depicted at a different stage of damage evolution. Again, the iso-error maps are obtained at low and high strain rates with the non-dimensional rate, $\mu\,\|\dot{\varepsilon}\|$, set respectively to 1 and 1000. For each non-dimensional rate, two values of rate-sensitivity parameter, ϵ, have been used: 10^0 and 10^{-1}. Once more for $\epsilon = 0$, the algorithm reproduces the rate-independent solution. By comparing Figs. 2 and 3, the dependency of the algorithm accuracy on the state of internal damage is made clear. The more damaged the material is, the more restricted an increment of trial stress must be to maintain the integration error within a prescribed limit. In other words, the accuracy of the algorithm deteriorates as damage increases. It is also possible to conclude that, in general, increasing (decreasing) rate-sensitivity and/or increasing (decreasing) strain rates tend to produce decreasing (increasing) integration errors. The largest errors are expected in the rate-independent limit.

5 Concluding Remarks

A computational model for elasto-viscoplastic solids, capable of handling regions of high rate-sensitivity to rate independent conditions in the presence of ductile damage, has been presented in this contribution. The material constitutive model is derived by coupling a power-law elasto-viscoplastic constitutive model, with a ductile damage evolution law. The evolution of the damage internal variable includes the important effect of micro-crack closure.

The model can describe different concurrent physical phenomena and implicitly contains, as limit cases, simpler constitutive descriptions. The resulting integration algorithm generalizes the standard *return mapping* procedures of the infinitesimal theory, and quite remarkably requires the solution of only *one scalar non-linear equation.*

References

1. Perić D (1993) On a class of constitutive equations in viscoplasticity: Formulation and computational issues. Int J Num Meth Engng 36:1365–1393
2. Ladevèze P, Lemaitre J (1984) Damage effective stress in quasi unilateral conditions. In: 16th Int. Congress Theor Appl Mech, Lyngby, Denmark
3. Andrade Pires FM , de Souza Neto EA, Owen DRJ (2004) On the finite element prediction of damage growth and fracture initiation in finitely deforming ductile materials. Comp Meth Appl Mech 193:5223–5256
4. Owen DRJ, Hinton E (1980) Finite Elements in Plasticity: Theory and Practice. Pineridge Press, Swansea
5. Bathe K-J (1996) Finite Element Procedures. Prentice-Hall, Englewood Cliffs, New Jersey
6. Crisfield MA (1991) Non-linear Finite Element Analysis of Solids and Structures. Vol.1: Essentials. John Wiley & Sons, Chichester
7. Crisfield MA (1997) Non-linear Finite Element Analysis of Solids and Structures. Vol.2: Advanced Topics. John Wiley & Sons, Chichester
8. Simo, JC, Hughes, TJR (1998) Computational Inelasticity. Springer-Verlag, New York
9. Belytschko T, Liu WK, Moran B (2000) Nonlinear Finite Elements for Continua and Structures. Wiley, New York
10. Zienkiewicz OC, Taylor, RL, Zhu, JZ (2005) The Finite Element Method – Its Basis & Fundamentals. Butterworth-Heinemann, 6th edition
11. Zienkiewicz OC, Taylor, RL (2005) The Finite Element Method for Solid and Structural Mechanics. Butterworth-Heinemann, 6th edition
12. Perzyna P (1971) Thermodynamic Theory of Viscoplasticity. Academic Press, New York

13. Ladèveze P (1983) On an anisotropic damage theory. In: J.P. Boehler, editor, Proc CNRS Int Coll 351 – Failure Criteria of Structured Media 355–363. Villars-de-Lans
14. Lemaitre J (1985) Coupled elasto-plasticity and damage constitutive equations. Comp Meth Appl Mech Engng 51:31–49
15. Odqvist FKG, Hult J (1961) Some aspects of creep rupture. Arkiv för fysik 19(26):379-382
16. Rabotnov YN (1969) Creep rupture. In: M. Hetényi and W.G. Vincenti (eds), Proc of the Twelfth International Congress of Applied Mechanics XII 342–349. Springer, Berlin, Germany
17. Lemaitre J, Chaboche JL (1990) Mechanics of Solid Materials. Cambridge Univ Press
18. Lemaitre J (1996) A Course on Damage Mechanics. Springer, 2nd edition
19. Lemaitre J (1983) A three-dimensional ductile damage model applied to deep-drawing forming limits. In: ICM 4 Stockholm Vol.2 1047–1053
20. Simo JC, Taylor RL (1985) Consistent tangent operators for rate-independent elastoplasticity. Comp Meth Appl Mech Engng 48:101–118
21. Ortiz M, Simo JC (1986) An analysis of a new class of integration algorithms for elastoplastic constitutive relations. Int J Num Meth Engng 23:353–366
22. Simo JC, Govindjee, S (1991) Non-linear b-stability and symmetry preserving return mapping algorithms for plasticity and viscoplasticity. Int J Num Meth Engng 31:151–176
23. Chadwick P, Ogden RW (1971) A theorem in tensor calculus and its application to isotropic elasticity. Arch Rat Mech Anal 44:54–68
24. Carlson DE, Hoger A (1986) The derivative of a tensor-valued function of a tensor. Quart Appl Math 44(3):409–423
25. Krieg RD, Krieg DB (1977) Accuracies of numerical solution methods for the elastic-perfectly plastic model. J Pressure Vessel Tech ASME 99:510–515
26. Ortiz M, Popov EP (1985) Accuracy and stability of integration algorithms for elastoplastic constitutive relations. Int J Num Meth Engng 21:1561–1576
27. Vaz Jr. M, Owen DRJ (2001) Aspects of ductile fracture and adaptive mesh refinement in damaged elasto-plastic materials. Int J Num Meth Engng 50:29–54, 2001.

On Multiscale Analysis of Heterogeneous Composite Materials: Implementation of Micro-to-Macro Transitions in the Finite Element Setting

D. Perić, E.A. de Souza Neto, A.J. Carneiro Molina and M. Partovi

Centre for Civil and Computational Engineering
School of Engineering, Swansea SA2 8PP, UK
d.peric@swansea.ac.uk

Summary. This paper describes a multiscale homogenization procedure required for computation of material response of non-linear microstructures undergoing small strains. Such procedures are important for computer modelling of heterogeneous materials when the length-scale of heterogeneities is small compared to the dimensions of the body.
The described homogenization procedure is based on the standard finite element discretisation of both macro- and micro-structure. The attention is restricted to two dimensional problem and the deformation-driven microstructures. Two classical types of boundary conditions are imposed over the unit cell: (a) linear displacements on the boundary, and (b) periodic displacements and antiperiodic tractions on the boundary. These boundary conditions satisfy the fundamental averaging condition, which equates microscopic and macroscopic virtual work. Numerical simulations, performed for an elasto-plastic material with micro-cavities, illustrate the scope and benefits of the described computational strategy.

1 Introduction

The ever increasing requirements in high-performance applications have provided a constant stimulus for the design of new materials. Often, this has been achieved by appropriately manipulating microstructure, for instance, by adding certain material component to the matrix phase, thus tailoring the overall material properties to specific applications. The added material phase is typically at a scale that is much smaller than the overall structural size, hence making the direct modelling of the material behaviour impractical. In many situations scales remain tightly coupled and the traditional phenomenological approach does not provide sufficiently general predictive modelling capability. Therefore a means of continuous interchange of information be-

Eugenio Oñate and Roger Owen (eds.), Computational Plasticity, 165–185.

tween scales is needed if predictive modelling of material behaviour is to be attempted.

Since the basic principles for the micro-macro modelling of heterogeneous materials were introduced (see Suquet [12, 13]), this technique has proved to be a very effective way to deal with arbitrary physically non-linear and time dependent material behaviour at micro-level. A number of recent works has been concerned with various approaches and techniques for the micro-macro simulation of heterogeneous materials. Among these we highlight the contributions by Moulinec and Suquet [10], Smit et al. [11], Miehe and co-workers [7, 8, 9], Kouznetsova et al. [5], Terada and Kikuchi [15] and Zohdi and Wriggers [16].

The present article discusses some issues related to computational strategy for homogenisation of microstructures with non-linear material behaviour undergoing small strains. Since the aim is to provide the basic ingredients of the computational strategy allowing for the concurrent simulation at different scales of the model, a simple model is considered comprising two scales arising, for instance, in modelling of heterogeneous composite materials. The focus of this article is on computational aspects; more specifically on the computational technique for prescribing the boundary conditions at the micro-scale and calculation of the macro-scale tangent moduli characterising relations between the macroscopic stress and strain tensors.

The attention is restricted to the deformation-driven microstructures, which have been proven to provide a convenient computational format [14]. Two types of boundary conditions are imposed over the unit micro-cell: (a) linear displacements on the boundary, and (b) periodic displacements and antiperiodic tractions on the boundary. These boundary conditions satisfy the fundamental Hill-Mandel averaging condition, which equates microscopic and macroscopic virtual work [4]. The resulting computational strategy is characterised by the Newton-Raphson solution of the discrete boundary value problem, and incorporates the appropriate tangent operators. Numerical examples of both micro-scale and two-scale finite element simulations are presented in order to illustrate the scope and the benefits of the described computational strategy.

2 Continuum Model at Small Strains

2.1 Preliminaries

A homogenized macro-continuum $\overline{\mathcal{B}} \subset \mathbb{R}^3$ with locally attached microstructures $\mathcal{B} \subset \mathbb{R}^3$ is visualized in Fig. 1. A point $\boldsymbol{x} \in \overline{\mathcal{B}}$ of the homogenized macromedium $\overline{\mathcal{B}} \subset \mathbb{R}^3$ is represented as a microstructure $\mathcal{B} \subset \mathbb{R}^3$. The tensors $\overline{\boldsymbol{\sigma}}$ and $\boldsymbol{\sigma}_\mu$ denote the macro and micro Cauchy stress tensor at $\boldsymbol{x} \in \overline{\mathcal{B}}$ and $\boldsymbol{y} \in \mathcal{B}$, respectively. The representative volume element (RVE) of the microstructure $\mathcal{V} \subset \mathbb{R}^3$ represents the part of the heterogeneous material consisting of the solid part $\mathcal{B}$ and the hole $\mathcal{H}$, i.e. $\mathcal{V} = \mathcal{B} \cup \mathcal{H}$ and $\partial\mathcal{B} = \partial\mathcal{V} \cup \partial\mathcal{H}$.

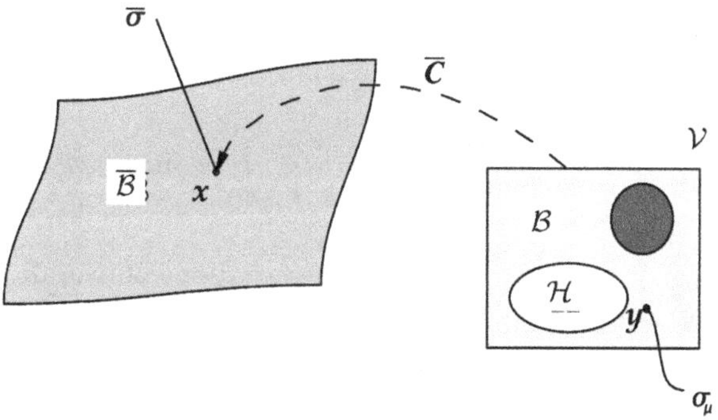

Fig. 1. Micro to macro transition

It is assumed that the traction field on the surface of the holes in the interior of RVE vanishes, i.e. $\boldsymbol{t}(\boldsymbol{y},t) = \boldsymbol{0}$ at $\boldsymbol{y} \in \partial\mathcal{H}$, where $\boldsymbol{t} \equiv \boldsymbol{\sigma}_\mu \boldsymbol{n}$ on $\partial\mathcal{B}$ is the traction field vector on the surface with outward normal $\boldsymbol{n}$ at $\boldsymbol{y} \in \partial\mathcal{B}$.

2.2 Basic Microvariables

Microscopic small strain tensor $\boldsymbol{\epsilon}_\mu$ is defined as the symmetric part of the displacement gradient tensor,

$$\boldsymbol{\epsilon}_\mu \equiv \mathrm{sym}\{\nabla(\boldsymbol{u})\} \tag{1}$$

where $\boldsymbol{u}$ is the displacement field at a material point $\boldsymbol{y} \in \mathcal{B}$.

Microequilibrium state is assumed in the presence of body forces per unit of mass $\boldsymbol{b}$,

$$\mathrm{div}(\boldsymbol{\sigma}_\mu) + \rho\, \boldsymbol{b} = \boldsymbol{0} \quad \text{in } \mathcal{B}\,, \tag{2}$$

where symmetric stress tensor $\boldsymbol{\sigma}_\mu$ is assumed to be related to the strain tensor $\boldsymbol{\epsilon}_\mu$ by some constitutive law

$$\boldsymbol{\sigma}_\mu = \hat{\boldsymbol{\sigma}}(\boldsymbol{\epsilon}_\mu\,; \boldsymbol{\alpha}\,;\ \boldsymbol{y}) \quad \text{in } \mathcal{B}, \tag{3}$$

where $\boldsymbol{\alpha}$ is the set of internal variables.

2.3 Basic Macrovariables and Averaging Theorem

Within the described homogenization technique no constitutive assumptions have been assumed at the macrolevel. Overall macrostress $\boldsymbol{\sigma}_\mathrm{M}$ of microstructure $\mathcal{V}$ is defined as an average of the microstresses over the unit cell. By applying the Gauss theorem and microquilibrium (2), $\boldsymbol{\sigma}_\mathrm{M}$ is given by the expression

$$\boldsymbol{\sigma}_\mathrm{M} \equiv \overline{\boldsymbol{\sigma}} = \frac{1}{|\mathcal{V}|}\int_\mathcal{V} \boldsymbol{\sigma}_\mu \,\mathrm{d}V = \frac{1}{|\mathcal{V}|}\int_{\partial\mathcal{V}} \mathrm{sym}[\boldsymbol{t}\otimes\boldsymbol{y}]\,\mathrm{d}A + \frac{1}{|\mathcal{V}|}\int_\mathcal{V} \rho\,\mathrm{sym}[\boldsymbol{b}\otimes\boldsymbol{y}]\,\mathrm{d}V \tag{4}$$

in terms of the traction $\boldsymbol{t}$ at $\boldsymbol{y} \in \partial\mathcal{V}$ and body force vector field $\boldsymbol{b}$ at $\boldsymbol{y} \in \mathcal{V}$.

By applying Green's Lemma in similar way, overall macrostrain $\boldsymbol{\epsilon}_{\mathrm{M}}$ is defined,

$$\boldsymbol{\epsilon}_{\mathrm{M}} \equiv \overline{\boldsymbol{\epsilon}} = \frac{1}{|\mathcal{V}|} \int_{\mathcal{V}} \boldsymbol{\epsilon}_{\mu} \, \mathrm{d}V = \frac{1}{|\mathcal{V}|} \int_{\partial\mathcal{V}} \mathrm{sym}[\boldsymbol{u} \otimes \boldsymbol{n}] \, \mathrm{d}A \tag{5}$$

in terms of the displacement $\boldsymbol{u}$ at $\boldsymbol{y} \in \partial\mathcal{V}$.

Overall tangent modulus $\overline{\boldsymbol{\mathcal{C}}}$ relates the variations of overall macrostress $\overline{\boldsymbol{\sigma}}$ and the macrostrain $\overline{\boldsymbol{\epsilon}}$ in the form

$$\overline{\boldsymbol{\mathcal{C}}} \equiv \frac{\mathrm{d}\overline{\boldsymbol{\sigma}}}{\mathrm{d}\overline{\boldsymbol{\epsilon}}} \tag{6}$$

The computation of these fourth-order tensors, in their discrete FE form, is an important aspect of this work.

The Hill-Mandel Principle

The Hill-Mandel principle or averaging theorem [4], demands that macroscopic stress work (or power) must equal the volume average of the microscopic stress work (or power) over the RVE associated with the macroscopic point, that is

$$\overline{\boldsymbol{\sigma}} : \overline{\boldsymbol{\epsilon}} = \frac{1}{|\mathcal{V}|} \int_{\mathcal{V}} \boldsymbol{\sigma} : \boldsymbol{\epsilon} \, \mathrm{d}V \tag{7}$$

Using (1) and integrating by parts the right hand side of (7) and then applying microequlibrium (2), the averaging theorem can be expressed in the following form

$$\overline{\boldsymbol{\sigma}} : \overline{\boldsymbol{\epsilon}} = \frac{1}{|\mathcal{V}|} \int_{\partial\mathcal{V}} \boldsymbol{t} \cdot \boldsymbol{u} \, \mathrm{d}A + \frac{1}{|\mathcal{V}|} \int_{\mathcal{V}} \rho \boldsymbol{b} \cdot \boldsymbol{u} \, \mathrm{d}V \tag{8}$$

2.4 Definition of the Boundary Conditions for the Small Scale

The boundary conditions for the displacement $\boldsymbol{u}$ and traction $\boldsymbol{t}$ at the microstructure, are chosen such that condition (7) is satisfied. Two classical types of boundary conditions that satisfy these conditions are prescribed on the unit cell: (a) linear displacements on the boundary, and (b) periodic displacements and antiperiodic tractions on the boundary. A crucial aspect is the formulation in deformation-driven context, where the macroscopic strain $\overline{\boldsymbol{\epsilon}}$ is prescribed as averaged over the microstructure. The deformation-driven format has proved more convenient than the stress driven format [14].

The displacement field is divided in two parts:

$$\boldsymbol{u}(\boldsymbol{y}) = \boldsymbol{u}^*(\boldsymbol{y}) + \widetilde{\boldsymbol{u}}(\boldsymbol{y}) = \overline{\boldsymbol{\epsilon}}\, \boldsymbol{y} + \widetilde{\boldsymbol{u}}(\boldsymbol{y}), \tag{9}$$

where $\boldsymbol{u}^*$ is the *Taylor displacement*, which defines a constant deformation $\overline{\boldsymbol{\epsilon}}$ over the unit cell as

$$\boldsymbol{u}^* \equiv \overline{\boldsymbol{\epsilon}}\, \boldsymbol{y}. \tag{10}$$

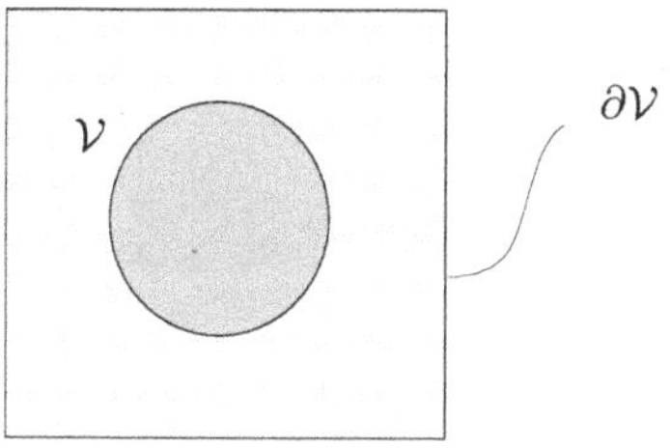

Fig. 2. Microstructure for linear b.c.

The component $\widetilde{\boldsymbol{u}}$ is known as the *displacement fluctuation*, which is considered to be unknown.

Insertion of (9) into the averaging theorem (8) yields in the following new form of the Hill-Mandel principle

$$\frac{1}{|\mathcal{V}|}\int_{\partial\mathcal{V}} \boldsymbol{t}\cdot\widetilde{\boldsymbol{u}}\,\mathrm{d}A + \frac{1}{|\mathcal{V}|}\int_{\mathcal{V}} \rho\,\boldsymbol{b}\cdot\widetilde{\boldsymbol{u}}\,\mathrm{d}V = \boldsymbol{0} \tag{11}$$

Linear displacements on the boundary

In this state, the definition of the linear deformation boundary constraint over the microstructure RVE shown in Fig. 2 assumes the following form

$$\widetilde{\boldsymbol{u}}(\boldsymbol{y}) \;=\; \boldsymbol{0} \qquad \text{at } \boldsymbol{y} \in \partial\mathcal{V}. \tag{12}$$

This condition defines a linear deformation on the boundary $\partial\mathcal{V}$ of the RVE. Insertion of the above condition (12) into the new averaging condition form (11) confirms that this model satisfies the averaging theorem only when body force effect is negligible.

Periodic deformation and antiperiodic traction on the boundary

Another possibility consists of applying periodic deformation and antiperiodic traction on the boundary of the RVE $\partial\mathcal{V}$, which can be represented as

$$\widetilde{\boldsymbol{u}}(\boldsymbol{y}^+) \;=\; \widetilde{\boldsymbol{u}}(\boldsymbol{y}^-) \quad \text{and} \quad \boldsymbol{t}(\boldsymbol{y}^+) = -\boldsymbol{t}(\boldsymbol{y}^-). \tag{13}$$

In order to apply these conditions the boundary of the unit cell is decomposed in two parts as indicated in Fig. 3. Thus $\partial\mathcal{V} = \partial\mathcal{V}^+ \cup \partial\mathcal{V}^-$ with outwards normals $\boldsymbol{n}^+ = -\boldsymbol{n}^-$ which are associated with the points $\boldsymbol{y}^+ \in \partial\mathcal{V}^+$ and $\boldsymbol{y}^- \in \partial\mathcal{V}^-$.

The body force effect is not taken into consideration so that this condition satisfies the averaging theorem. This can be proved easily by inserting $(13)_1$ and $(13)_2$ into (11).

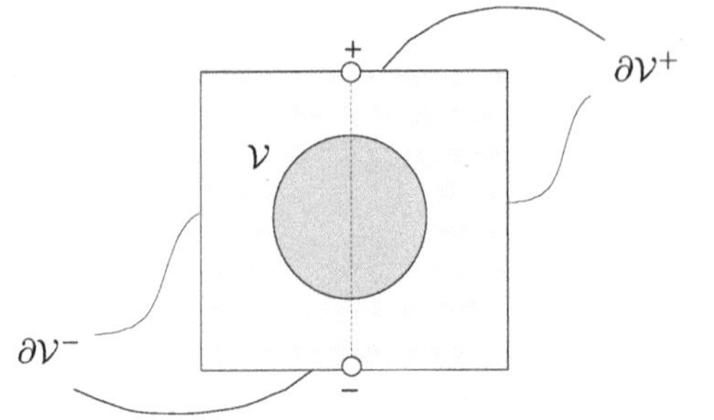

Fig. 3. Microstructure for periodic b.c.

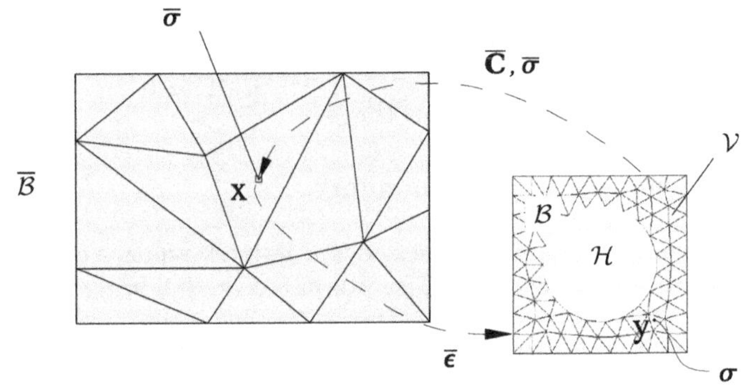

Fig. 4. Micro to macro transition

3 Discretised Model at Small Strains

3.1 Introduction

The discretisation of the continuum multi-scale problem described in Section 2 is based on finite element formulation. A representative finite element dicretisations of the macro and microstructure are depicted in Fig. 4. One can notice that at every integration Gauss point of the macrostructure, a discrete RVE microstructure is considered as representation of a Gauss point at the macro-level.

The present approach is based on the deformation-driven microstructures, in which the value of the overall macroscopic deformation $\overline{\epsilon}$ is prescribed on the discretised RVE. The goal is then to develop a numerical procedure for computing macroscopic average stress $\overline{\boldsymbol{\sigma}}$ and the overall tangent moduli $\overline{\mathcal{C}}$ at each macroscopic integration point with locally attached microstructure.

3.2 Displacement Field Partition and Matrix Notation

Following the previous Section 2.4, where the continuum displacement field partition (9) was established, the displacement field is divided in two parts:

$$\boldsymbol{u} = \boldsymbol{u}^* + \widetilde{\boldsymbol{u}} \tag{14}$$

where the *Taylor displacement* $\boldsymbol{u}^*$ (previously defined in the continuum form (10)) is expressed in its discrete form

$$\boldsymbol{u}_j^* \equiv \overline{\boldsymbol{\epsilon}}\, \boldsymbol{y}_j \quad j = 1 \cdots n \tag{15}$$

for the n nodes of the discretised microstructure RVE. The *displacement fluctuation* $\widetilde{\boldsymbol{u}}$ is the unknown for every node of the discretised microstructure.

In what follows the standard finite element matrix notation will be used, where the tensor entities so far used, can be expressed in the matrix form as

$$\overline{\boldsymbol{\epsilon}} \equiv \begin{Bmatrix} \bar{\varepsilon}_{11} \\ \bar{\varepsilon}_{22} \\ 2\,\bar{\varepsilon}_{12} \end{Bmatrix}, \quad \boldsymbol{u}_j \equiv \begin{Bmatrix} u_1 \\ u_2 \end{Bmatrix}_j, \quad \overline{\boldsymbol{\sigma}} \equiv \begin{Bmatrix} \bar{\sigma}_{11} \\ \bar{\sigma}_{22} \\ \bar{\sigma}_{12} \end{Bmatrix} \text{ and } \boldsymbol{f}_j \equiv \begin{Bmatrix} f_1 \\ f_2 \end{Bmatrix}_j . \tag{16}$$

Here $\overline{\boldsymbol{\epsilon}}$ is the matrix representation of the macrostrain tensor, $\boldsymbol{u}_j$ is the displacement field at the node j of the discretised unit cell $\mathcal{V}$, $\overline{\boldsymbol{\sigma}}$ is the averaged stress field while $\boldsymbol{f}_j$ denotes the force vector associated with the microcell node j.

The *Taylor displacement* $\boldsymbol{u}_j^*$ of the node j is computed in the following matrix form

$$\boldsymbol{u}_j^* = \mathbb{D}_j^{\mathrm{T}}\, \overline{\boldsymbol{\epsilon}}\,, \quad j = 1 \cdots n. \tag{17}$$

where $\mathbb{D}_j$ is the *coordinate matrix* at node j of the microstructure defined in [9].

3.3 Discretised Micro-equilibrium State and Solution Procedure

Following standard procedure, the *discrete boundary value problem* is formulated as follows: Find the *nodal displacements global vector* $\boldsymbol{u}$, such that

$$\boldsymbol{r}(\boldsymbol{u}) \equiv \boldsymbol{f}^{int}(\boldsymbol{u}) - \boldsymbol{f}^{ext} = \boldsymbol{0} \tag{18}$$

where $\boldsymbol{f}^{int}$ and $\boldsymbol{f}^{ext}$ are, respectively, the *internal* and *external global force vectors*, and $\boldsymbol{r}$ is the *residual* (or *out-of-balance force*) vector.

An iterative Newton-type procedure for the solution of the Non-linear microscopic equilibrium (18) is considered here. Each iteration determines the current fluctuation field assuming frozen macroscopic strain $\overline{\boldsymbol{\epsilon}}$. At the end of the procedure, when microequilibrium is reached, the averaged macroscopic stress over the microstructure RVE can be updated.

3.4 General Average Stress and Overall Tangent Modulus Computation

Average Stress Computation

Assuming no body forces in the expression for the average stress (4), in the discrete setting, $\boldsymbol{t}\,\mathrm{d}A \rightarrow \boldsymbol{f}_j^{ext}$, that is the infinitesimal force $\boldsymbol{t}\,\mathrm{d}A$ becomes the finite force $\boldsymbol{f}_j^{ext}$ at position $\boldsymbol{y}_j$ on the boundary $\partial\mathcal{V}$. Therefore (4) degenerates into the discrete sum

$$\overline{\boldsymbol{\sigma}} = \frac{1}{|\mathcal{V}|} \sum_{j=1}^{n_\mathrm{b}} \mathrm{sym}[\boldsymbol{f}_j^{ext} \otimes \boldsymbol{y}_j] \tag{19}$$

where n_b is the number of nodes on the boundary $\partial\mathcal{V}$. Using matrix representation this expression becomes

$$\overline{\boldsymbol{\sigma}} = \frac{1}{|\mathcal{V}|} \sum_{j=1}^{n_\mathrm{b}} \mathbb{D}_j\, \boldsymbol{f}_j^{ext} \tag{20}$$

where $\mathbb{D}_j$ is the coordinate matrix evaluated at node j on the boundary of the discretised microstructure RVE. The above expression can be rearranged in the following global expression

$$\overline{\boldsymbol{\sigma}} = \frac{1}{|\mathcal{V}|} \mathbb{D}_\mathrm{b}\, \boldsymbol{f}_\mathrm{b}^{ext}\,, \tag{21}$$

where $\boldsymbol{f}_\mathrm{b}^{ext}$ is the external nodal force vector of the boundary nodes, and $\mathbb{D}_\mathrm{b}$ is the *boundary coordinate matrix* defined by:

$$\mathbb{D}_\mathrm{b} \equiv \left[\, \mathbb{D}_1^\mathrm{b}\, \mathbb{D}_2^\mathrm{b} \ldots \mathbb{D}_{n_\mathrm{b}}^\mathrm{b} \right]. \tag{22}$$

Overall Tangent Modulus Computation

In the computational homogenization approach no explicit form of the constitutive behavior on the macrolevel is assumed a priori, so that the tangent modulus has to be determined numerically by relations between variations of the macroscopic stress and variations of the macroscopic strain at such integration point. This can be accomplished by numerical differentiation of the numerical macroscopic stress strain relation, for instance, by using forward difference approximations as suggested in [8]. An alternative approach is to condense the microstructural stiffness matrix to the macroscopic matrix tangent modulus. This task is achieved by reducing the total RVE system of equations to the relation between the forces acting on the boundary $\partial\mathcal{V}$ and the displacement on the boundary. This procedure has been used in [5], [6], and also in [9], in combination with the Lagrange multiplier method to impose the boundary constraints. A similar scheme have been used in this work whereby the direct condensation is employed to obtain a relation between the forces acting on the boundary $\partial\mathcal{V}$ and the Taylor displacement on the boundary nodes array $\boldsymbol{u}^*$ which depends linearly on the macroscopic strain $\bar{\boldsymbol{\epsilon}}$.

The total microstructural system of equations that gives the relation between the iterative nodal displacement $\mathrm{d}\boldsymbol{u}$ and iterative nodal external force vectors is

$$\boldsymbol{K}\ \mathrm{d}\boldsymbol{u} = \mathrm{d}\boldsymbol{f}^{ext} \tag{23}$$

With the displacement partition (14) the system can be rearranged as

$$\Rightarrow\ \boldsymbol{K}\ \mathrm{d}\boldsymbol{u}^* + \boldsymbol{K}\ \mathrm{d}\widetilde{\boldsymbol{u}} = \mathrm{d}\boldsymbol{f}^{ext} \Rightarrow\ \boldsymbol{K}\ \mathrm{d}\widetilde{\boldsymbol{u}} = \mathrm{d}\boldsymbol{f}^{ext} - \boldsymbol{K}\ \mathrm{d}\boldsymbol{u}^* \tag{24}$$

The boundary constraints are then applied to this system in the following sections to condense the system. This procedure gives the expression that relates the variation of boundary external forces $\mathrm{d}\boldsymbol{f}_{\mathrm{b}}^{ext}$ against the variation of the Taylor displacement $\mathrm{d}\boldsymbol{u}^*$.

3.5 Linear Displacements on the Boundary Assumption

In view of the discrete formulation of the boundary conditions outlined before in section 2.4, the nodes of the mesh are partitioned into those on the surface $\partial\mathcal{V}$ of RV and those in the interior of $\mathcal{V}$. In this mesh n_{b} boundary nodes and n_{i} internal nodes are distinguished. More details of the discrete form of this linear constraint are given in [1].

Partitioning of Algebraic Equations

Partitioning of the current nodal displacements and internal forces is given as

$$\boldsymbol{u} = \begin{Bmatrix} \boldsymbol{u}_{\mathrm{i}} \\ \boldsymbol{u}_{\mathrm{b}} \end{Bmatrix} \equiv \begin{Bmatrix} \mathbb{L}_{\mathrm{i}}\,\boldsymbol{u} \\ \mathbb{L}_{\mathrm{b}}\,\boldsymbol{u} \end{Bmatrix} \quad \text{and} \quad \boldsymbol{f} = \begin{Bmatrix} \boldsymbol{f}_{\mathrm{i}} \\ \boldsymbol{f}_{\mathrm{b}} \end{Bmatrix} \equiv \begin{Bmatrix} \mathbb{L}_{\mathrm{i}}\,\boldsymbol{f} \\ \mathbb{L}_{\mathrm{b}}\,\boldsymbol{f} \end{Bmatrix}. \tag{25}$$

Here $\mathbb{L}_{\mathrm{i}}$ and $\mathbb{L}_{\mathrm{b}}$ are the *connectivity matrices* [2], which define the contributions of the interior and boundary nodes, respectively. These are Boolean matrices, i.e. they consist of integers 0 and 1. Displacements $\boldsymbol{u}_{\mathrm{i}}$ and $\boldsymbol{u}_{\mathrm{b}}$ are arranged as shown in (25).

In line with (25) the tangent stiffness matrix is rearranged as

$$\boldsymbol{K} = \frac{\mathrm{d}\boldsymbol{f}^{int}}{\mathrm{d}\boldsymbol{u}} = \begin{bmatrix} \boldsymbol{k}_{\mathrm{ii}} & \boldsymbol{k}_{\mathrm{ib}} \\ \boldsymbol{k}_{\mathrm{bi}} & \boldsymbol{k}_{\mathrm{bb}} \end{bmatrix} \equiv \begin{bmatrix} \mathbb{L}_{\mathrm{i}}\,\boldsymbol{K}\,\mathbb{L}_{\mathrm{i}}^{\mathrm{T}} & \mathbb{L}_{\mathrm{i}}\,\boldsymbol{K}\,\mathbb{L}_{\mathrm{b}}^{\mathrm{T}} \\ \mathbb{L}_{\mathrm{b}}\,\boldsymbol{K}\,\mathbb{L}_{\mathrm{i}}^{\mathrm{T}} & \mathbb{L}_{\mathrm{b}}\,\boldsymbol{K}\,\mathbb{L}_{\mathrm{b}}^{\mathrm{T}} \end{bmatrix} \tag{26}$$

into contributions associated with internal nodes and nodes on the surface of the RVE.

Linear Displacement

At each node j of the boundary $\partial\mathcal{V}$ the condition (12) induces the discrete constraint

$$\widetilde{\boldsymbol{u}}_j = \boldsymbol{0}, \quad j = 1 \cdots n_{\mathrm{b}}. \tag{27}$$

According to the matrix notation introduced in Section 3.2, we define the *global coordinate matrix*

$$\mathbb{D}_{\mathrm{global},l} \equiv \left[\, \mathbb{D}_{\mathrm{i}}\; \mathbb{D}_{\mathrm{b},l} \right], \tag{28}$$

where $\mathbb{D}_{\mathrm{i}}$ and $\mathbb{D}_{\mathrm{b}}$ are the *interior coordinate matrix* and the *boundary coordinate matrix*, respectively, given as

$$\mathbb{D}_{\mathrm{i}} \equiv \left[\, \mathbb{D}^{\mathrm{i}}_1\; \mathbb{D}^{\mathrm{i}}_2 \ldots \mathbb{D}^{\mathrm{i}}_{n_{\mathrm{i}}} \right] \quad \text{and} \quad \mathbb{D}_{\mathrm{b},l} \equiv \left[\, \mathbb{D}^{\mathrm{b}}_1\; \mathbb{D}^{\mathrm{b}}_2 \ldots \mathbb{D}^{\mathrm{b}}_{n_{\mathrm{b}}} \right]. \tag{29}$$

Matrices $\mathbb{D}_{\mathrm{i}}$ and $\mathbb{D}_{\mathrm{b}}$ are defined in terms of the node coordinate matrices of the interior and boundary nodes, as discussed in Section 3.2. The Taylor displacement $\boldsymbol{u}^*$ defined in (17), is now represented in global form as $\boldsymbol{u}^* = \mathbb{D}^{\mathrm{T}}_{\mathrm{global},l}\, \bar{\boldsymbol{\epsilon}}$, where $\bar{\boldsymbol{\epsilon}}$ is the matrix representation of the prescribed macroscopic strain (16). In this model the variation of the Taylor displacement vector $\mathrm{d}\boldsymbol{u}^*$ is represented as

$$\mathrm{d}\boldsymbol{u}^* = \mathbb{D}^{\mathrm{T}}_{\mathrm{global},l}\, \mathrm{d}\bar{\boldsymbol{\epsilon}}, \tag{30}$$

that is, as a function of the variation of the macroscopic average strain vector $\mathrm{d}\bar{\boldsymbol{\epsilon}}$.

Tangent Modulus of Linear Displacements on the Boundary Constraint

Using Partitioning of the algebraic equations (25) and (26), the system (23) can be rewritten for the case when $\mathrm{d}\boldsymbol{f}^{ext}_{\mathrm{i}} = \boldsymbol{0}$.

The general procedure explained in Section 3.4, where the rearranged system (24) was obtained, is followed. By applying linear displacement constraint in discrete form (27) and by using the Taylor displacement variation $\mathrm{d}\boldsymbol{u}^*$ given by (30), after some algebraic manipulation, a variation for the external nodal force vector at the boundary nodes is obtained as

$$\mathrm{d}\boldsymbol{f}^{ext}_{\mathrm{b}} = \boldsymbol{K}^{\mathrm{B}}_{\mathrm{lin}}\, \mathbb{D}^{\mathrm{T}}_{\mathrm{global},l}\, \mathrm{d}\bar{\boldsymbol{\epsilon}} \tag{31}$$

in terms of the global coordinate matrix (28) and the overall averaged macrostrain.

The overall tangent moduli defined in (6), can be computed in its discretised matrix form, using previous averaged stress expression (21), in the following way

$$\bar{\boldsymbol{\mathcal{C}}}_l = \frac{\mathrm{d}\bar{\boldsymbol{\sigma}}}{\mathrm{d}\bar{\boldsymbol{\epsilon}}} = \frac{1}{|\mathcal{V}|}\, \mathbb{D}_{\mathrm{b}}\, \frac{\mathrm{d}\boldsymbol{f}^{ext}_{\mathrm{b}}}{\mathrm{d}\bar{\boldsymbol{\epsilon}}} \tag{32}$$

Substituting (31) into (32), the overall tangent modulus representation is obtained as

$$\boxed{\overline{\mathcal{C}}_l = \frac{1}{|\mathcal{V}|} \, \mathbb{D}_{\mathrm{b},l} \, \boldsymbol{K}^{\mathrm{B}}_{\mathrm{lin}} \, \mathbb{D}^{\mathrm{T}}_{\mathrm{global},l}} \,. \tag{33}$$

Clearly the modulus $\overline{\mathcal{C}}_l$ is given as a function of the boundary coordinate matrix $\mathbb{D}_{\mathrm{b},l}$ defined in (29), the *condensed stiffness matrix* $\boldsymbol{K}^{\mathrm{B}}_{\mathrm{lin}}$ and the *global coordinate matrix* $\mathbb{D}^{\mathrm{T}}_{\mathrm{global},l}$ outlined in (28). Finally we remark that using (33) the tangent moduli can be computed for heterogeneous material with arbitrary microstructures. When using this tangent modulus the quadratic rate of convergence is attained at the macroscopic level.

3.6 Periodic Displacements and Antiperiodic Traction on the Boundary

In order to discretise the continuum model of the periodic boundary conditions described in 2.4, the nodes of the mesh are partitioned in four groups: 1) n_{i} interior nodes, 2) n_{p} positive boundary nodes which are located at the top and right of the microstructure boundary $\partial\mathcal{V}$ of the RVE, 3) n_{p} negative boundary nodes which are located at the bottom and left of the microstructure boundary $\partial\mathcal{V}$ of the RVE, and 4) $n_{\mathrm{c}} = 4$ node at the corners. More details on these discrete constraints are given in [1].

Partitioning of Algebraic Equations

The partition of the nodal displacements and internal forces for the periodic boundary condition is as follows

$$\boldsymbol{u} = \begin{Bmatrix} \boldsymbol{u}_{\mathrm{i}} \\ \boldsymbol{u}_{\mathrm{p}} \\ \boldsymbol{u}_{n} \\ \boldsymbol{u}_{\mathrm{c}} \end{Bmatrix} \equiv \begin{Bmatrix} \mathbb{L}_{\mathrm{i}}\,\boldsymbol{u} \\ \mathbb{L}_{\mathrm{p}}\,\boldsymbol{u} \\ \mathbb{L}_{n}\,\boldsymbol{u} \\ \mathbb{L}_{\mathrm{c}}\,\boldsymbol{u} \end{Bmatrix} \quad \text{and} \quad \boldsymbol{f} = \begin{Bmatrix} \boldsymbol{f}_{\mathrm{i}} \\ \boldsymbol{f}_{\mathrm{p}} \\ \boldsymbol{f}_{n} \\ \boldsymbol{f}_{\mathrm{c}} \end{Bmatrix} \equiv \begin{Bmatrix} \mathbb{L}_{\mathrm{i}}\,\boldsymbol{f} \\ \mathbb{L}_{\mathrm{p}}\,\boldsymbol{f} \\ \mathbb{L}_{n}\,\boldsymbol{f} \\ \mathbb{L}_{\mathrm{c}}\,\boldsymbol{f} \end{Bmatrix} \tag{34}$$

Here $\mathbb{L}_{\mathrm{i}}$, $\mathbb{L}_{\mathrm{p}}$, $\mathbb{L}_{n}$ and $\mathbb{L}_{\mathrm{c}}$ are the *connectivity matrices* which define respectively: the interior contribution, the contribution of positive boundary nodes, the one from their corresponding negative boundary nodes, and finally the contribution from the nodes at the corners. In correspondence to (34), the tangent stiffness matrix is partitioned in the following way

$$\boldsymbol{K} = \frac{\mathrm{d}\boldsymbol{f}^{int}}{\mathrm{d}\boldsymbol{u}} = \begin{bmatrix} \boldsymbol{k}_{\mathrm{ii}} & \boldsymbol{k}_{\mathrm{ip}} & \boldsymbol{k}_{\mathrm{in}} & \boldsymbol{k}_{\mathrm{ic}} \\ \boldsymbol{k}_{\mathrm{pi}} & \boldsymbol{k}_{\mathrm{pp}} & \boldsymbol{k}_{\mathrm{pn}} & \boldsymbol{k}_{\mathrm{pc}} \\ \boldsymbol{k}_{\mathrm{ni}} & \boldsymbol{k}_{\mathrm{np}} & \boldsymbol{k}_{\mathrm{nn}} & \boldsymbol{k}_{\mathrm{nc}} \\ \boldsymbol{k}_{\mathrm{ci}} & \boldsymbol{k}_{\mathrm{cp}} & \boldsymbol{k}_{\mathrm{cn}} & \boldsymbol{k}_{\mathrm{cc}} \end{bmatrix} \equiv \begin{bmatrix} \mathbb{L}_{\mathrm{i}}\,\boldsymbol{K}\,\mathbb{L}_{\mathrm{i}}^{\mathrm{T}} & \mathbb{L}_{\mathrm{i}}\,\boldsymbol{K}\,\mathbb{L}_{\mathrm{p}}^{\mathrm{T}} & \mathbb{L}_{\mathrm{i}}\,\boldsymbol{K}\,\mathbb{L}_{n}^{\mathrm{T}} & \mathbb{L}_{\mathrm{i}}\,\boldsymbol{K}\,\mathbb{L}_{\mathrm{c}}^{\mathrm{T}} \\ \mathbb{L}_{\mathrm{p}}\,\boldsymbol{K}\,\mathbb{L}_{\mathrm{i}}^{\mathrm{T}} & \mathbb{L}_{\mathrm{p}}\,\boldsymbol{K}\,\mathbb{L}_{\mathrm{p}}^{\mathrm{T}} & \mathbb{L}_{\mathrm{p}}\,\boldsymbol{K}\,\mathbb{L}_{n}^{\mathrm{T}} & \mathbb{L}_{\mathrm{p}}\,\boldsymbol{K}\,\mathbb{L}_{\mathrm{c}}^{\mathrm{T}} \\ \mathbb{L}_{n}\,\boldsymbol{K}\,\mathbb{L}_{\mathrm{i}}^{\mathrm{T}} & \mathbb{L}_{n}\,\boldsymbol{K}\,\mathbb{L}_{\mathrm{p}}^{\mathrm{T}} & \mathbb{L}_{n}\,\boldsymbol{K}\,\mathbb{L}_{n}^{\mathrm{T}} & \mathbb{L}_{n}\,\boldsymbol{K}\,\mathbb{L}_{\mathrm{c}}^{\mathrm{T}} \\ \mathbb{L}_{\mathrm{c}}\,\boldsymbol{K}\,\mathbb{L}_{\mathrm{i}}^{\mathrm{T}} & \mathbb{L}_{\mathrm{c}}\,\boldsymbol{K}\,\mathbb{L}_{\mathrm{p}}^{\mathrm{T}} & \mathbb{L}_{\mathrm{c}}\,\boldsymbol{K}\,\mathbb{L}_{n}^{\mathrm{T}} & \mathbb{L}_{\mathrm{c}}\,\boldsymbol{K}\,\mathbb{L}_{\mathrm{c}}^{\mathrm{T}} \end{bmatrix} . \tag{35}$$

Periodic Displacements and Antiperiodic Tractions

At each node pair j on the boundary $\partial\mathcal{V}^+ \cup \partial\mathcal{V}^-$, the continuum conditions $(13)_1$ and $(13)_2$ induce discrete constraints at boundary nodes of the discretised RVE. The displacement fluctuation at the corners is prescribed to zero in order to avoid the solid rigid body motion.

Using the matrix notation introduced in Section 3.2, we redefine the *global coordinate matrix* for the periodic b.c. assumption as

$$\mathbb{D}_{\text{global},p} \equiv \left[\, \mathbb{D}_{\text{i}} \; \mathbb{D}_{\text{b},p} \right] \tag{36}$$

where $\mathbb{D}_{\text{i}}$ is the *interior coordinate matrix* defined in (29) and the $\mathbb{D}_{\text{b},p}$ is the *boundary coordinate matrix* for periodic assumption defined this time as

$$\mathbb{D}_{\text{b},p} = \left[\, \mathbb{D}_{\text{p}} \; \mathbb{D}_n \; \mathbb{D}_{\text{c}} \right] \tag{37}$$

where $\mathbb{D}_{\text{p}}$, $\mathbb{D}_n$, $\mathbb{D}_{\text{c}}$ are the *positive boundary coordinate matrix*, *negative boundary coordinate matrix* and *corner coordinate matrix*, respectively.

The Taylor displacement $\boldsymbol{u}^*$ defined as a constant for each node in (17), is given in a compact form as $\boldsymbol{u}^* = \mathbb{D}^{\text{T}}_{\text{global},p}\,\overline{\boldsymbol{\epsilon}}$, where $\mathbb{D}_{\text{global},p}$ is the global coordinate matrix for periodic assumption and $\overline{\boldsymbol{\epsilon}}$ is the matrix representation of the prescribed macroscopic strain tensor. In this model the variation of the Taylor displacement vector $\text{d}\boldsymbol{u}^*$ is considered as follows

$$\text{d}\boldsymbol{u}^* = \mathbb{D}^{\text{T}}_{\text{global},p}\,\text{d}\overline{\boldsymbol{\epsilon}} \,, \tag{38}$$

i.e. the displacement $\text{d}\boldsymbol{u}^*$ is a function of the variation of the macroscopic average strain vector $\text{d}\overline{\boldsymbol{\epsilon}}$.

Tangent Modulus of Periodic Displacements and Antiperiodic Traction on the Boundary Constraints

After rearranging the displacement nodal vector $\boldsymbol{u}$, the external nodal force vector $\boldsymbol{f}^{ext}$ and the stiffness matrix $\boldsymbol{K}$, as defined in (34) and (35), respectively, the general system (23) that relates the variations $\text{d}\boldsymbol{u}$ and $\text{d}\boldsymbol{f}^{ext}$ can be obtained.

Again the general procedure of Section 3.4 is followed to rearrange the system in the way described in (24). The variation of the Taylor displacement $\text{d}\boldsymbol{u}^*$ is given by (38). In this system the displacement fluctuation variation vector $\text{d}\tilde{\boldsymbol{u}}$ is considered as unknown. The application of the periodic displacement and antiperiodic external force in its discrete form, after some algebraic operations, gives the variation of the external force vector as

$$\text{d}\boldsymbol{f}^{ext}_{\text{b}} = \boldsymbol{K}^{\text{B}}_{\text{per}} \; \mathbb{D}^{\text{T}}_{\text{global},p}\,\text{d}\overline{\boldsymbol{\epsilon}} \tag{39}$$

where the Taylor displacement variation (38) has been inserted into the above equation (39). Therefore the desired expression is obtained as

$$\frac{\mathrm{d}\boldsymbol{f}_{\mathrm{b}}^{ext}}{\mathrm{d}\overline{\boldsymbol{\epsilon}}} = \boldsymbol{K}_{\mathrm{per}}^{\mathrm{B}}\ \mathbb{D}_{\mathrm{global},p}^{\mathrm{T}} \tag{40}$$

which gives the sensitivity of external boundary force vector $\mathrm{d}\boldsymbol{f}_{\mathrm{b}}^{ext}$ in terms of the macroscopic average strain matrix $\mathrm{d}\overline{\boldsymbol{\epsilon}}$.

The overall tangent moduli defined in (6), can be computed in its discretised F.E. matrix form, using previous averaged stress expression (21), in the following way

$$\overline{\mathcal{C}}_p = \frac{\mathrm{d}\overline{\boldsymbol{\sigma}}}{\mathrm{d}\overline{\boldsymbol{\epsilon}}} = \frac{1}{|\mathcal{V}|}\ \mathbb{D}_{\mathrm{b},p}\ \frac{\mathrm{d}\boldsymbol{f}_{\mathrm{b}}^{ext}}{\mathrm{d}\overline{\boldsymbol{\epsilon}}} \tag{41}$$

Inserting (40) into (41), the overall tangent modulus matrix form for periodic deformation and antiperiodic traction on the boundary of RVE is finally obtained as

$$\boxed{\overline{\mathcal{C}}_p = \frac{1}{|\mathcal{V}|}\ \mathbb{D}_{\mathrm{b},p}\ \boldsymbol{K}_{\mathrm{per}}^{\mathrm{B}}\ \mathbb{D}_{\mathrm{global},p}^{\mathrm{T}}}\ . \tag{42}$$

Clearly the modulus $\overline{\mathcal{C}}_p$ is a function of the boundary coordinate matrix $\mathbb{D}_{\mathrm{b},p}$ defined in (37), the *condensed periodic stiffness matrix* $\boldsymbol{K}_{\mathrm{per}}^{\mathrm{B}}$ and the global coordinate matrix $\mathbb{D}_{\mathrm{global},p}$ outlined in (36). Finally we remark that with the above expression (42), the tangent moduli can be computed for heterogeneous materials with arbitrary microstructures of the RVE. This results in the desired *quadratic rate of convergence* of the Newton-type solution procedure applied to solve the homogenized nonlinear macrostructure under *periodic deformation* and *antiperiodic traction* on the boundary of the RVE.

4 Numerical Examples

In this section numerical examples are presented in order to illustrate the scope and benefits of the described computational strategy. First set of numerical simulations focuses on microstructure simulations and discusses some important issues regarding numerical analysis at the micro-level such as the effect of boundary conditions, topology and distribution of heterogeneities, etc. Second numerical example considers a full two-scale simulation of a boundary value problem and incorporates all computational ingredients described in this paper. This example also includes a comparison with a detailed single scale analysis.

4.1 Study of the Effect of Topology of Cavities on the Properties of the RVE

Problem Specifications

A square unit cell is considered representing an RVE at the micro-level. The cell is composed of an elasto-plastic material with heterogeneity being induced

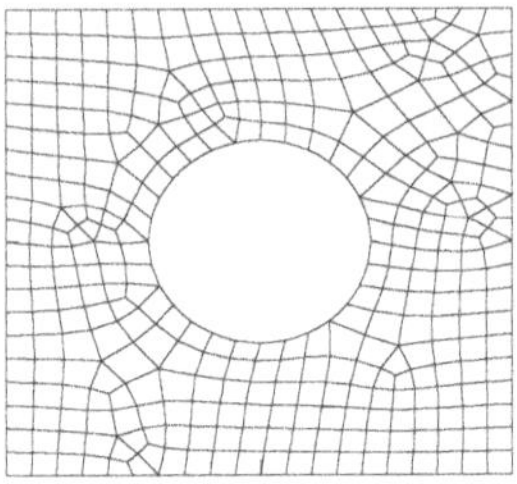

Fig. 5. Regular cavity model

by cavities. Two models are considered: (i) a regular cell with a single circular hole embedded in a soft matrix depicted in Fig. 5, (ii) randomly generated distribution of cavities surrounded by soft matrix given in Fig. 7. For both models the void volume fraction of the unit cell is taken as 15%.

Two types of finite elements are employed: linear 3-noded triangle element and 8-noded quadrilateral element with 4-Gauss points. The matrix in all models is assumed to be composed of the von Mises elasto-plastic material with linear strain hardening. The material properties assigned are: Young's modulus $E = 70\,GPa$, Poisson's ratio $\nu = 0.2$, the initial yield stress $\sigma_{Y_0} = 0.243\,GPa$ and the strain hardening modulus $H = 0.2\,GPa$.

Analysis Approach

All simulations in this section have been performed by employing the computational homogenisation under the plane-stress assumption in small strain regime. The average stress is obtained by imposing the macro-strain over the unit cell and solving the problem for defined boundary condition over the RVE. The generic imposed macro-strain tensor is expressed by:

$$[\bar{\varepsilon}_{11}, \bar{\varepsilon}_{22}, 2\bar{\varepsilon}_{12}] = [0.001, 0.001, 0.0034]\,.$$

To obtain the load step at each load increment, the generic strain tensor is multiplied by the relevant load factor. The analysis is performed under two different boundary conditions: (i) linear displacement boundary condition, and (ii) periodic displacement boundary condition.

Study of the Regular Cavity Model

An 8-node quadrilateral element with 4-Gauss points is employed in this simulation. Figure 5 depicts a finite element mesh containing 350 elements and 1158 nodes.

Figures 6 (a) and (b) show, respectively, the deformed mesh and the equivalent plastic strain distribution for the linear displacement boundary condition. This plastic zone is clearly positioned along the diagonal side of the unit

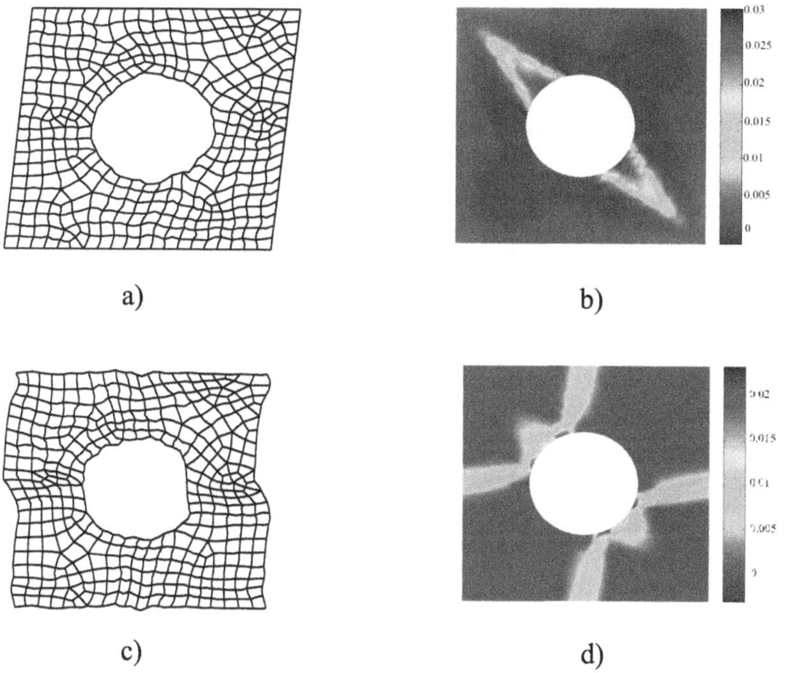

Fig. 6. Regular cavity model under linear displacement boundary condition and periodic condition. a) Deformed mesh. b)Effective plastic strain contour plot

cell in direction of the imposed shear. The corresponding results for the periodic boundary condition are given in Figs. 6 (c) and (d). From Fig. 6, it can be seen that the plastic zone has a distinctively different pattern under periodic boundary condition.

The overall stress-strain response is presented in terms of the Euclidean norm of the average stress and strain, given, respectively as

$$\|\bar{\sigma}\| = \sqrt{\bar{\sigma}_{11}^2 + \bar{\sigma}_{22}^2 + \bar{\sigma}_{12}^2}, \quad \|\bar{\varepsilon}\| = \sqrt{\bar{\varepsilon}_{11}^2 + \bar{\varepsilon}_{22}^2 + \bar{\varepsilon}_{12}^2}.$$

Figure 9 shows the resulting average stress - strain curves for this model. The obtained results show that under linear displacement boundary condition the overall response of the regular cavity model shows significantly stiffer behaviour with respect to the overall response under periodic boundary condition.

The RVE with Randomly Generated Voids

In this study a unit cell at the micro-level with a randomly generated distribution of void placements and sizes is considered (see Fig. 7). A standard 3-node linear triangular element is employed in this simulation. Again plane-stress conditions are prescribed and two boundary conditions at the micro-

Fig. 7. Unit cell with randomly generated voids

level are considered: (i) linear displacement boundary condition, and (ii) periodic boundary condition. The imposed macro-strain and material properties for this model are identical to the previous examples in this section.

Figure 8 shows the equivalent plastic strain distribution for both linear displacement boundary condition and periodic boundary conditions. The occurrence of localised bands with significant plastic straining can be observed on both contour plots. Significantly, unlike in the case of the single cavity model both boundary conditions give similar distribution of the plastic strain indicating the convergence of the results at the micro-level with the increase of the statistical sample of heterogeneities.

Figure 9 shows the average stress - strain curves for this model. It can be observed that the micro-cell with randomly generated void distribution results in the stress-strain behaviour that shows small difference between the two different boundary conditions imposed at the micro-level. This clearly indicates the convergence of the average properties with the increase of the statistical sample representing the heterogeneities at the micro-level.

4.2 Two-scale Analysis of Stretching of an Elasto-plastic Perforated Plate

In this section a full two-scale analysis of a perforated plate is performed. This is a classical example often used as a verification problem in computational plasticity. The plate is composed of an elasto-plastic material and contains regularly distributed voids. The plate has width $10\,mm$, length $18\,mm$ and

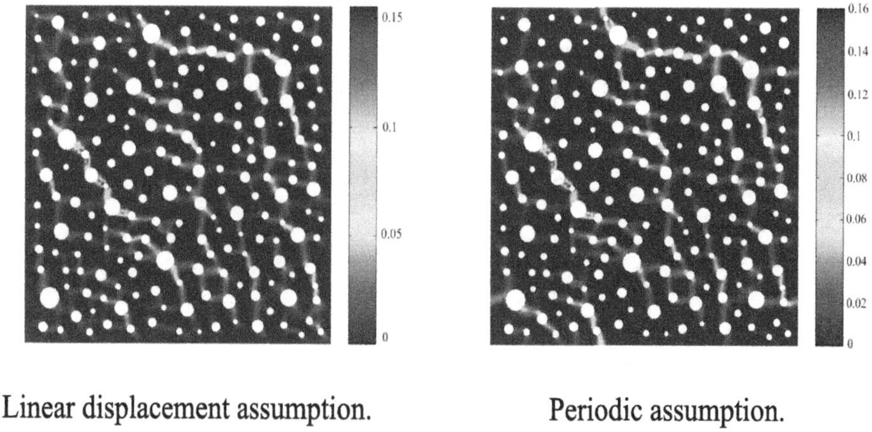

Fig. 8. Effective plastic strain contours for the unit cell with randomly generated voids under two boundary conditions

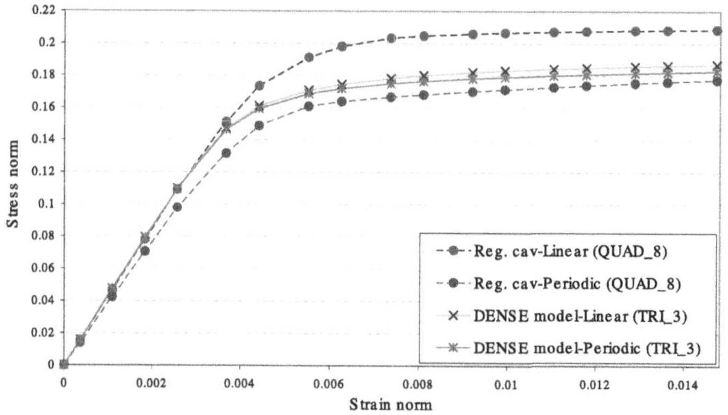

Fig. 9. Stress-Strain norm curves for dense model under two boundary conditions

uniform thickness of $1\,mm$ (see Fig. 10). For obvious symmetry reasons only one-quarter of the specimen is considered (see Fig. 10). The simulation is performed by imposing uniform displacement along the upper boundary. The elasto-plastic material is assumed to follow the standard von Mises model with linear isotropic hardening. Material properties are: Young modulus $E = 70\,GPa$, Poisson's ration $\nu = 0.2$, yield stress $\sigma_{Y_0} = 0.243\,GPa$ and hardening modulus $H = 0.2\,GPa$. Both two-scale analysis and a single scale analysis of this problem are performed.

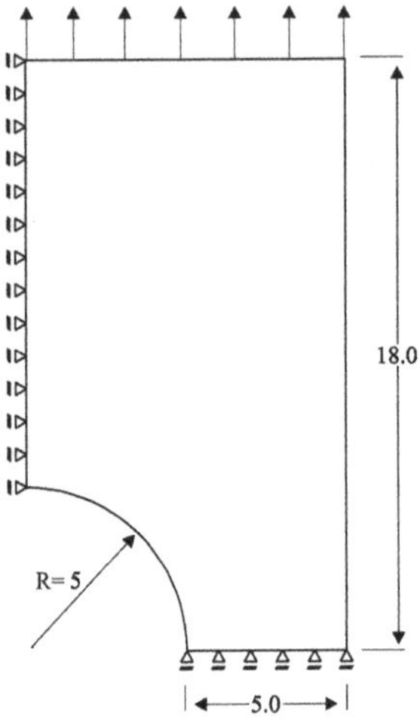

Fig. 10. Plane-stress strip with a circular hole. Geometry and boundary conditions

Single-scale Analysis

Single scale analysis is used for comparative purposes and is performed on a detailed finite element mesh of the problem given in Fig. 11 (a). The mesh is composed of 11216 4-node quadrilateral elements and 12147 nodes. Figure 11 (b) illustrates distribution of an equivalent plastic strain at latter stages of the simulation.

Two-scale Analysis

For the multi-scale finite element analysis the perforated plate is defined as a homogeneous structure at the macro-level, while at the micro-level a unit cell is defined with side length equal to 1 mm and a single void in the centre of the micro-cell giving the volume fraction of 50%. Linear 3-noded triangle element is employed at both macro and micro-level (see Fig. 12). The mesh at the macro-level number is composed of 25 elements and 21 nodes, while at the micro-level the FE mesh is composed of 603 elements and 352 nodes.

Multi-scale analysis has been performed under three different boundary conditions at the micro-level: (i) Taylor assumption, (ii) linear displacement boundary condition and (iii) periodic boundary condition. As can be seen from Fig. 13, which gives reaction force against the prescribed displacement, different boundary conditions result in markedly different force-displacement diagrams. As expected, the results obtained for the Taylor assumption show substantially stiffer behaviour with comparison to the other two boundary assumptions. Periodic boundary assumption generates the softest response, and significantly the resulting overall behaviour shows very good correspondence with the results obtained by the detailed single-scale analysis.

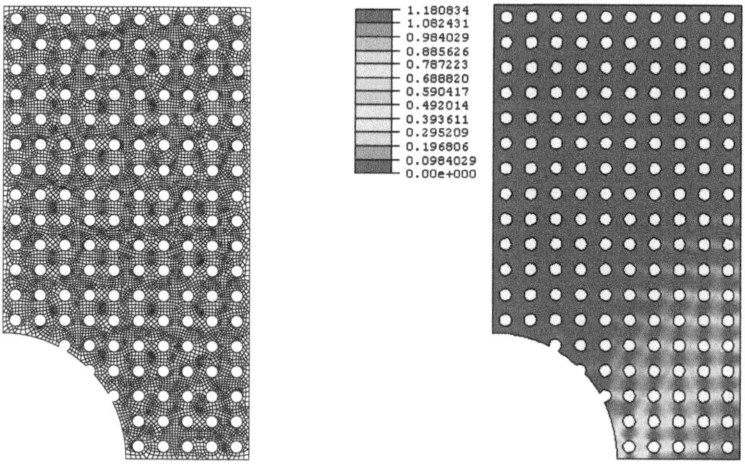

a) b)

Fig. 11. Single-scale analysis of an elasto-plastic perforated plate: (a) Finite element mesh, and (b) distribution of equivalent plastic strain

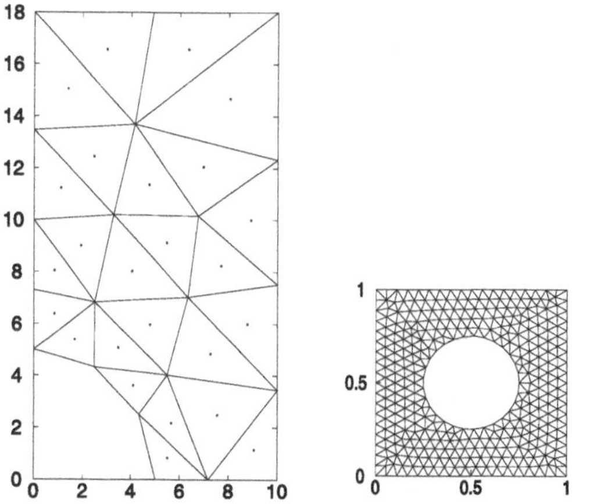

Macro-structure Micro-structure

Fig. 12. FE meshes at macro- and micro-level for multi-scale analysis

5 Conclusions

A multiscale computational strategy for homogenisation of material behaviour of heterogeneous composites has been described. The presented numerical tests have confirmed the successful implementation of the computational procedure and efficient solution of the discrete multiscale problem.

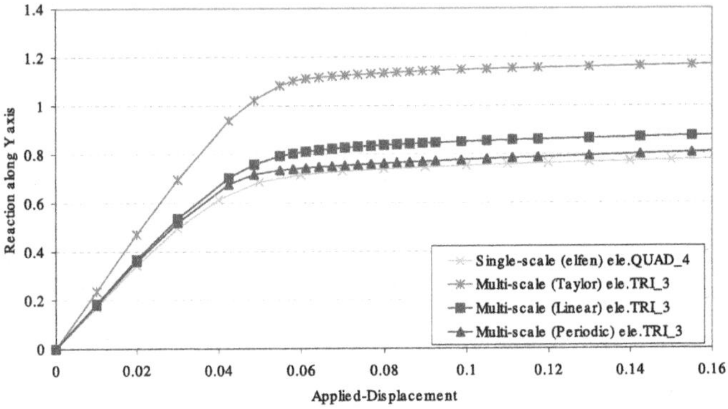

Fig. 13. Reaction along Y direction against the applied displacement

The ongoing research is concerned with the analysis of more general non-linear material behaviour at the microscale and incorporation of the finite strain kinematics. This work will be reported in future publications.

References

1. Carneiro Molina AJ, de Souza Neto EA and Perić D (2005) Homogenized tangent moduli for heterogeneous materials. In: Crouch P (ed) Proceedings of the 13th ACME Conference. Sheffield UP, Sheffield
2. Belytschko T, Liu WK, Moran B (2000) Nonlinear Finite Element for Continua and Structures. Wiley, New York
3. de Souza Neto EA, Perić D, Owen DRJ (to be published) Computational Plasticity: Small and Large Strain Finite Element Analisys of Elastic and Inelastic Solids.
4. Hill R (1972) On constitutive macro-variables for heterogeneous solids at finite strain. Proc Roy Soc London 326:131-147
5. Kouznetsova VG, Brekelmans WAM, Baaijens FPT (2001) An Approach to micro-macro modelling of heterogeneous materials. Comput Mech 27:37-48
6. Kouznetsova VG (2002) Computational homogenization for multiscale analysis of multi-phase materials PhD Thesis, TU University Eindhoven, Eindhoven
7. Miehe C, Schotte J and Schröder J (1999) Computational micro-macro transitions and overall moduli in the analysis of polycrystals at large strains. Comput Material Sci 16:372-382
8. Miehe C, Schröder J and Schotte J (1999) Computational homogenization analysis in finite plasticity - Simulation of texture development in polycrystalline materials. Comp Meth Appl Mech Engng 171:387-418
9. Miehe C and Koch A (2002) Computational micro-to-macro transitions of discretized microstructures undergoing small strains. Arch Appl Mech 72:300-317
10. Moulinec H and Suquet P (1998) A numerical method for computing the overall response of nonlinear composites with complex microstructure. Comp Meth Appl Mech Engng 157:69-94

11. Smit RJM, Brekelmans WAM and Meijer HEH (1998) Prediction of the mechanichal behaviour of nonlinear heterogeneous syatems by multi-level finite element modeling. Comp Meth Appl Mech Engng 155:181-192
12. Suquet PM (1985) Local and global aspects in the mathematical theory of plasticity. In: Sawczuk A (ed) Plasticity today: modelling methods and applications. Elsevier Applied Science Publishers, Amsterdam
13. Suquet PM (1987) Elements of homogenization for inelastic solid mechanics. In: Sanchez-Palencia E and Zaoui A (eds) Homogenization Techniques for Composite Media. Springer-Verlag, Berlin
14. Swan CC (1994) Techniques for stress-controlled and strain-controlled homogenization of inelastic periodic composites. Comp Meth Appl Mech Engng 117:249-267
15. Terada K and Kikuchi N (2001) A class of general algorithms for multi-scale analyses of heterogeneous media. Comp Meth Appl Mech Engng 190:5427-5464
16. Zohdi TI and Wriggers P (2005) Introduction to Computational Micromechanics. Springer, Berlin

Assessment of Protection Systems for Gravel-Buried Pipelines Considering Impact and Recurrent Shear Loading Caused by Thermal Deformations of the Pipe

B. Pichler, Ch. Hellmich, St. Scheiner, J. Eberhardsteiner and H.A. Mang

Institute for Mechanics of Materials and Structures
Vienna University of Technology
Karlsplatz 13/202, A-1040 Vienna, Austria
e-mail: Herbert.Mang@tuwien.ac.at

Summary. Two safety topics in pipeline engineering are considered: (1) rockfall onto gravel-buried steel pipes and (2) protection of the outer anti-corrosion coating of soil-covered steel pipelines. For both cases, effective protection systems are identified, based on the results of non-linear elasto-plastic Finite Element analyses.

Introduction

Two safety topics in pipeline engineering will be treated: (i) rockfall onto gravel-buried steel pipes and (ii) protection of the outer anti-corrosion coating of soil-covered steel pipelines. In both cases non-linear elasto-plastic Finite Element (FE) analyses provide insight into the structural behavior, as needed for the design of effective protection systems. For rockfall, a two-component protection system is recommended. It consists of an impact damping layer and of a buried load-distributing and load-carrying structure. As regards wear of the anti-corrosion coating, two well-established means of protection are considered to be most effective: (i) burying pipelines by sand and (ii) covering pipelines by (fibre-)reinforced concrete.

1 Protection Systems for Gravel-Buried Pipelines Subjected to Rockfall

Recent increase of rockfall activities in the European Alps has raised the need for designing impact protection systems for pipelines in Alpine valleys. This section deals with the assessment of different protection systems including

Eugenio Oñate and Roger Owen (eds.), Computational Plasticity, 187–206.

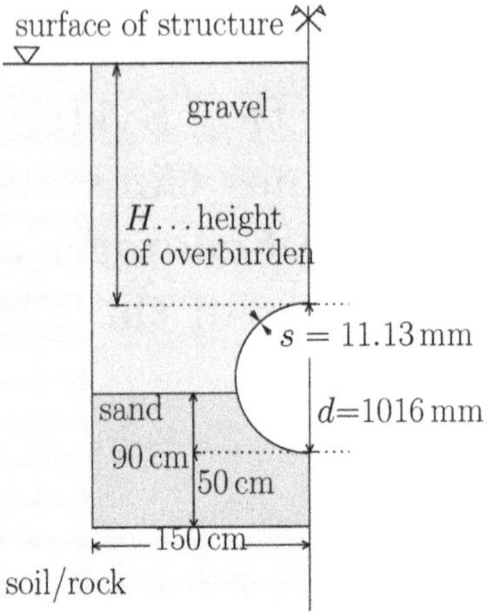

Fig. 1. Geometric dimensions of the gravel-buried steel pipeline in a section perpendicular to the axis of the pipe

sandy gravel as an energy-absorbing and load-distributing structural component. In Subsection 1.1, the development of a FE model allowing for prediction of the loading of a gravel-buried steel pipe subjected to rockfall is described. In Subsection 1.2, results from a real-scale impact experiment of a boulder onto a buried steel pipe are used for an assessment of this model. To ensure the significance of this validation procedure, the identification of the model parameters is based on experiments which are *independent* of the real-scale impact test. Finally, in Subsection 1.3 and 1.4, respectively, the structural model is used to study the performance of two different rockfall-protection systems covering a steel pipe.

1.1 Development of a Structural Model

Geometric Dimensions of the Considered Problem

A pipeline with an outer diameter $d = 1016\,\text{mm}$ and a wall thickness $s = 11.13\,\text{mm}$ is considered. It is buried in the middle of a trench of 3 m width, resting on a 50 cm thick layer of sand. The tube is laterally buried up to a height of 40 cm by sand. The rest of the trench is filled by sandy gravel, see Fig. 1.

Impact Scenario and Mode of Analysis

Single boulders impacting vertically, directly above the axis of the steel pipe, are considered as typical rockfall events. Making use of symmetries, it is sufficient to discretize one fourth of the entire structure, see Fig. 2 (a). The material beside and beneath the trench is represented by a Winkler foundation modeled by bar elements representing linear springs, see Figs. 1 and 2 (a).

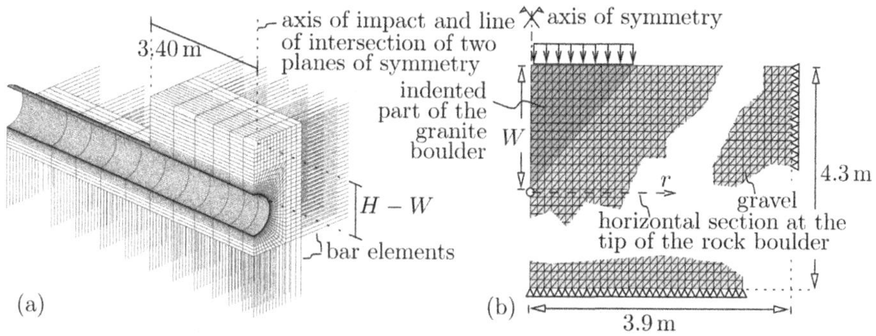

Fig. 2. (a) FE discretization used for the validation of the structural model; (b) axisymmetric analysis performed to obtain the distribution of surface loads applied to the 3D FE model

Dead load results in stresses in the pipe, which are by two orders of magnitude smaller than the stresses caused by the investigated types of impact. Therefore, dead load is not taken into account and, hence, the overburden of the pipe is only discretized to simulate the load distribution. Since the impact stresses in the backfill material *above* the pipe occur only in the vicinity of the impact axis, the discretization of the overburden of the pipe is restricted to a longitudinal distance of 3.4 m from the impact axis, see Fig. 2 (a). Outside this domain, only the backfill material beside and beneath the tube is discretized.

The penetration process is not modeled in detail. Instead, the height of the overburden H is reduced by the penetration depth W at which the maximum impact force occurs, see Fig. 2 (a). Both dynamic and static 2D plain strain analyses delivered approximately the same maximum stresses in the pipe [18]. Therefore, the maximum impact force is applied quasistatically onto the surface of the FE model. This mode of modeling requires estimates of (i) the maximum impact force and the penetration depth at maximum impact force, both as a function of the mass m and the impact velocity v_0 of an impacting boulder, and (ii) the distribution of stresses corresponding to the maximum impact force, which are prescribed as stress boundary conditions for the FE analysis.

Estimates of the Maximum Impact Force and the Penetration Depth at Maximum Impact Force

The final penetration depth X, i.e. the penetration depth after the end of the impact, is estimated based on the dimensionless formula [16]

$$\frac{X}{d} = \sqrt{\frac{1 + k\,\pi/4\,N}{(1 + I/N)}\,\frac{4k}{\pi}\,I} \quad \text{for} \quad \frac{X}{d} \leq k\,, \quad \text{with} \quad I = \frac{m\,v_0^2}{d^3\,R}\,, \tag{1}$$

where d denotes the characteristic size of the boulder, N is a geometry function characterizing the “sharpness” of the boulder nose, k stands for the dimensionless depth of a surface crater, and R denotes the indentation resistance of the hit material. For cubic granite boulders with a volume V and a mass density equal to $2700\,\mathrm{kg/m^3}$, impacting with a tip onto gravel with a mass density equal to $1800\,\mathrm{kg/m^3}$, the following relationships were obtained based on a hybrid experimental-analytical approach [16]

$$N = 2.385\,, \quad k = 1.257\,, \quad d = 1.050\,\sqrt[3]{V}\,, \quad \text{and} \quad R = 9.22\,\mathrm{MPa}\,. \qquad (2)$$

Relations between X and the maximum impact force F, and between X and the penetration depth at maximum impact force, W, were derived on the basis of a model for the impact kinematics which was deduced from experimental acceleration measurements [16, 17] as

$$\frac{X}{d} = \frac{m\,v_0^2}{F\,d} = \frac{W}{d}\left(\frac{4\,\pi^2}{3\,\pi^2+4}\right)\,. \qquad (3)$$

Computation of the Distribution of Stresses Corresponding to the Maximum Impact Force

In order to perform a 3D FE analysis of rockfall onto gravel as described previously, the stress distribution corresponding to the maximum impact force must be specified. Herein, this stress distribution is computed by means of an axisymmetric linear-elastic FE model. It comprises the gravel and the tip of the granite boulder approximated as a conical indenter at a penetration depth W, see Fig. 2 (b). The maximum impact force F is applied to the conical indenter as a spatially constant surface stress σ. The material parameters for gravel and granite are taken from Table 1. The vertical stresses obtained for the axisymmetric FE model in a horizontal section through the model at the tip of the boulder, see Fig. 2 (b), serve as surface loads for the 3D FE model.

Material Modeling of Steel, Gravel and Sand

The material behavior of steel is modeled by small-strain von Mises elasto-plasticity, see e.g. [10, 22] and Table 1 for Young's modulus E, Poisson's ratio ν, and the uniaxial yield strength σ_y.

The material behavior of gravel is represented by the elasto-plastic Cap Model [3, 5, 19]. The elastic domain follows the isotropic generalized Hooke's law [12]: $\boldsymbol{\sigma} = \mathbf{C} : (\boldsymbol{\varepsilon} - \boldsymbol{\varepsilon}^p)$, where $\boldsymbol{\sigma}$ denotes the Cauchy stress tensor, and $\boldsymbol{\varepsilon}$ and $\boldsymbol{\varepsilon}^p$ stand for the linearized strain tensor and the plastic strain tensor, respectively. $\mathbf{C}$ represents the isotropic constitutive elasticity tensor which can be expressed as a function of the bulk modulus K and the shear modulus G, reading $\mathbf{C} = K\,\mathbb{I}\otimes\mathbb{I} + 2G(\mathbb{II} - \frac{1}{3}\mathbb{I}\otimes\mathbb{I})$, where $\mathbb{I}$ and $\mathbb{II}$ denote the 2nd-order unity tensor and the 4th-order unity tensor, respectively. In the principal stress space the elastic domain is bounded by three surfaces, (i) a tension cut-off accounting for tensile failure, (ii) a Drucker-Prager surface defining shear failure under pronounced deviatoric stress states, and (iii) an ellipsoidal cap

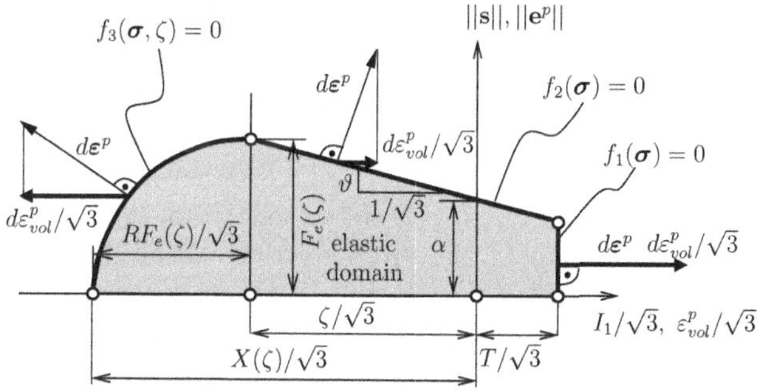

Fig. 3. Cap model for gravel: elastic domain and direction of plastic flow, respectively, in a meridional plane of the principal stress and the plastic-strain space, respectively

representing the hardening of the material associated with compaction, see Fig. 3. In mathematical terms, these functions read [8, 9]

$$\begin{aligned} f_1(\boldsymbol{\sigma}) &= I_1 - T = 0\,, \\ f_2(\boldsymbol{\sigma}) &= ||\mathbf{s}|| - F_e(I_1) = 0 && \text{for} \quad T \geq I_1 \geq \zeta\,, \\ f_3(\boldsymbol{\sigma},\zeta) &= F_c(||\mathbf{s}||, I_1, \zeta) - F_e(\zeta) = 0 && \text{for} \quad \zeta \geq I_1 \geq X(\zeta) \quad \text{with} \end{aligned} \tag{4}$$

$$\begin{aligned} F_e(\xi) &= \alpha - \vartheta\,\xi\,, \\ F_c(||\mathbf{s}||, I_1, \zeta) &= \sqrt{||\mathbf{s}||^2 + ([I_1 - L(\zeta)]/R)^2} \end{aligned} \quad \text{and} \quad L(\zeta) = \begin{cases} \zeta & \text{if} \quad \zeta < 0\,, \\ 0 & \text{if} \quad \zeta \geq 0\,, \end{cases}$$

where I_1 and $\mathbf{s}$ denote the first invariant of the stress tensor and the deviatoric stress tensor, respectively. The direction of plastic flow is given by the associated flow rule [7]

$$d\boldsymbol{\varepsilon}^p = \sum_{\alpha \in J_{act}} d\lambda_\alpha \frac{\partial f_\alpha}{\partial \boldsymbol{\sigma}}\,, \tag{5}$$

where the consistency parameters are denoted as $d\lambda_\alpha$, whereas J_{act} stands for the set of active yield surfaces, defined as $J_{act} = \{\alpha \in [1,2,3] \mid f_\alpha(\boldsymbol{\sigma},\zeta) = 0\}$. Activation of the cap mode leads to compaction, whereas activation of the failure-surface mode or the tension cut-off mode results in plastic volume dilatation, see Fig. 3. The tension cut-off and the Drucker-Prager surface are fixed in the stress space. The ellipsoidal cap, however, expands if it is activated, reflecting material compaction described by

$$\varepsilon^p_{vol} = -W\,(1 - \exp[D\,X(\zeta)])\,, \quad \text{with} \quad X(\zeta) = \zeta + R\,F_e(\zeta)\,, \tag{6}$$

where ζ denotes the hardening state variable. Algorithmic issues of this model as well as the expressions for the consistent tangent moduli can be found in [5, 8, 9, 22].

Parameter Identification

The parameters K and G of the backfill material are identified on the basis of the theory of elastic waves from a series of small strain acoustic experiments on unloaded sandy gravel [15], see Table 1. The tested material (mass density of 1800 kg/m^3) consisted to 60–80 % of particles with diameters < 70 mm, and to 20–40 % of edged stones with diameters from 70 mm – 200 mm. The loading surfaces of the Cap Model are defined by seven material parameters (see Fig. 2 (b)): T, α, ϑ, W, D, R, and ζ_{ini}, denoting the initial value of the hardening state variable ζ. Identification of the parameters α, ϑ, W, and D was based on triaxial tests on gravel [14, 15, 18], where cylindrical specimens were subjected to hydrostatic compression up to different confining pressures p. While p was subsequently kept constant, the specimens were subjected to additional uniaxial loading in the direction of the cylinder axis, up to shear failure. The parameters α and ϑ (Table 1) were identified from the stress states at failure. W and D were identified such that measured relationships between the mean normal stress and the volumetric strain are well predicted for $p \leq 0.9$ MPa, see (Table 1). The initial size and the shape of the cap were determined from constant volume tests, $R = 4.4$ and $\zeta_{ini} = 0$ kPa, see [8]. For numerical reasons, the material parameter referring to tensile failure of gravel, T, was set equal to a very small positive value (100 Pa). Since, in a first approximation, material parameters of dense sand are comparable to the material parameters of the investigated type of gravel [20], no distinction between sand and gravel was made in the numerical simulation.

Table 1. Material parameters of steel, gravel, sand, granite, and concrete

Material parameters of steel: $E = 210\,\text{GPa}$, $\nu = 0.3$, $\sigma_y = 514\,\text{MPa}$
Material parameters of sandy gravel (also valid for dense sand)
$K = 244\,\text{MPa}$, $G = 72\,\text{MPa}$, $\alpha = 149\,\text{kPa}$, $\vartheta = 0.40$, $W = 0.28$, $D = 0.05\,\text{MPa}^{-1}$, $R = 4.4$, $\zeta_{ini} = 0\,\text{kPa}$, $T = 0.1\,\text{kPa} \approx 0$
Material parameters of granite: $E = 50\,\text{GPa}$, $\nu = 0.17$
Material parameters of concrete
$E = 26.2\,\text{GPa}$, $\nu = 0.3$, $\alpha = 21.74\,\text{MPa}$, $\vartheta = 0.099$, $T = 0.1\,\text{kPa} \approx 0$

1.2 Validation of the Developed Structural Model

Real-scale Impact Experiment

The assessment of the developed structural model is based on a real-scale impact experiment, which is independent from all experiments related to the development of the structural model. In an area of a quarry, used for disposal of wet clay, a 22 m long steel pipe was buried with sandy gravel as illustrated in Fig. 1. The height of the overburden H was equal to 2 m. A granite boulder of approximately cubic shape with a mass $m = 18260\,\mathrm{kg}$ ($V = 6.76\,\mathrm{m}^3$, $d = 1.99\,\mathrm{m}$) was dropped from a height $h_f = 18.85\,\mathrm{m}$ ($v_0 = 19.23\,\mathrm{m/s}$) onto the buried pipeline such that it impacted with a tip. The impact of the boulder caused a crater with a depth of $X = 0.85\,\mathrm{m}$.

The cross-section of the pipe containing the impact axis was equipped with four strain gauge rosettes. The strain histories measured during the impact were converted into stress histories by means of ideal elasto-plasticity of von Mises type, see e. g. [10], assuming plane-stress conditions. The obtained maximum von Mises stresses at the locations of the strain gauge rosettes are listed in Table 2.

Table 2. Von Mises stresses and second invariant of the strain tensor referring to the maximum loading of the pipe subjected to rockfall: experimental results, numerical results, and relative errors

position	experimental result	FE result	relative error*
$12h$ ○	$\sigma_{vM}^{exp} = 514\,\mathrm{MPa} = \sigma_y$	$\sigma_{vM}^{FE} = 514\,\mathrm{MPa} = \sigma_y$	0.0 %
$3h$ ○	$\sigma_{vM}^{exp} = 5386\,\mathrm{MPa}$	$\sigma_{vM}^{FE} = 357\,\mathrm{MPa}$	7.5 %
$6h$ ○	$\sigma_{vM}^{exp} = 5204\,\mathrm{MPa}$	$\sigma_{vM}^{FE} = 305\,\mathrm{MPa}$	49.5 %
$12h$ ○	$J_{2,\varepsilon}^{exp} = -3.4761 \cdot 10^{-6}$	$J_{2,\varepsilon}^{FE} = -3.1754 \cdot 10^{-6}$	8.7 %

* evaluated as $|\sigma_{vM}^{exp} - \sigma_{vM}^{FE}|/\sigma_{vM}^{exp}$ and $|J_{2,\varepsilon}^{exp} - J_{2,\varepsilon}^{FE}|/J_{2,\varepsilon}^{exp}$, respectively

Comparison between Model-predicted and Experimentally Determined Stress Distribution in the Steel Pipe

The real-scale impact experiment was simulated by means of the developed 3D FE model. The maximum impact force and the penetration depth at maximum impact force were calculated from Eqs. (3) as $F = 7.94\,\mathrm{MN}$, and $W = 0.72\,\mathrm{m}$. The values for F and W served as input for modeling of the impact as described in Subsection 1.1. The coefficient for the sub-grade reaction, k_s, characterizing the elastic foundation of the trench, was set equal to $18\,\mathrm{MN/m^3}$, which is the mean value of the interval $[12\,\mathrm{MN/m^3}; 24\,\mathrm{MN/m^3}]$ recommended for soft clay in [6]. Finite elements with trilinear shape functions for the displacements were used. The nonlinear elasto-plastic FE simulation was performed in an incremental-iterative manner, based on consistent tangent moduli [5, 8, 9].

The computed stresses in the pipe were compared to the corresponding experimental results, see Table 2. For a graphical representation of the obtained von Mises stresses along the inner surface of the pipe, see the thick solid line in Fig. 4 (a). In general, the simulated behavior of the steel pipe reflects the experimentally observed behavior of the tube satisfactorily, both qualitatively and quantitatively. At the positions $12h$ and $3h$, good agreement between the numerical predictions and the experimental results is observed. Since the steel

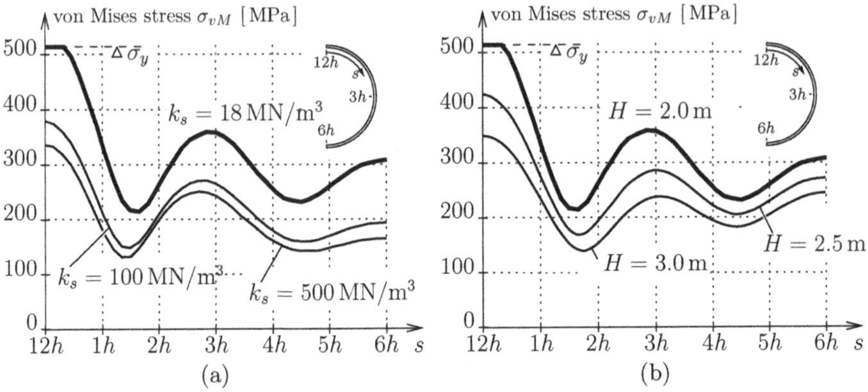

Fig. 4. Prognoses of the distribution of the von Mises stress along the inner surface of the pipe in the cross-section beneath the impact as a function of (a) the coefficient of the sub-grade reaction and (b) the height of overburden

yielded at the top of the pipe, also strain measurements and FE-predictions are compared at the position $12h$, based on the second invariant of the strain deviator,

$$J_2^{\varepsilon} = \frac{1}{6}\left[(\varepsilon_{xx} - \varepsilon_{\vartheta\vartheta})^2 + (\varepsilon_{\vartheta\vartheta} - \varepsilon_{rr})^2 + (\varepsilon_{rr} - \varepsilon_{xx})^2\right] + \varepsilon_{x\vartheta}^2 + \varepsilon_{\vartheta r}^2 + \varepsilon_{rx}^2 \, . \quad (7)$$

Also this comparison yields a discrepancy of less than 10 %, see Table 2. The largest relative error between the numerical simulation and the experiment is obtained at the position $6h$. There, the numerically predicted loading of the pipe is 49.5 % greater than the experimentally obtained value. Nevertheless, in regions where the highest loading of the steel pipe occurs, the developed structural model yields satisfactory results. Consequently, this model possesses predictive capabilities, i. e., the model is successfully validated.

1.3 Prognoses of Structural Behavior

Prognoses Considering a Change of the Boundary Conditions

The trench in which pipelines in Alpine valleys are buried by gravel is usually surrounded by rock rather than by soft clay. This provides the motivation to study the influence of different coefficients of sub-grade reaction on the loading of the pipe. In order to model the behavior of weathered rock, k_s is set equal to $100\,\mathrm{MN/m^3}$, as recommended for dense sand in [6]. Secondly, k_s is set equal to $500\,\mathrm{MN/m^3}$ in order to investigate an (almost) rigid bedding, as encountered with unweathered rock. In both cases, the height of the overburden and the intensity of the impact are the same as in the FE simulation described in Subsection 1.2. For increased values of the coefficient of sub-grade reaction, the FE-predicted maximum von Mises stress of the pipe does not reach the yield stress, see Fig. 4 (a). An increase of k_s from 18 MN/m^3 to 100 MN/m^3 and 500 MN/m^3 results in a reduction of the loading of the pipe by approximately 25 % and 30 %, respectively.

Prognoses Considering a Change of the Structural Dimensions

In order to assess the capability of gravel layers to serve as a protection system for steel pipes endangered by rockfall, the influence of different heights of overburden ($H = 2.5$ m and 3.0 m) on the loading of the pipe is studied. In these two simulations, the coefficient of sub-grade reaction and the intensity of the impact are the same as in the FE simulation which allowed for validation of the model, see the Subsection 1.2. For increased heights of the overburden, the maximum von Mises stress of the pipe, predicted by the FE simulation, does not reach the yield stress, see Fig. 4 (b). An increase of H from 2.0 m to 2.5 m and 3.5 m results in a reduction of the loading by approximately 20 % and 30 %, respectively. However, this decrease is less than linear. Therefore, in order to effectively increase the safety of a gravel-buried pipeline subjected to rockfall, the burying depth must be increased significantly.

Prognoses Considering a Change of the Boundary Conditions and of the Structural Dimensions

In the two previous parameter studies the influence of the increasing height of the overburden and the one of the increasing coefficient of sub-grade reaction on the loading of the pipe were investigated separately. In order to investigate the synergism of both modifications, H and k_s are increased simultaneously in an additional numerical analysis.

The height of overburden is set equal to 3 m and the coefficient of sub-grade reaction equal to 500 MN/m^3, whereas the intensity of the impact remains unchanged. Results from this FE analysis show a further reduction of the loading of the pipe (compare the thick line in Fig. 6 with the results illustrated in Figs. 4 (a) and (b)). The maximum von Mises stress encountered along the inner surface of the pipe is equal to 33.5 % of the yield stress.

1.4 Assessment of an Enhanced Protection System Consisting of Gravel and, Additionally, of Buried Load-Carrying Structural Elements

A protection system for a steel pipe subjected to rockfall must satisfy two requirements: (i) damping of the impact, in order to keep the forces arising from rockfall reasonably small and (ii) load distribution and load-carrying capacity, in order to reduce the loading of the steel pipe. As for the protection system investigated so far, both tasks are accomplished by the layer of gravel. However, the flexibility of gravel required for the damping of the impact is opposed to the stiffness of the material required for distribution and carrying of the load.

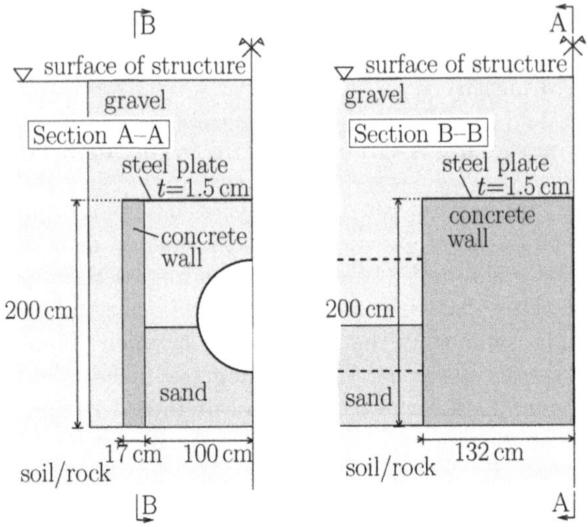

Fig. 5. Alternative protection system consisting of gravel as an energy-absorbing and impact-damping system and a buried steel plate supported by concrete walls as a load-carrying structural component

In order to further improve the effectiveness of a protection system for a steel pipe subjected to rockfall, the aforementioned two tasks should be performed by two separate structural elements. Such an enhanced protection system is investigated subsequently.

Buried Steel Plate Resting on Concrete Walls

The investigated protection system consists of (i) gravel as an energy-absorbing and impact-damping system and (ii) a buried steel plate supported by concrete walls, representing a load-carrying structural component, as illustrated in Figs. 5 (a) and (b). Length, width, and thickness of the steel plate are equal to 264 cm, 234 cm, and 1.5 cm, respectively. It consists of the same steel as used for the pipe (Table 1). The boundary conditions at the interface between the steel plate and the concrete walls are chosen such that only compressive forces can be transmitted to the concrete walls.

For the description of the material behavior of concrete, the Cap Model is employed, with the ellipsoidal failure surface being removed. The considered type of concrete is characterized by the uniaxial compressive strength $f_{cu} = -30.3\,\text{MPa}$ ($\sigma_{11} = f_{cu}$, $\sigma_{22} = \sigma_{33} = \sigma_{12} = \sigma_{23} = \sigma_{31} = 0$) and the biaxial compressive strength $f_{cb} = -35.15\,\text{MPa}$ ($\sigma_{11} = \sigma_{22} = f_{cb}$, $\sigma_{33} = \sigma_{12} = \sigma_{23} = \sigma_{31} = 0$). These two strength values are the basis for calculating the material parameters α and ϑ of the Drucker-Prager failure criterion

$$\vartheta = \sqrt{\frac{2}{3}}\left(\frac{f_{cu} - f_{cb}}{f_{cu} - 2\,f_{cb}}\right) = 0.099\,, \quad \alpha = \sqrt{\frac{2}{3}}\left(\frac{f_{cu}\,f_{cb}}{f_{cu} - 2\,f_{cb}}\right) = 21.74\,\text{MPa}\,. \quad (8)$$

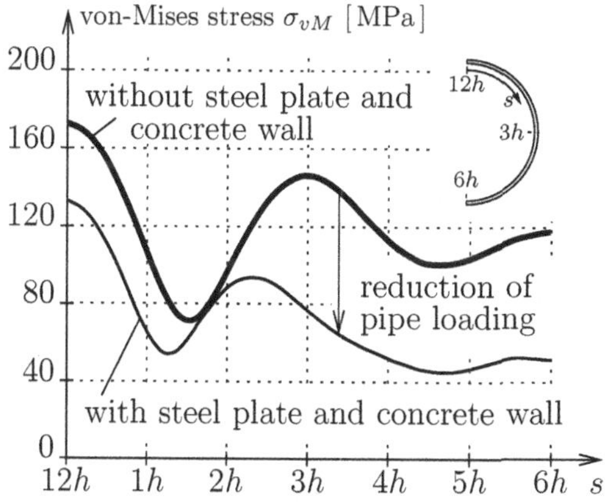

Fig. 6. Prognoses of the distribution of the von Mises stress along the inner surface of the pipe in the cross-section beneath the impact for a height of overburden $H = 3\,\mathrm{m}$ and a coefficient of sub-grade reaction $k_s = 500\,\mathrm{MN/m^3}$ with and without the additional construction consisting of a steel plate and concrete walls

Young's modulus E, Poisson's ratio ν and the parameter T referring to the tension cut-off are listed in Table 1.

Assessment of the Enhanced Protection System

The performance of the enhanced protection system is assessed considering the same rockfall scenario as previously investigated, i.e., the mass and the height of fall of the boulder are equal to 18260 kg and 18.85 m, respectively, the coefficient of sub-grade reaction is set equal to 500 MN/m^3, the height of overburden equal to 3 m, and F and W again equal to 7.94 MN and 0.72 m, respectively. Results from the FE analysis show that the additional structural elements result in a further significant reduction of the loading of the pipe, see Fig. 6.

1.5 Conclusions

It was shown that a protection system for a steel pipe subjected to rockfall must satisfy two requirements: (i) damping of the impact, in order to keep the forces arising from rockfall reasonably small and (ii) load distribution and load-carrying capacity, in order to reduce the loading of the steel pipe. In case of a purely gravel-based protection system, both tasks must be performed by gravel. Thereby, the flexibility of gravel required for the damping of the impact is opposed to the stiffness of the material required for the task of load distribution and carrying of the load.

This was the motivation to investigate an enhanced protection system consisting of (i) gravel as an energy-absorbing and impact-damping system and (ii) a buried steel plate supported by concrete walls as a load-carrying structural component. This analysis showed that buried load-carrying structural elements can significantly increase the safety of pipelines endangered by rockfall.

2 Protection Systems Against Abrasive Shear Loading Caused by Thermal Deformation of Soil-Covered Pipelines

Oil and gas pipelines are commonly bedded in and covered with sand. The transport of sand to the construction site may be expensive. Therefore, the replacement of sand by a (coarse) filling material which is directly available at the construction site (e.g. possibly sharp-edged stones with dimensions up to $\approx 10 \times 10 \times 10\,\text{cm}$) could prove profitable. However, such a construction method would result in larger contact forces exerted from relatively large, sharp-edged stones onto the pipelines. This is particularly disadvantageous in combination with motions of the pipeline relative to the backfill material, resulting from temperature-induced displacements of the pipeline in longitudinal direction: Recurrent abrasive shear loading by tips of stones may damage the anti-corrosion film at the outer surface of the pipeline.

This section deals with the protection of the anti-corrosion film against such shear forces, discussing the role of geosynthetics and steel-fibre reinforced concrete.

2.1 Assessment of Static Forces Exerted by Single Stone Tips onto Soil-Covered Pipelines

Loading Scenario

The magnitude of forces acting onto soil-covered pipelines depends on the height of the cover, on the mechanical properties and the specific weight of the filling material, and on the settlements of the pipeline and the filling material. If no settlements occur, the force acting onto the pipeline is equal to the weight of the filling material directly above the pipeline. If, however, the settlements of the filling material beside the pipeline are greater than the ones of the pipeline, the forces acting onto the pipeline increase because of significant load re-arrangement processes in the soil [21]. The extent of this re-arrangement depends on the shear resistance of the filling material. It increases with decreasing shear resistance.

Structural Model

The relevant force acting on a soil-covered pipeline follows from the combination of a filling material of low shear resistance, piled up to a standard maximum cover height of 1.5 m, with an adverse settlement scenario. We consider

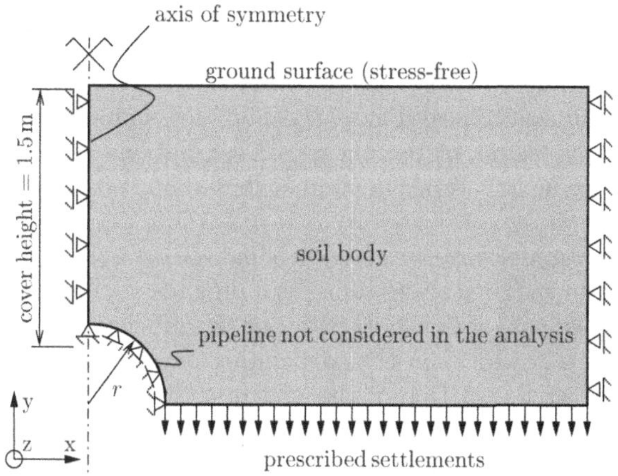

Fig. 7. Structural model of the soil body above and lateral of the pipeline

uniform settlements of the filling material on both sides of a fixed pipeline. Consequently, no longitudinal displacements occur and displacements in a plane orthogonal to the pipe axis do not change along the longitudinal direction. This corresponds to a plane state of strain, and allows for the use of a plane structural model [12] (see Fig. 7).

Steel pipes for oil and gas pipelines are so stiff that the displacements resulting from a soil cover of 1.5 m height are negligibly small as compared to typical soil settlements of several centimeters. Hence, the pipe does not have to be considered explicitly in the structural model. Instead, the contact surface between the pipe and the filling material is spatially fixed in the structural model. Since both the structure and the settlements are symmetric, investigation of half the structure is sufficient. Thereby, the axis of symmetry (left boundary in Fig. 7) and the boundary opposite to the axis of symmetry (right boundary in Fig. 7) are fixed in the x-direction. The upper boundary, corresponding to the ground surface, is stress-free. At the lower boundary, settlements, i.e. displacements in the y-direction, are prescribed (see Fig. 7). The boundary opposite to the axis of symmetry is sufficiently far from this axis so to exert no influence on the stress state around the pipe.

Determination of this stress state requires a realistic consideration of the nonlinear material behavior of the filling material. Therefore, the Cap Model described in Subsection 1.1 is adopted. Aiming at determination of the relevant forces that are acting on a soil-covered pipeline embedded in a material with low shear resistance, the subsequent calculations are based on a *cohesionless* material, i.e. we use the parameters of sandy gravel listed in Table 1, setting, however, α equal to 0.5 kPa.

Finite Element Simulation

The structural problem, involving elasto-plastic material behavior, is solved numerically for three standard diameters of oil or gas pipelines ($d = 500$, 1000 and 1500 mm), by means of nonlinear FE simulations [21], which are performed in an incremental-iterative manner, based on consistent tangent moduli [5, 8, 9].

For determination of forces exerted by single stone tips onto a pipeline, the distribution of the radial stresses along the pipeline perimeter must be computed. Prior to settlements on both sides of the pipeline, radial compression stresses amount to 25 kPa – 35 kPa, depending on the pipeline diameter [see Fig. 8(a)–(c)]. After completion of the re-arrangement of the loading, caused by the settlements, release occurs in the vicinity of the vertex of the pipeline. In the lateral regions of the upper half of the pipeline the radial stresses are increasing significantly (see Figs. 8). The increase of the maximum load is the larger, the smaller the pipeline diameter [see Fig. 8(d)].

Relevant Forces of Tips of Single Stone Acting on Soil-covered Pipelines, as a Function of the Pipe Diameter

These forces can be determined through integration of the maximum radial stresses over parts of the surface associated with a characteristic distribution of stone tips. For the sake of simplicity, a quadratic grid of stone tips with distances a is assumed (see Fig. 9). Thus, forces exerted by one stone tip read as

$$F = a^2 \cdot \max \sigma_{rr,\Delta s} \quad \text{with} \quad \max \sigma_{rr,\Delta s} = \begin{cases} -106\,\text{kPa} & \text{for } d = 500\,\text{mm}, \\ -69\,\text{kPa} & \text{for } d = 1000\,\text{mm}, \\ -63.4\,\text{kPa} & \text{for } d = 1500\,\text{mm}. \end{cases} \tag{9}$$

Assuming a uniformly granulated filling material, the stone tip distance a can be set equal to the grain size of the filling material. Thus, following from Eq. (9_1), the force exerted by one tip of a stone for a given pipeline diameter is a quadratic function of the grain size of the filling material (see Fig. 10).

2.2 Identification of Wear Protection Strategies

Archard's Wear Law

Using Archard's wear law [2], the maximum contact forces determined in Subsection 2.1 allow for calculation of the service life of the protection layer serving as an envelope of the pipe. In order to carry out this calculation, the wear resistance must be determined experimentally (see Fig. 11). Archard's wear law reads as [2]

$$\dot{V} = K_A \frac{F\,v}{H} \,. \tag{10}$$

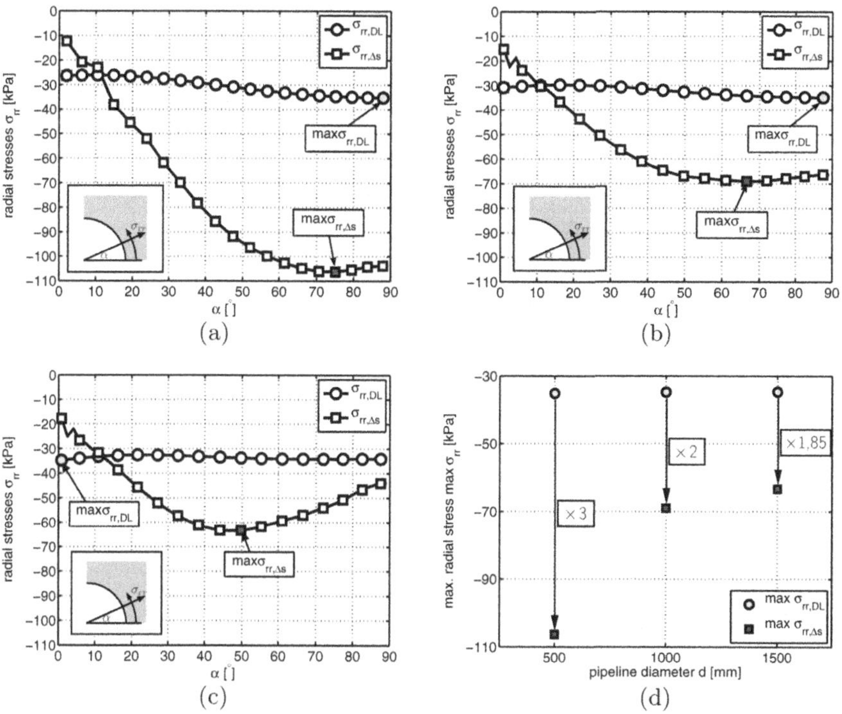

Fig. 8. Distribution of radial stresses resulting from the dead load of the filling material (specific weight γ of covering soil $20.0\,\mathrm{kN/m^3}$ and cover height 1.5 m) along the perimeter of a soil-covered pipeline *before* (○, $\sigma_{rr,DL}$) and *after* (□, $\sigma_{rr,\Delta s}$) settlement-induced load re-arrangement for (a) $d = 500\,\mathrm{mm}$, (b) $d = 1000\,\mathrm{mm}$, and (c) $d = 1500\,\mathrm{mm}$; and (d) settlement-induced increase of maximum radial stresses in the pipe

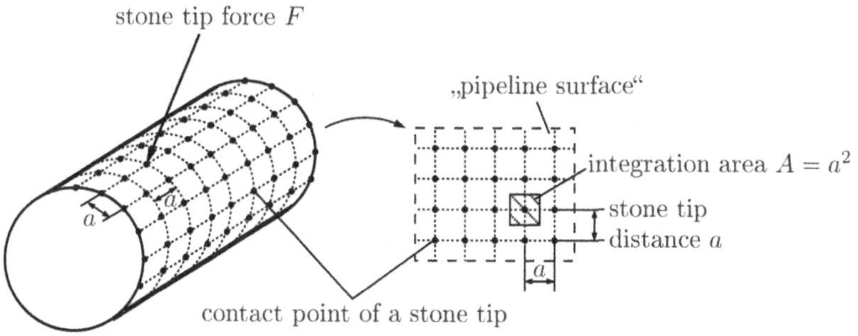

Fig. 9. Definition of distance a between stone tips and stress integration area A of a single stone tip: stone tips exert contact force F in a regular grid ($a \times a$) on the pipeline surface

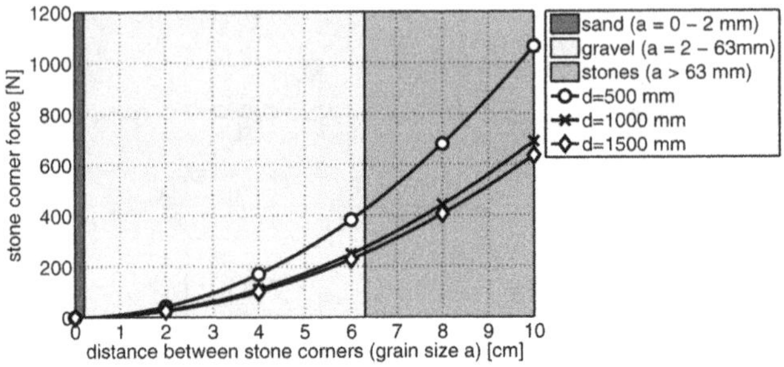

Fig. 10. Relevant forces of single stone tips as function of the distance between stone tips, and of pipeline diameter (specific weight γ of covering soil $20.0\,\mathrm{kN/m^3}$, cover height 1.5 m)

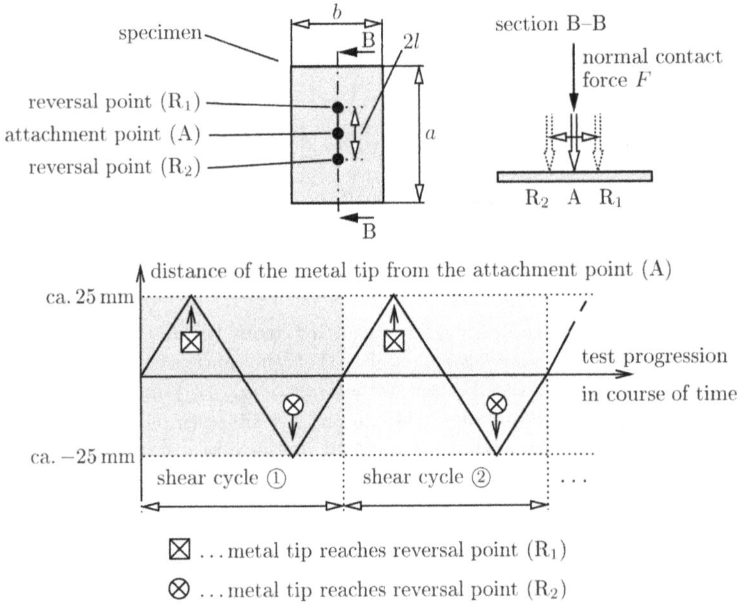

Fig. 11. Definition of the term 'shear cycle' in context with wear tests conducted on specimens representing the protection layer of the outer anti-corrosion layer of soil-covered steel pipelines: per shear cycle, twice the shear distance, $2l$, is covered

In Eq. (10), $\dot{V}$ is the abrasion rate (abrased volume per time), F is the compressive force, v denotes the velocity of the abrasion-inducing body, H represents the hardness of the loaded body, and K_A is a dimensionless parameter, related to the probability of (microscopic) abrasion phenomena depending on characteristics of the abrased and the abrasive body (e.g. mineralogical hardness, viscosity, stiffness) [11, 13].

Since time-dependent effects play no role in the considered wear tests, the integration of Eq. (10) is trivial,

$$V = K_A \frac{F\,s}{H}\,, \tag{11}$$

where V is the abrased volume along the total shear path length s. V can be expressed as the product of the tip width w of the abrasion-inducing body, the shear distance l, and the cutting depth t. The total shear path length can be expressed as the product of the shear cycle number n and twice the shear distance covered during one shear cycle, $2l$ (see Fig. 11), $s = n \cdot 2l$. Thus, following from Eq. (11),

$$w \cdot l \cdot t = K_A \frac{F \cdot n \cdot 2l}{H} \quad \Rightarrow \quad t = \frac{2K_A}{w\,H} \cdot F \cdot n\,. \tag{12}$$

According to Eq. (12), the cutting depth t depends on the compressive force F, on the number of conducted shear cycles n, on K_A, on the metal tip width w, and on the hardness H of the specimen. Notably, the shear distance l has no influence on the cutting depth.

Protection Performance of Geosynthetics

The wear resistance of common geosynthetics was determined by means of shear tests (see Fig. 11) on 6 mm thick geosynthetic specimens of the type *Polyfelt AR 20* [4] ($a = 21$ cm, $b = 14.85$ cm). For this purpose, a metal tip was used, similar to the cone-shaped tip described in [1]. It was attached to the geosynthetic specimen, loaded by a compressive force of approximately 50 N, acting perpendicular to the specimen. In two independent tests the specimens were completely transected after ≈ 50 shear cycles.

Evaluation of Eq. (12) for the tests ($t = 6$ mm; $n = 50$; $F = 50$ N) yields $(2K_A)/(wH) = 2.4 \cdot 10^{-6}$ m/N. This value, together with Eq. (12), allows for estimation of the cutting depth t into a geosynthetic specimen of the aforementioned type, as a function of the compressive force F and the shear cycle number n.

The order of magnitude of the shear cycle number which leads to a transection of a geosynthetic of 6 mm thickness can be assessed through insertion of $t = 6$ mm, $(2K_A)/(wH) = 2.4 \cdot 10^{-6}$m/N, and Eq. (9) into Eq. (12),

$$n_{d=500/1000/1500\,\mathrm{mm}} = \begin{cases} 60/90/100 & \text{for } a = 2\,\mathrm{cm}\,, \\ 7/10/11 & \text{for } a = 6\,\mathrm{cm}\,, \\ 2/4/4 & \text{for } a = 10\,\mathrm{cm}\,. \end{cases} \tag{13}$$

For common pipe diameters (500 – 1500 mm) and grain sizes (distance between stone tips) of 2 – 10 cm, the 6 mm geosynthetic investigated herein resists 2 – 100 shear cycles, until being completely transected.

Given a typical service life of $n = 5000$ shear cycles for oil and gas pipelines, the above results [see Eq. (13)] reveal that a 6 mm geosynthetic of the considered type protects the outer anti-corrosion film of the pipeline for only $\approx 1/2500$ to $1/50$ of the service life.

The stone tip force which results in transection of the 6 mm geosynthetic of the type *Polyfelt AR 20* after 5000 shear cycles can be assessed through insertion of $t = 6\,\mathrm{mm}$, $(2K_A)/(wH) = 2.4 \cdot 10^{-6}\mathrm{m/N}$, and $n = 5000$ into Eq. (12). This force amounts to

$$F(n = 5000,\ t_{AR\,20}^{Polyfelt} = 6\,\mathrm{mm}) = 0.5\,\mathrm{N}\,. \tag{14}$$

The maximum admissible distance a between stone tips exerting a force of $F = 0.5\,\mathrm{N}$, can be determined by means of Eqs. (9) as a function of the outer pipe diameter,

$$a = \begin{cases} 2.2\,\mathrm{mm} & \text{for } d = 500\,\mathrm{mm}\,, \\ 2.7\,\mathrm{mm} & \text{for } d = 1000\,\mathrm{mm}\,, \\ 2.8\,\mathrm{mm} & \text{for } d = 1500\,\mathrm{mm}\,. \end{cases} \tag{15}$$

Hence, when covered by *sand grains*, the 6 mm geosynthetic investigated herein resists 5000 shear cycles before being transected.

Effective Means of Protection

The results of the previous paragraph suggest two strategies for providing a sufficient protection of the outer anti-corrosion film of soil-covered pipelines for oil and gas transport against abrasive shear loading exerted by coarse filling material, caused by temperature-induced displacements of the pipeline in the longitudinal direction:

(a) Increase of the hardness H and the thickness t, respectively, of the protective layer subjected to wear: e.g. by coating the pipeline by steel-fibre reinforced concrete (SFRC) [see Fig. 12(a)].
(b) Decrease of the abrasion force by reducing the distance a between the stone tips, as the pipeline is embedded in and buried by sand. Thereby, leaching of sand can be avoided by enveloping the sand body with a filtration layer which separates the sand body and the coarse filling material from each other. Preferably, geosynthetics are attached as filtration layer [see Fig. 12(b)].

2.3 Conclusions

The resistivity of geosynthetics of a type similar to ones tested herein against abrasion through sharp-edged stones larger than 2 mm, is too low as to provide, during the entire service life, sufficient protection of the outer anti-corrosion film of the pipeline. Hence, sand cannot be replaced by coarser

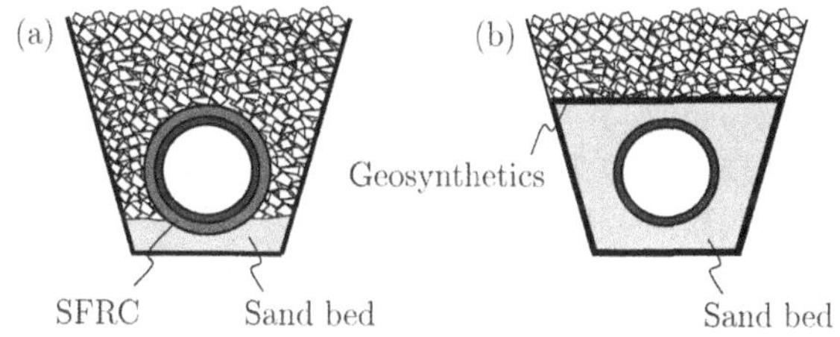

Fig. 12. Strategies for protecting the outer anti-corrosion film of soil-covered steel pipelines in case a mixture of sandy gravel and square-edged stones is used as filling material: (a) increase of hardness and thickness of the protective layer, e.g. made of steel-fibre reinforced concrete (SFRC) or (b) decrease of the abrasion force through use of sandy material around the pipe

filling material together with an envelope of the steel pipe by geosynthetics comparable to ones investigated herein. However, when subjected to shear loading exerted by sand grains, failure (transection) of *Polyfelt AR 20*-type geosynthetics can indeed be excluded prior to 5000 shear cycles, i.e. within the entire service life of such a pipe. Hence, two strategies can be recommended for protecting the outer anti-corrosion film of soil-covered steel pipelines against abrasive shear loading: On the one hand, the wear resistance of the protection layer can be improved by increasing its hardness and thickness, e.g. through the use of steel-fibre reinforced concrete. On the other hand, encasing the pipeline with sand minimizes the abrasion force, which can be withstood by the outer anti-corrosion film of the pipeline without any additional means of protection. Leaching of sand can be avoided by an envelope of the sand body by geosynthetics.

References

1. ÖNORM EN 14574 (2005) Geokunststoffe – Bestimmung des Pyramidendurchdrückwiderstandes von Geokunststoffen auf harter Unterlage [Geosynthetics – Determination of the Pyramid Puncture Resistance of Geosynthetics on Hard Underlay]. Österreichisches Normungsinstitut In German.
2. Archard JF (1953) Contact and rubbing of flat surfaces. Journal of Applied Physics 24(8):981–988
3. DiMaggio FL, Sandler IS (1971) Material models for granular soils. Journal of Engineering Mechanics (ASCE) 97(3):935–950
4. Polyfelt Geosynthetics (2006) Technical data polyfelt.AR – 500 917 | 02.2002. *www.polyfelt.com.*
5. Hofstetter G, Simo JC, Taylor RL (1993) A modified cap model: closest point solution algorithms. Computers & Structures 46(2):203–214
6. JDAAF (1983) Soils and Geology Procedures for Foundation Design of Buildings and Other Structures (Except Hydraulic Structures): Technical Manual 5-818-

1, Air Force Manual 88-3. Joint Departments of the Army and the Air Force, Washington, DC, USA
7. Koiter WT (1960) General theorems for elastic-plastic solids, volume I, Chapter IV, pp 167–218. North-Holland Publishing Company, Amsterdam
8. Kropik Ch (1994) Three-dimensional elasto-viscoplastic finite element analysis of deformations and stresses resulting from the excavation of shallow tunnels. PhD thesis, Vienna University of Technology, Vienna, Austria
9. Kropik Ch, Mang HA (1996) Computational mechanics of the excavation of tunnels. Engineering Computations 13(7):49–69
10. Lubliner J (1990) Plasticity Theory. Macmillan Publishing Company, New York
11. Magneé A (1993) Modelization of damage by abrasion. Wear 162–164:848–855
12. Mang H, Hofstetter G (2000) Festigkeitslehre [Strength of Materials]. Springer, Wien, NewYork. In German
13. Olofsson U, Telliskivi T (2003) Wear, plastic deformation and friction of two rail steels – a full scale test and a laboratory study. Wear 254:80–93
14. Penumadu D, Zhao R (1999) Triaxial compression behavior of sand and gravel using artificial neural networks (ANN). Computers & Geotechnics 24:207–230
15. Pichler B (2003) Parameter identification as the basis for prognoses in geotechnical engineering. PhD thesis, Vienna University of Technology, Vienna, Austria
16. Pichler B, Hellmich Ch, Mang HA (2005) Impact of rocks onto gravel - design and evaluation of experiments. International Journal of Impact Engineering 31:559–578. Available online at www.sciencedirect.com.
17. Pichler B, Hellmich Ch, Mang HA, Eberhardsteiner J (2005) Assessment of protection systems for buried steel pipelines endangered by rockfall. Computer-Aided Civil and Infrastructure Engineering 20:331–342
18. Pichler B, Hellmich Ch, Mang HA, Eberhardsteiner J (2006) Gravel-buried steel pipe subjected to rockfall: Development and verification of a structural model. Journal of Geotechnical and Geoenvironmental Engineering (ASCE) 132(11):1465–1473
19. Sandler IS, Rubin D (1979) An algorithm and a modular subroutine for the cap model. International Journal for Numerical and Analytical Methods in Geomechanics 3:173–186
20. Sawicki A, Świdziński W (1998) Elastic moduli of non-cohesive particulate materials. Powder Technology 96:24–32
21. Scheiner St, Pichler B, Hellmich Ch, Eberhardsteiner J (2006) Loading of soil-covered oil and gas pipelines due to adverse soil settlements – protection against thermal dilatation-induced wear, involving geosynthetics. Computers and Geotechnics 33(8):371–380
22. Simo JC, Taylor RL (1985) Consistent tangent operators for rate independent elasto–plasticity. Computer Methods in Applied Mechanics and Engineering 48:101–118

Enriched Free Mesh Method: An Accuracy Improvement for Node-based FEM

Genki Yagawa[1] and Hitoshi Matsubara[2]

[1] Center for Computational Mechanics Research, Toyo University, 2-36-5, Hakusan, Bunkyo-ku, Tokyo, Japan, 112-8611
yagawa@eng.toyo.ac.jp
[2] Center for Computational Science and Engineering, Japan Atomic Energy Agency, 6-9-3 Higashi-Ueno, Taito-ku, Tokyo, Japan, 110-0015
matsubara.hitoshi@jaea.go.jp

Summary. In the present paper, we discuss the accuracy improvement for the free mesh method: a node based finite element technique. We propose here a scheme where the strain field is defined over clustered local elements in addition to the standard finite element displacement field. In order to determine the unknown parameter, the least square method or the Hellinger-Reissner Principle is employed. Through some bench mark examples, the proposed technique has shown excellent performances.

1 Introduction

Recent advances in computer technology have enabled a number of complicated natural phenomena to be accurately simulated, which were ever only observed by experiments. Among various computer simulation techniques, the finite element method (hereinafter referred to as "FEM") has been most widely used due to the capability of analyzing an arbitrary domain, and results, accurate enough for engineering purposes, are obtainable at reasonable cost[1][2]. However, mesh generation for finite element analysis becomes very difficult and time consuming if the degree of freedom of the analysis model is extremely large, for example exceeding 100-million, and the geometries of the model are complex. In order to overcome the above shortcoming of the standard FEM, the so called mesh-free methods[3][4] have been studied. The Element-Free Galerkin Method (EFGM)[5][6] is among them with the use of integration by background-cells instead of by elements, based on the moving least square and diffuses element methods. The Reproducing Kernel Particle Method (RKPM)[7][8] is another mesh-free scheme, which is based on a particle method and wavelets. The general feature of these mesh-free methods is that, contrary to the standard FEM, the connectivity information between

Eugenio Oñate and Roger Owen (eds.), Computational Plasticity, 207–219.

nodes and elements is not required explicitly, since the evaluation of the total stiffness matrix is performed generally by the node-wise calculations instead of the element-wise calculations.

On the other hand, a virtually mesh-free approach called the free mesh method (hereinafter referred to as "FMM")[9][10] is based on the usual FEM, having a cluster of local meshes and equations constructed in a node-by-node manner. In other word, the FMM is a node-based FEM, which still keeps the well-known excellent features of the standard FEM. Through the node-wise manner of the FMM, a seamless flow in simulation procedures from local mesh generation to visualization of the results without user's consciousness is realized. The method has been applied to solid/fluid dynamics [11], crack problems[12], concrete problems[13], and so on. In addition, in order to achieve a high accuracy, the FMM with vertex rotations has been studied[14][15].

In this paper, we discuss another high accurate FMM: the Enriched FMM (hereinafter referred to as "EFMM"). In the following section, the fundamental concept of the original FMM is reviewed, and the third section deals with two EFMMs, one is "EFMM based on the localized least square method" and the other "EFMM based on the Hellinger-Reissner principle". In the fourth section, some numerical examples are presented, and concluding remarks are given in the final section.

2 Basic Concept of Free Mesh Method (FMM)

The FMM starts with only the nodes distributed in the analysis domain (Ω), without the global mesh data, as following equation.

$$p_i(x_i, y_i, r_i) \ \forall_i \in \{1, 2, \cdots, m\} \tag{1}$$

where m is the number of node, $p_i(x_i, y_i)$ are the Cartesian coordinates, and r_i is the nodal density information, which is used to generate appropriate nodes as illustrated in Fig. 1(a). From above nodal information, a node is selected as a central node and nodes within a certain distance from the central node are selected as candidate nodes. This distance is usually decided from the prescribed density of the distribution of nodes. Then, satellite nodes are selected from the candidate nodes, which generate the local elements around the central node (shown in Fig. 1(b)). For each local element, the element stiffness matrix is constructed in the same way as the FEM, however in FMM, only the row vector of stiffness matrix for each local element is necessary. The local stiffness matrix of each temporary element is given by

$$\mathbf{k}_{e_i} = [\, \mathbf{k}_{p_i} \ \mathbf{k}_{S_j} \ \mathbf{k}_{S_k} \,] \tag{2}$$

where $\mathbf{k}_{e_i}$ is the row vector of the stiffness matrix for element e_i and $\mathbf{k}_{p_i}$, $\mathbf{k}_{S_i}$ and $\mathbf{k}_{S_k}$ are components for node of p_i, S_i and S_k (j and k are number

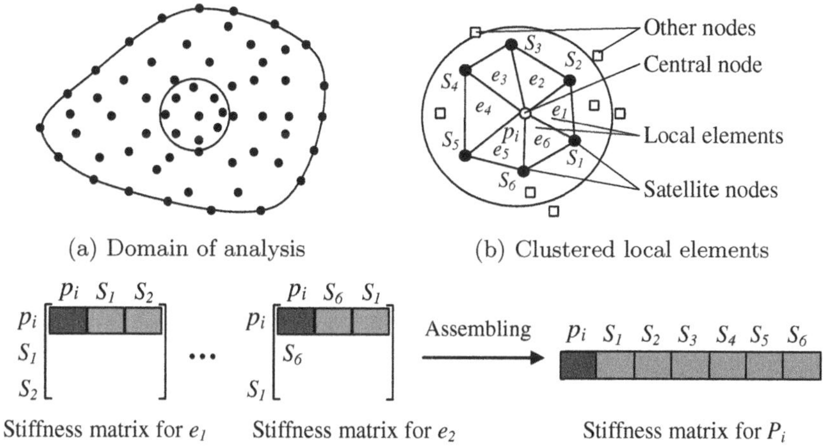

(c) Stiffness matrix for central node

Fig. 1. Concept of Free Mesh Method

of current satellite nodes). Through the above procedures are carried out for all local elements, the stiffness matrix for a central node is given by

$$\mathbf{k}_{p_i} = \sum_{i=1}^{n_e} \mathbf{k}_{e_i} \tag{3}$$

where $\mathbf{k}_{p_i}$ is the stiffness matrix for central node p_i, and n_e the number of local elements. Through the above procedures for all nodes is carried out, the global stiffness matrix is given by assembling $\mathbf{k}_{p_i}$ which are computed by node-wise manner:

$$\mathbf{K} = \begin{bmatrix} \mathbf{k}_{p_1} \\ \mathbf{k}_{p_2} \\ \vdots \\ \mathbf{k}_{p_m} \end{bmatrix} \tag{4}$$

Brief of the nodal stiffness matrix is shown in Fig. 1(c). After the construction of the global stiffness matrix, a derivation of the solution is processed. The great advantage of the FMM is that the global stiffness matrix can be evaluated in parallel with respect to each node through the node-wise manner, and only satellite node information is required with each nodal calculation. Finally, a derivation of the solution is performed as the usual FEM. Thus, the FMM is a node-wise FEM, which still keeps the well-known excellent features of the usual FEM. The features of FMM are summarized as follows,

(1) Easy to generate a large-scale mesh automatically
(2) Processed without being conscious of mesh generation
(3) The result being equivalent to that of the FEM

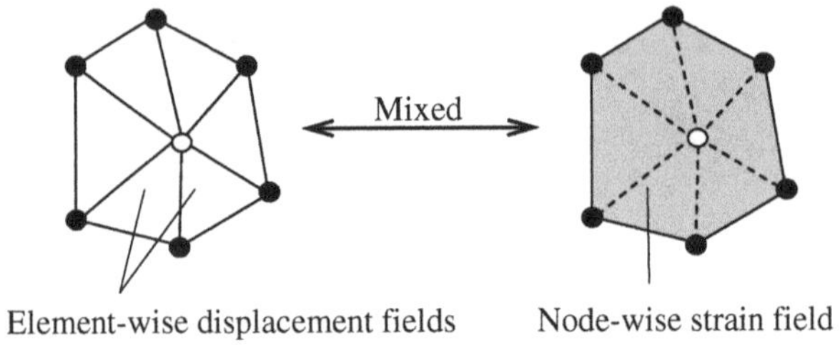

Fig. 2. Concept of enriched free mesh method

3 Enriched Free Mesh Method (EFMM)

3.1 Outline of EFMM

"Assumed strain on the clustered local elements" is the concept of EFMM as shown in Fig. 2. In the EFMM, the strain field on the clustered local elements and the displacement field of each local element are assumed independently. Relating these independent fields, we propose here two approaches, one is the localized least square method and the other is the method based on the Hellinger-Reissner principle.

3.2 EFMM Based on the Localized Least Square Method

The EFMM based on the localized least square method (hereinafter referred to as "EFMM-LS") assumes the strain field on the clustered local elements as

$$\{\varepsilon(\mathbf{x})\} = [\mathbf{N}^{\varepsilon}]\{\mathbf{a}\} \tag{5}$$

where $\{\varepsilon(\mathbf{x})\} = \{\varepsilon_{xx}, \varepsilon_{yy}, \gamma_{xy}\}$ is the strain field defined on the clustered local elements and each component of $\{\varepsilon_{xx}, \varepsilon_{yy}, \gamma_{xy}\}$ is assumed independently, and $[\mathbf{N}^{\varepsilon}]$ is a matrix, which consists of arbitrary polynomials as follows,

$$[\mathbf{N}^{\varepsilon}] = \begin{bmatrix} p^t(\mathbf{x}) & 0 & 0 \\ 0 & p^t(\mathbf{x}) & 0 \\ 0 & 0 & p^t(\mathbf{x}) \end{bmatrix} \tag{6}$$

where $p^t(x)$ is given on the clustered local elements as

$$\begin{array}{ll} p^t(\mathbf{x}) = \begin{bmatrix} 1 \; x \; y \end{bmatrix} & \textit{linear basis} \\ p^t(\mathbf{x}) = \begin{bmatrix} 1 \; x \; y \; x^2 \; xy \; y^2 \end{bmatrix} & \textit{quadratic basis} \\ p^t(\mathbf{x}) = \begin{bmatrix} 1 \; x \; y \; x^2 \; xy \; y^2 \; x^3 \; x^2y \; xy^2 \; y^3 \end{bmatrix} & \textit{cubic basis} \\ \cdots & \end{array} \tag{7}$$

In this paper, $p^t(x)$ is assumed to be linear or quadratic basis polynomial. The coefficients vector $\{\mathbf{a}\}$ in Eq. (5) is determined by minimizing the discrete L_2 norm as follows,

$$J = \sum_{c=1}^{n_e} \sum_{i=1}^{p} \left[\{\varepsilon(\mathbf{x})\} - \{\varepsilon_i^c\}\right]^2 \tag{8}$$

where n_e is the number of local elements with $c(= 1, 2, \cdots, n_e)$ being current local element, p the number of points, which are called as the "strain monitoring points" on the clustered local elements with $i(= 1, 2, \cdots, p)$ being the current strain monitoring point and $\{\varepsilon_i^c\}$the strain vector of i-th strain monitoring point on the c-th local element, which is called as the "mother element". The stationary condition of Eq. (8) is

$$\delta J = 2\{\mathbf{a}\}^T \sum_{c=1}^{n_e} \sum_{i=1}^{p} \left[[\mathbf{N}_i^\varepsilon]^T [\mathbf{N}_i^\varepsilon]\{\mathbf{a}\} - [\mathbf{N}_i^\varepsilon]^T \{\varepsilon_i^c\}\right] = 0 \tag{9}$$

which yields the coefficients vector $\{\mathbf{a}\}$ as follows,

$$\{\mathbf{a}\} = \sum_{c=1}^{n_e} \sum_{i=1}^{p} \left[\left[[\mathbf{N}_i^\varepsilon]^T [\mathbf{N}_i^\varepsilon]\right]^{-1} [\mathbf{N}_i^\varepsilon]^T \{\varepsilon_i^c\}\right] \tag{10}$$

Let us consider a simple Constant Strain Triangle as the mother element in which the displacement field of each local element is defined by

$$\{\mathbf{u}\} = \sum_{i=1}^{3} \{\mathbf{u}_i\}\zeta_i \tag{11}$$

where $\{\mathbf{u}\}$ is the displacement field of the local element, $\{\mathbf{u}_i\}$ is the nodal displacement, and ζ_i is the area-coordinate[16]. Thus, the strain value on the strain monitoring points is given by

$$\{\varepsilon_i^c\} = [\mathbf{B}_i^c]\{\mathbf{u}_i\} \tag{12}$$

where

$$\begin{aligned} &[\mathbf{B}_i^c] = \left[\,[\mathbf{B}_1]\ [\mathbf{B}_2]\ [\mathbf{B}_3]\,\right] \\ &\textit{with} \\ &[\mathbf{B}_j] = \begin{bmatrix} \partial\zeta_j/\partial x & 0 \\ 0 & \partial\zeta_j/\partial y \\ \partial\zeta_j/\partial y & \partial\zeta_j/\partial x \end{bmatrix} \quad , \quad j = 1, 2, 3 \end{aligned} \tag{13}$$

By substituting Eq. (12) into Eq. (10), the unknown coefficient $\{\mathbf{a}\}$ is determined as

$$\{\mathbf{a}\} = \sum_{c=1}^{n_e} \sum_{i=1}^{p} \left[\left[[\mathbf{N}_i^\varepsilon]^T [\mathbf{N}_i^\varepsilon]\right]^{-1} [\mathbf{N}_i^\varepsilon]^T [\mathbf{B}_i^c]\{\mathbf{u}_i\}\right] \tag{14}$$

Substituting Eq. (14) into Eq. (15), we obtain

$$\{\varepsilon(\mathbf{x})\} = [\mathbf{N}^{\varepsilon}]\sum_{e=1}^{n_e}\sum_{i=1}^{p} [[[\mathbf{N}_i^{\varepsilon}]^T[\mathbf{N}_i^{\varepsilon}]]^{-1}[\mathbf{N}_i^{\varepsilon}]^T[\mathbf{B}_i^{c}]\{\mathbf{u}_i\}] = [\mathbf{A}]\{\mathbf{u}_i\} \tag{15}$$

where

$$[\mathbf{A}] = [\mathbf{N}^{\varepsilon}]\sum_{e=1}^{n_e}\sum_{i=1}^{p} [[[\mathbf{N}_i^{\varepsilon}]^T[\mathbf{N}_i^{\varepsilon}]]^{-1}[\mathbf{N}_i^{\varepsilon}]^T[\mathbf{B}_i^{c}]] \tag{16}$$

In the elasticity problem, the stress vector $\{\sigma\}$ and the strain vector $\{\varepsilon\}$ have the relation as follows,

$$\{\sigma\} = [\mathbf{D}]\{\varepsilon\} \tag{17}$$

where $[\mathbf{D}]$ is a symmetric matrix of material stiffness. With $[\mathbf{D}]$ given by Eq. (16), the stiffness matrix based on the localized least square method is computed on the clustered local elements as

$$[\mathbf{k}_{LS}] = \int_{\Omega} [\mathbf{A}]^T[\mathbf{D}][\mathbf{A}]\mathbf{d}\Omega \tag{18}$$

where Ω is area of the clustered local elements. It is important to say that the above stiffness matrix is computed in a node-wise manner.

It is noted that the present EFMM-LS is closely related to the superconvergent patch recovery proposed by Zienkiewicz and Zhu[17][18]. In an adaptive finite element method[19][20], the Z-Z error estimator has been most widely used to estimate the error. The error estimator requires an exact solution, but generally it is impossible to compute the exact value because the exact solution is not available in general. The Z-Z technique then obtains the recovered solution in a post processing stage. The clustered local elements in the present method are equivalent to the superconvergent patch used in the Z-Z technique. The difference lies in that the recovering procedure in the EFMM-LS is in a main process stage when computing element stiffness matrices. The use of the assumed strain is therefore, in some sense, equivalent to the "post-process" of the Z-Z superconvergent patch recovery.

3.3 EFMM Based on Hellinger-Reissner Principle

In the EFMM based on the Hellinger-Reissner principle [1][21] (hereinafter referred to as "EFMM-HR"), the Hellinger-Reissner (hereinafter referred to as "HR") principle is employed to obtain better accuracy. Let the HR principle of a linear elastic body be defined on the clustered local elements by

$$\begin{aligned}\prod(\varepsilon,\mathbf{u}) = \int_{\Omega}\{\varepsilon\}^{\mathbf{T}}[\mathbf{D}]\{\partial\mathbf{u}\}d\Omega - \frac{1}{2}\int_{\Omega}\{\varepsilon\}^{\mathbf{T}}[\mathbf{D}]\{\varepsilon\}d\Omega \\ - \int_{\Omega}\{\mathbf{u}\}^{\mathbf{T}}\{\mathbf{b}\}d\Omega - \int_{S_\sigma}\{\mathbf{u}\}^{\mathbf{T}}\{\tilde{\mathbf{t}}\}dS\end{aligned} \tag{19}$$

where

$$\{\partial \mathbf{u}\} = [\mathbf{B}]\{\bar{\mathbf{u}}\} \ , \ \{\varepsilon\} = [\mathbf{N}^{\varepsilon}]\{\bar{\varepsilon}\} \tag{20}$$

with $\{\mathbf{b}\}$ being the applied body force per unit mass, and $\{\tilde{t}\}$ the applied traction on boundary S_σ. $\{\bar{\mathbf{u}}\}$ is the unknown nodal displacement and $\{\bar{\varepsilon}\}$ the unknown nodal strain. The unknown values $(\bar{\mathbf{u}}, \bar{\varepsilon})$ of the HR principle satisfy the following equations in a weak manner,

$$\int_{\Omega} \delta\{\varepsilon\}^{\mathbf{T}}[\mathbf{D}]\left([\mathbf{B}]\{\bar{\mathbf{u}}\} - [\mathbf{N}]\{\bar{\varepsilon}\}\right) d\Omega = 0 \tag{21}$$

$$\int_{\Omega} \delta\{\mathbf{u}\}^{\mathbf{T}}[\mathbf{B}]^{\mathbf{T}}[\mathbf{D}][\mathbf{N}]\{\bar{\mathbf{u}}\} d\Omega - \int_{\Omega} \delta\{\mathbf{u}\}^{\mathbf{T}}\{\mathbf{b}\} d\Omega - \int_{S_\sigma} \delta\{\mathbf{u}\}^{\mathbf{T}}\{\tilde{\mathbf{t}}\} dS = 0 \tag{22}$$

It is noted here that the strain field is defined on the clustered local elements by node-wise manner, where the displacement field is defined on each element by element-wise manner. Equations (21) and (22) yields the following linear matrix equation,

$$\begin{bmatrix} -\mathbf{A} & \mathbf{C} \\ \mathbf{C}^{\mathbf{T}} & \mathbf{0} \end{bmatrix} \begin{Bmatrix} \bar{\varepsilon} \\ \bar{\mathbf{u}} \end{Bmatrix} = \begin{Bmatrix} \mathbf{f}_1 \\ \mathbf{f}_2 \end{Bmatrix} \tag{23}$$

where

$$\begin{cases} \mathbf{A} = \int_{\Omega} [\mathbf{N}^{\varepsilon}]^{\mathbf{T}}[\mathbf{D}][\mathbf{N}^{\varepsilon}] d\Omega \\ \mathbf{C} = \int_{\Omega} [\mathbf{N}^{\varepsilon}]^{\mathbf{T}}[\mathbf{D}][\mathbf{B}] d\Omega \\ \mathbf{f}_1 = 0 \\ \mathbf{f}_2 = \int_{\Omega} [\mathbf{N}^{u}]^{\mathbf{T}}\{\mathbf{b}\} d\Omega + \int_{\Gamma} [\mathbf{N}^{u}]^{\mathbf{T}}\{\tilde{t}\} d\Gamma \end{cases} \tag{24}$$

By condensing the coefficient matrix of Eq. (23), we obtain the following equation,

$$\mathbf{C}^{T}\left(\mathbf{A}^{-1}\mathbf{C}\bar{\mathbf{u}}\right) = \mathbf{f}_2 \tag{25}$$

where the condensation should be executed on the clustered local elements. Thus, the stiffness matrix based on the HR principle is computed on the clustered local elements as follows,

$$[\mathbf{k}_{HR}] = \mathbf{C}^{T}\mathbf{A}^{-1}\mathbf{C} \tag{26}$$

It is noted here that we can obtain the enriched stiffness matrix without increasing the number of nodal degrees of freedom.

4 Examples

4.1 Convergence Study: Displacement

To study the convergence characters of the present methods, a cantilever model is solved as shown Fig. 3, where the three different mesh patterns are prepared and the mesh division in the x direction is varied. As shown in the figure, a beam of length $L = 10$, height $D = 1$ and thickness $t = 1$

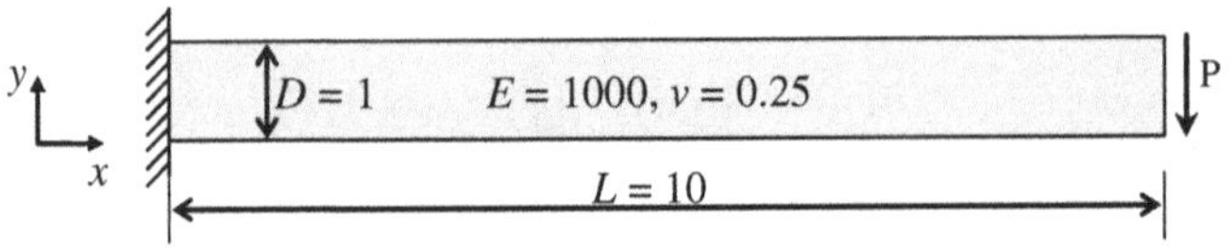

(a) Cantilever beam model

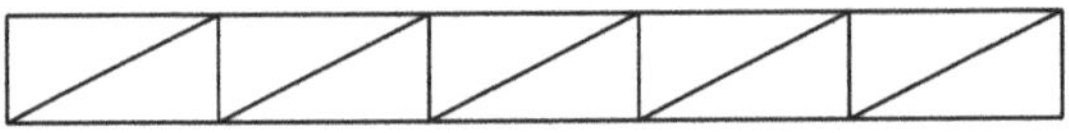

(b) One layer mesh

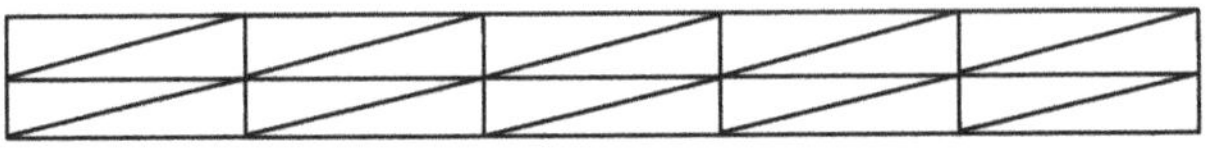

(c) Two layers mesh

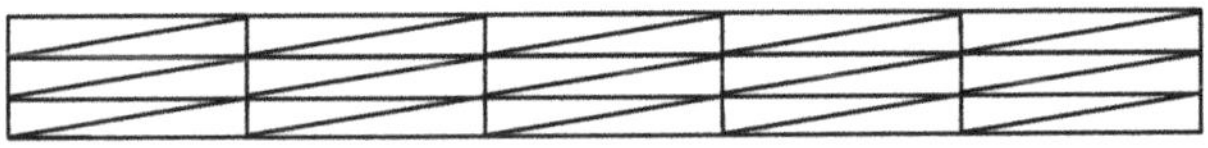

(d) Three layers mesh

Fig. 3. Mesh patterns for cantilever beam model

is subjected to a shear load in plane stress condition. The material parameters are given as the Young's modulus $E = 1000.0$ and the Poisson's ratio $\nu = 0.25$. The displacements at the loaded edge normalized by the exact value are plotted against the degrees of freedom (see Fig. 4). From the comparison of displacement results among the six different solutions, it can be observed that

(a) The accuracy of the FEM with the three noded linear element of constant strain is the worst, whereas that with the six noded quadratic element is the best irrespective of the mesh patterns.
(b) As the number of layers in the thickness direction increase, the accuracy of EFMMs approaches that of the quadratic FEM.
(c) Regarding the comparisons among the EFMMs, the accuracy of the EFMM-HR and the EFMM-LS with the linear strain field are the best, whereas, for the finer meshes (see Fig. 4(c)), the results of EFMMs with the quadratic strain field are almost equivalent to those of the formers.

4.2 Convergence Study: Error Norms

As another convergence measures, two kinds of error norms for the beam problems as shown Fig. 5 [22] are employed, which are, respectively, given as

$$\|E\|_2 = \left[\int_\Omega \left(\mathbf{u} - \mathbf{u}^{exact}\right)^T \left(\mathbf{u} - \mathbf{u}^{exact}\right) d\Omega \right]^{1/2} \tag{27}$$

(a) One layer mesh

(b) Two layer mesh

(c) Three layer mesh

—○— FEM (linear)	—●— FEM (quadratic)
—△— EFMM-LS (linear)	—▲— EFMM-LS (quadratic)
—▱— EFMM-HR (linear)	—■— EFMM-HR (quadratic)

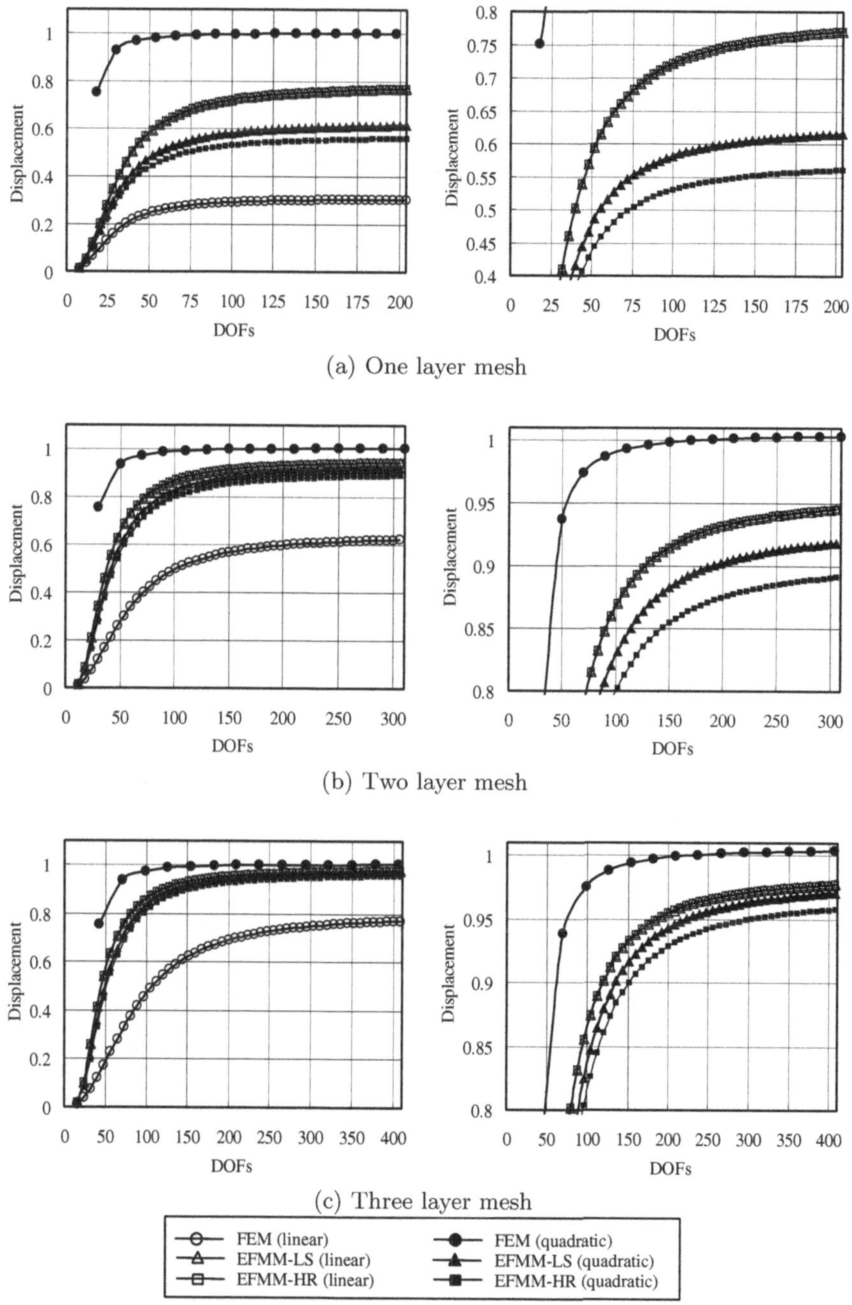

Fig. 4. Normalized displacements at the loaded edge vs. DOFs (The figures in the right hand side are zoomed ones)

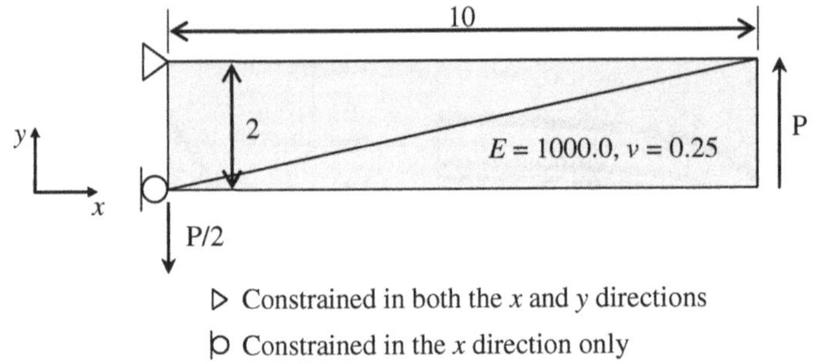

Fig. 5. Cantilever beam model for error norm study; only the case of mesh is shown as examples

$$\|E\|_e = \left[\int_\Omega \frac{1}{2}\left(\varepsilon - \varepsilon^{exact}\right)^T \left(\sigma - \sigma^{exact}\right) d\Omega\right]^{1/2} \tag{28}$$

where $\|E\|_2$ is the displacement error norm and $\|E\|_e$ that of the energy error. $\mathbf{u}$, ε and σ are, respectively, the numerical results of displacement, strain and stress, whereas $\mathbf{u}^{exact}$, $\varepsilon^{\mathbf{exact}}$ and $\sigma^{\mathbf{exact}}$ are the exact solutions. A beam of length $L = 10$, height $D = 2$ and thickness $t = 1$ is subjected to a shear load in plane stress condition. The material parameters are given by the Young's modulus $E = 1000.0$ and the Poisson's ratio $\nu = 0.25$. The above displacement and energy convergence norms are plotted against the DOFs in Figs. 6 and 7, respectively, where the meshes are, respectively, 1×1, 2×2, 4×4, 16×16, 32×32, and 64×64. It can be seen from these figures that

(a) Again, the error norms of the displacement of the EFMMs are between those of the linear and the quadratic FEMs (see Fig. 6). However, the convergence slopes of the EFMMs are almost equal to that of the quadratic FEM.
(b) The error norms of the energy of the quadratic EFMMs are almost the same as that of the quadratic FEM and those of the linear EFMMs are between the linear and the quadratic FEMs.

4.3 Patch Test

The patch test is performed using the three models of patch as shown in Fig. 8, where the displacement field

$$\begin{Bmatrix} u(x) \\ v(y) \end{Bmatrix} = \begin{Bmatrix} 0.2x \\ -0.6y \end{Bmatrix} \tag{29}$$

is applied at the boundary. Table 1 shows the test results for the FEMs and the EFMMs. As illustrated in the table, all the method passes the patch test for the Model A, which is a regular mesh division model. However, for the Model B and C, which are irregular ones, the EFMM-LSs do not pass the

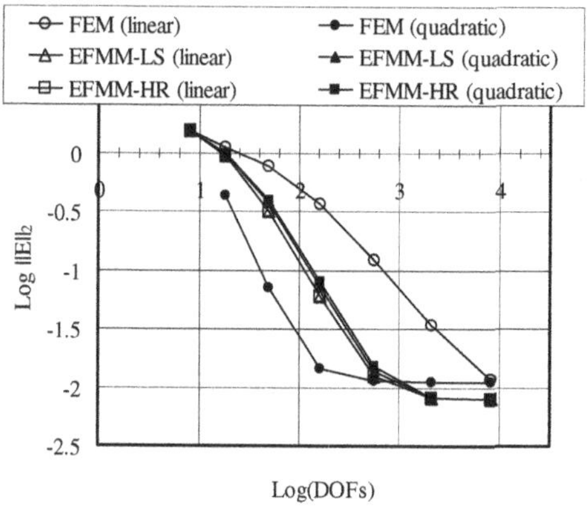

Fig. 6. L_2 error norms of displacement vs. DOFs

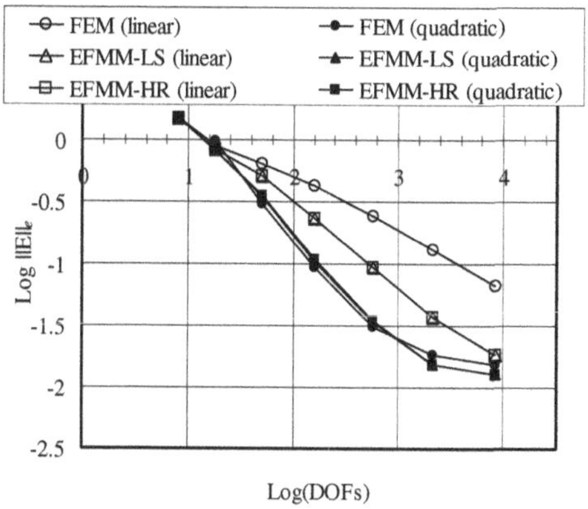

Fig. 7. Energy error norms vs. DOFs

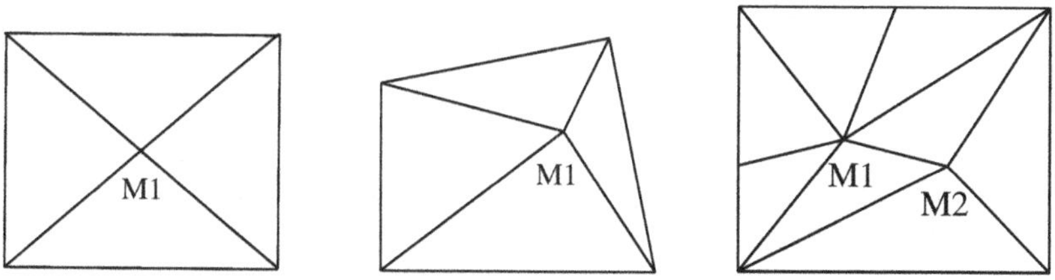

(a) Model A (b) Model B (c) Model C

Fig. 8. Models for patch tests

test. Here, "Pass" means that the displacement of the internal node ($M1$ or $M2$) satisfies Eq. (29). This means that the EFMM-LSs are nonconforming for irregular mesh, which is an open question.

Table 1. Displacements at internal nodes $M1$ and $M2$

	Model A		Model B		Model C			
	$u(M1)$	$v(M1)$	$u(M1)$	$v(M1)$	$u(M1)$	$v(M1)$	$u(M2)$	$v(M2)$
FEM(linear)	0.900	-2.400	1.200	-3.000	0.600	-2.400	1.300	-1.800
FEM(quadratic)	0.900	-2.400	1.200	-3.000	0.600	-2.400	1.300	-1.800
EFMM-LS(linear)	0.900	-2.400	1.198	-3.170	0.611	-2.372	1.285	-1.899
EFMM-LS(quadratic)	0.900	-2.400	1.198	-3.141	0.611	-2.368	1.290	-1.859
EFMM-HR(linear)	0.900	-2.400	1.200	-3.000	0.600	-2.400	1.300	-1.800
EFMM-HR(quadratic)	0.900	-2.400	1.200	-3.000	0.600	-2.400	1.300	-1.800
Exact	0.900	-2.400	1.200	-3.000	0.600	-2.400	1.300	-1.800

5 Concluding remarks

A new Free Mesh Method called "Enriched Free Mesh Method" is proposed in this paper, in which a high accuracy can be obtained without explicitly increasing the degree of freedoms. The work is summarized as follows,

(1) The key idea of the proposed method is that the strain field is assumed on clustered local elements, in addition to the usual displacement field on each element. To relate the above two fields, the localized least square method or the Hellinger-Reissner principle are, respectively, employed.

(2) The convergence characteristics of the displacement L2 error norms in the cantilever problem are between that of the FEM with the linear displacement field and that with the quadratic one, whereas that of the energy error norms with the quadratic strain field for the clustered elements is equivalent to that of the FEM with the quadratic displacement field.

(3) The EFMM based on the Hellinger-Reissner principle passes the patch test, whereas, for irregular nodal arrangements, the EFMM based on the localized least square method does not. This would be an open question and there is a room for future research.

References

1. Zienkiewicz OC, Taylor RL (2000) The finite element method. Fifth edition, Butterworth Heinemann
2. Cook RD, Malkus DS, Plesha ME (1989) Concepts and applications of finite element analysis. Third edition, Wiley
3. Barth TJ, Griebel M, Keyes DE, Nieminen RM, Roose D, Schlick T (2003) Meshfree methods for partial differential equation. Springer-Verlag Berlin Heidelberg
4. Liu GR, Liu MB (2003) Smoothed Particle Hydrodynamics a meshfree particle method. World Scientific publishing
5. Belytschko T, Lu YY, Gu L (1994) Element-free Galerkin methods. Int J Num Meth Engng 37:229–256
6. Lu YY, Belytschko T, Gu L (1994) A new implementation of the element-free Galerkin Method. Comput Meth Appl Mech Engng 113:397–414

7. Liu WK, Jun S, Adee J, Belytischko T (1995) Reproducing kernel particle methods for structural dynamics. Int J Num Meth Engng 38:1655–1680
8. Liu WK, Li S, Belytschko T (1997) Moving least square kernel particle method Part 1: methodology and convergence. Comput Meth Appl Mech Engng 143:113–154
9. Yagawa G, Yamada T (1996) Free mesh method: a new meshless finite element method. Computational Mechanics 18:383–386
10. Yagawa G, Furukawa T (2000) Recent developments approaches for accurate free mesh method. Int J Num Meth Engng 47:1445–1462
11. Fujisawa T, Inaba M, Yagawa G (2003) Parallel computing of high-speed compressible flows using a node-based finite-element method. Int J Num Meth Engng 58:481–511
12. J. Imasato, Y. Sakai (2002) Application of 2-dimentional crack propagation problem using FMM. Advances in Meshfree and X-FEM Methods, Liu GR, editor, World-Scientific
13. Matsubara H, Iraha S, Tomiyama J, Yagawa G (2002) Application of 3D free mesh method to fracture analysis of concrete. Advances in Meshfree and X-FEM Methods, Liu GR, editor, World-Scientific
14. Matsubara H, Iraha S, Tomiyama J, Yamashiro T, Yagawa G (2004) Free mesh method using a tetrahedral element including vertex rotations. JSCE J of Structural Mechanics and Earthquake Engineering 766(I-68):97–107
15. Tian R, Matsubara H, Yagawa G, Iraha S, Tomiyama J (2004) Accuracy improvement on free mesh method: a high performance quadratic tetrahedral/triangular element with only corners. Proc of the 2004 Sixth World Congress on Computational Mechanics (WCCM VI)
16. O.C. Zienkiewicz, K. Morgan (1983), Finite element and approximation, John Wiley & Sons
17. Zienkiewicz OC, Zhu JZ (1992) The superconvergent patch recovery and a posteriori error estimates. PART 1: The recovery technique. Int J Num Meth Engng 33:1331–1364
18. Zienkiewicz OC, Zhu JZ (1992) The superconvergent patch recovery and a posteriori error estimates. PART 2: Error estimates and adaptivity. Int J Num Meth Engng 33:1365–1382
19. Zienkiewicz OC, Zhu JZ (1992) The superconvergent patch recovery (SPR) and adaptive finite element refinement. Comput Meth Appl Mech Engng 101:207–224
20. Babuska I, Rheinboldt WC (1978) A-posteriori error estimates for the finite element method. Int J Num Meth Engng 12:1597–1615
21. Washizu K (1968) Variational methods in elasticity and plasticity. Pergamon Press, New York
22. Timoshenko SP, Goodier JN (1987) Theory of elasticity. McGraw-Hill

Modelling of Metal Forming Processes and Multi-Physic Coupling

J.-L. Chenot and F. Bay

Center for Material Forming (CEMEF)
Ecole des Mines de Paris - UMR CNRS
7635 BP 207, F-06904 Sophia-Antipolis Cedex, France
jean-loup.chenot@ensmp.fr

Summary. The main physical phenomena which can be coupled with the mechanical computation of metal forming processes are analyzed. Recalling the classical thermal coupling, it is shown that a stronger numerical coupling is necessary when localization of the deformation occurs. Several situations where we have mechanical solid and liquid interactions with thermal coupling are briefly mentioned. A more complete description of electro magnetic and thermal coupling is given in view of induction heating. Finally the case of multi scale coupling for metallurgic microstructure evolution is introduced.

Key words: Computational Plasticity, Forming Process, Thermal Coupling, Induction Heating, Micro-Structure.

1 Introduction

Since about thirty years ago, numerical modeling of metal forming processes was starting to be developed in laboratories [1] to [4] and later in software companies as well, using mostly the finite element method. To-day, with the help of powerful modern computers and parallel computation, it is possible to predict accurately several important technical parameters, for complex 3-D forming processes such as forging, rolling or extrusion.

On the other hand, several physical phenomena still remain very approximately known, either during the forming process itself, or during the heating or cooling process, and in heat treatment operations. It is an industrial demand to predict with the same software the material evolution during the whole process. This particularly important if one wants to predict the microstructure evolution.

The objective of the paper is to review some of the most important physical evolutions, which are coupled with the mechanical working process, keeping in mind that the final microstructure is responsible for the final properties of the work-piece.

Eugenio Oñate and Roger Owen (eds.), Computational Plasticity, 221–238.

2 Mechanical Approach of Metal Forming Processes

For a detailed account of the finite element modeling of metal forming processes, the reader is referred to [5].

2.1 Constitutive Modeling

Despite rigid plastic or rigid viscoplastic laws are still widely used in many research teams, the elastic plastic or viscoplastic laws must be preferred. The later ones are intrinsically more realistic; they allow the user to model properly unloading after forming, to predict spring back and the associated residual stresses. The elastic viscoplastic laws are also compulsory when small deformation processes are simulated such as during cooling, heat treatment, stretching, etc. In the literature, several degrees of sophistication can be found.

The most widely used approximation, which reveals accurate enough in most forming processes is the additive decomposition of the strain rate tensor $\dot{\boldsymbol{\varepsilon}}$ into the elastic and plastic parts. It is usually written:

$$\dot{\boldsymbol{\varepsilon}} = \dot{\boldsymbol{\varepsilon}}^e + \dot{\boldsymbol{\varepsilon}}^p \tag{1}$$

Introduction of an objective derivative of the stress tensor; which is necessary when large rotations take place:

$$\frac{d_j \boldsymbol{\sigma}}{dt} = \lambda^e trace(\dot{\boldsymbol{\varepsilon}}^e) + 2\mu^e \dot{\boldsymbol{\varepsilon}}^e \tag{2}$$

The relatively general Perzyna formalism is often retained for the plastic component of the strain rate tensor:

$$\dot{\boldsymbol{\varepsilon}}^p = \frac{1}{k} \left\langle \frac{\sigma_{eq} - R}{K} \right\rangle^{\frac{1}{m}-1} \boldsymbol{\sigma}' \tag{3}$$

where $\boldsymbol{\sigma}'$ is the deviatoric stress tensor, σ_{eq} the equivalent stress, R the yield stress, K the consistency and m the strain rate sensitivity index. At the interface between part and tool the friction shear stress can be modelled by :

$$\boldsymbol{\tau} = \alpha_f(\sigma_n) K |\Delta v|^{q-1} \Delta v \tag{4}$$

where α_f is the friction coefficient, σ_n function of the normal stress and Δv is the tangential velocity.

For an incompressible or quasi incompressible flow, it is desirable to utilize a mixed formulation. In the domain Ω of the part, this formulation is written for any virtual velocity and pressure fields v^*, p^* as:

$$\int_\Omega \dot{\boldsymbol{\sigma}}' : \dot{\boldsymbol{\varepsilon}}^* dV - \int_{\partial\Omega_c} \dot{\tau} \cdot v^* dS - \int_\Omega \dot{p} \mathrm{div}(v^*) dV = 0 \tag{5}$$

$$\int_\Omega (\mathrm{div}(v) + \dot{p}/\kappa) p^* dV = 0 \tag{6}$$

2.2 Finite Element Approximation

Many different finite element formulations were proposed, and developed at the laboratory level, but it is now realized that the discretization scheme must be compatible with other numerical and computational constraints, among which we can quote:

- Remeshing and adaptive remeshing,
- Unilateral contact analysis,
- Iterative solving of non linear and linear systems;
- Domain decomposition and parallel computing;
- Easy transfer of physical internal parameters, when multi physic coupling must be taken into account.

Today a satisfactory compromise is based on a mixed velocity and pressure formulation using tetrahedral elements, and a bubble function to stabilize the solution for incompressible or quasi incompressible materials. The velocity field is discretized with shape functions N_n, in term of nodal velocity vectors $\mathbf{V}_n$:

$$\mathbf{v} = \sum_n \mathbf{V}_n N_n(\xi) \tag{7}$$

Using isoparametric elements the mapping between the physical space with coordinates $\mathbf{x}$ and the reference space, with coordinate ξ is:

$$\mathbf{x} = \sum_n \mathbf{X}_n N_n(\xi) \tag{8}$$

The strain rate tensor can be computed with the help of the conventional $\mathbf{B}$ discretized linear operator:

$$\dot{\boldsymbol{\varepsilon}} = \sum_n \mathbf{V}_n \mathbf{B}_n(\xi) \tag{9}$$

The discretized mixed integral formulation for the mechanical problem is:

$$\begin{aligned} R_n^V = & \int_\Omega \rho\gamma N_n dV + \int_\Omega dK(\sqrt{3}\dot{\bar{\varepsilon}})^{m-1}\dot{\varepsilon} : \mathbf{B}_n dV \\ & + \int_{\partial\Omega_c} \alpha_f K|\Delta v|^{p-1}\Delta v N_n dS - \int_\Omega p tr(B_n) dV = 0 \end{aligned} \tag{10}$$

$$R_m^P = \int_\Omega M_m(\mathrm{div}(v) + 3\alpha_d \dot{T}) dV = 0 \tag{11}$$

The most widely used integration method remains the simple Euler one step scheme. It was recognized that a two levels Runge and Kutta method leads to a considerable improvement of the solution, which is particularly desirable when the work-piece undergoes a large rotation, such as in ring rolling.

2.3 Numerical Issues

The non-linear equations resulting from the mechanical behaviour are linearized with the Newton-Raphson method. The resulting linear systems are often solved now with iterative methods, which appear faster and require much less CPU memory than the direct ones.

Prediction of possible formation of folding defects during forging is based on the analysis of the contact of the part with itself, so providing a similar problem like the coupling with tools.

Automatic dynamic remeshing during the simulation of the whole forming process is almost always necessary, as elements undergo very high strain which could produce degeneracy. Before this catastrophic event, decrease of element quality must be evaluated and a remeshing module must be launched periodically to recover a satisfactory element quality. The global mesh can be completely regenerated, using a Delaunay or any front tracing method, but the method of iterative improvement of the mesh, with a possible *local change* of element structure and connectivity, seems to be much more effective.

For industrial, complicated applications with short delays, the computing time can be decreased dramatically using several or several tens of processors. This requires to use an iterative solver and to define a partition of the domain, each sub domain being associated with a processor. But the parallelization is made more complex due to remeshing and the remeshing process itself must be parallelized.

In order to avoid the necessity for the user to perform several computations, with different meshes to check the accuracy, an error estimation can be developed using for example the generalization of the method proposed by Zienckiewicz and Zhu. Then, if the rate of convergence of the computation is known, the local mesh refinement necessary to achieve a prescribed tolerance can be computed, and the meshing modules are improved to be able to respect the refinement when generating the new mesh.

3 Thermal and Fluid-Solid Coupling

3.1 Classical Thermal and Mechanical Coupling

The classical heat equation for deformable bodies is written simply:

$$\rho c \frac{dT}{dt} = \mathrm{div}(\mathrm{kgrad}(T)) + f_w(\sqrt{3}\dot{\bar{\varepsilon}})^{m+1} \tag{12}$$

where the last term is a fraction f_w of the viscoplastic heat dissipation. The constitutive law depends on temperature. For example we have the equations:

$$K = K_0(\varepsilon_0 + \bar{\varepsilon})^n \exp(\beta/T), \quad m = m_0 + m_1 T \tag{13}$$

The radiation condition on the free surface $\partial\Omega_s$ is written:

$$-k\frac{\partial T}{\partial n} = \varepsilon_r\sigma_r(T^4 - T_0^4) \tag{14}$$

where ε_r is the emissivity parameter, σ_r the Stephan constant and T_0 the outside temperature.

The conduction with the tools on $\partial\Omega_c$, is modeled by:

$$-k\frac{\partial T}{\partial n} = h_{cd}(T - T_{\text{tool}}) + \frac{b}{b + b_{\text{tool}}}\alpha_f K|\Delta v|^{q+1} \tag{15}$$

where T_{tool} is the tool temperature, b and b_{tool} are the effuvisity parameters of the part and of the tool respectively. The thermal integral formulation can be written, for any test function w:

$$\int_{\Omega}\rho c\frac{dT}{dt}wdV + \int_{\Omega}k\mathbf{grad}(T)\cdot\mathbf{grad}(w)dV - \int_{\Omega}\dot{q}_V wdV + \int_{\partial\Omega_f}\varepsilon_r\sigma_r(T^4 - T_0^4)wdS +$$
$$+\int_{\partial\Omega_c}\left(h_{cd}(T - T_{\text{tool}}) + \frac{b}{b + b_{\text{tool}}}\alpha K|\Delta\mathbf{v}|^{q+1}\right)wdS = 0 \tag{16}$$

The previous equations are discretized with finite element, generally similar to those for the mechanical problem:

$$T = \sum_n T_n N_n(\xi) \tag{17}$$

so that the discretized form of the equations is:

$$\int_{\Omega}\rho c\dot{T}N_n dV + \int_{\Omega}k\mathbf{grad}(T)\cdot\mathbf{grad}(N_n)dV - \int_{\Omega}\dot{q}_V wN_n dV$$
$$+\int_{\partial\Omega_f}\varepsilon_r\sigma_r(T^4 - T_0^4)N_n dS + \int_{\partial\Omega_c}\left(h_{cd}(T - T_{\text{tool}}) - \frac{b}{b + b_{\text{tool}}}\tau_f\Delta\mathbf{v}\right)N_n dS = 0 \tag{18}$$

Different time discretization schemes can be used, which allow to compute the temperature field at time $t + \Delta t$, knowing it at time t and $t - \Delta t$, using an explicit scheme which avoids solving a non linear thermal equations. In most codes, coupling between mechanical and thermal equations is not performed at the increment level, as the scheme consists in computing first the thermal field, then the mechanical field (without a fixed point iteration algorithm).

3.2 Localization

When the thermal and mechanical coupling is strong enough, such as in processes where localization in narrow shear bands can occur (e. g. in high speed machining), the previous method is not satisfactory and a fixed point algorithm may not converge. In this case it is desirable to solve simultaneously the mechanical and thermal equations. Therefore we have a non linear problem with the following unknowns nodal unknowns at time $t+\Delta t$: temperature $\mathbf{T}^{t+\Delta t}$, velocity $\mathbf{V}^{t+\Delta t}$, and pressure $\mathbf{P}^{t+\Delta t}$ when a mixed formulation is used. A Newton-Raphson method must be utilized, where the unknowns are the same and the stiffness matrix must contain all the derivatives of the residual equation, including the coupling terms.

3.3 Fluid Solid Coupling During Heating or Heat Treatment

Fluid solid coupling is an important issue in several industrial processes. During heating in a furnace, we must take into account the flow of the surrounding gas and heat exchange between the gas and the preforms, in order to determine precisely the temperature field inside the work-pieces which will be formed. The problem is even more complicated when one wants to predict the quenching process with water which will be vaporized at the contact of a hot work piece. The prediction of the temperature evolution in the part is again important, as it will be responsible for the geometric change of the part and the microstructure evolution.

An analogous situation arises in casting of large work pieces during the cooling process, when a solid fraction interacts with a moving fluid with complicated thermal and mechanical evolutions.

An efficient frame for this coupling is to use an ALE formulation for the liquid phase where the material velocity v is different from the mesh velocity $\mathbf{v}_g$, which will be defined by a smoothing operator, such as:

$$\mathrm{div}(\mathbf{grad}(\mathbf{v}_g)) = 0 \tag{19}$$

and boundary conditions preserving the geometry of the free surface: $(\mathbf{v}_g - \mathbf{v}) \cdot \mathbf{n} = 0$. With the ALE scheme, the grid derivative of any parameter a will be evaluated in term if the material derivative according to:

$$\frac{d_g a}{dt} = \frac{da}{dt} + (\mathbf{v}_g - \mathbf{v}) \cdot \mathbf{grad}(a) \tag{20}$$

4 Electro Magnetic Coupling

4.1 The Induction Heating Process

Many industrial processes are based on an efficient use of coupling between electromagnetic, thermal and mechanical phenomena. They generally use direct or induced currents to generate heat inside a work piece in order to get

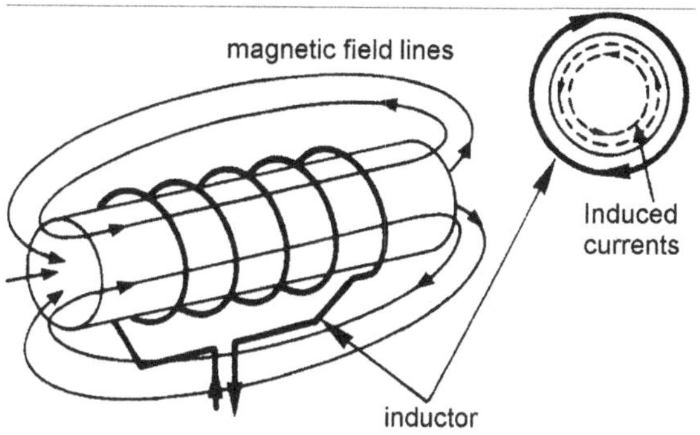

Fig. 1. Induction heating setup

either a prescribed temperature field or some given mechanical or metallurgical properties through an accurate control of temperature evolution with respect to time.

The basic induction setup [6] consists of one or several inductors and metallic work pieces to be heated (see Fig. 1). The inductors are supplied with alternating current with frequencies ranging from fifty to several hundred thousand cycles per second. A rapidly oscillating magnetic field is generated and in turn induces eddy currents in the work piece due to the Joule effect. These currents generate ohmic heat losses inside the work piece. Moreover, for ferromagnetic materials, alternating magnetization and hysteresis effect also contribute to heat generation.

Most of the heat is produced in a thin layer under the surface of the work piece; the skin depth - defined as the depth at which the magnitude of the field drops to a value of e^{-1} of its surface value:

$$\delta = \sqrt{\frac{1}{\pi f \sigma \mu}} \tag{21}$$

where f is the frequency, σ the electrical conductivity and μ the magnetic permeability. High frequencies are used to achieve surface heating, while low frequencies generate a more uniform heating.

4.2 The Direct Electro-Thermal Formulation

The electromagnetic model is classically based on the set of Maxwell equations:

$$div(\mathbf{B}) = 0 \tag{22}$$

$$div(\mathbf{D}) = 0 \tag{23}$$

$$\mathbf{rot}(\mathbf{E}) = -\frac{\partial \mathbf{B}}{\partial t} \tag{24}$$

$$\mathbf{rot}(\mathbf{H}) = \mathbf{J} + \frac{\partial \mathbf{D}}{\partial t} \tag{25}$$

where **H** is the magnetic field, **B** the magnetic induction, **E** the electric field, **D** the electric flux density, and **J** the electric current density associated with free charges.

We also have the following relations which for the intrinsic material properties:

$$\mathbf{D} = \varepsilon \mathbf{E} \tag{26}$$

$$\mathbf{B} = \mu(|H|)\mathbf{H} \tag{27}$$

$$\mathbf{J} = \sigma \mathbf{E} \tag{28}$$

where ε is the dielectric constant, μ the magnetic permeability, and σ the electrical conductivity. They all depend on temperature and the magnetic permeability μ depends also on **H**. The range of frequencies dealt with in induction heating (less than 10^6 Hz) enables us to neglect the displacement currents in the Maxwell-Ampere equation (magneto-quasi-static approximation). A combination of the previous relations leads us to the following equation where the unknown is the electric field.

$$\sigma \frac{\partial \mathbf{E}}{\partial t} + \mathbf{rot}\left(\frac{1}{\mu}\mathbf{rot}(\mathbf{E})\right) = -\frac{\partial \mathbf{J}_S}{\partial t} \tag{29}$$

with $\sigma = \sigma(T)$ and $\mu = \mu(T, \mathbf{H})$.

We deal here with axisymmetrical cases, in which the electric field will only have a non-zero component in the θ direction:

$$\mathbf{E} = (0, E_\theta(r,z), 0) \tag{30}$$

Temperature evolution in the work piece is governed by the classical heat transfer equation:

$$\rho C \frac{\partial T}{\partial t} - div(k\mathbf{grad}\ (T)) = \dot{Q}_{\mathrm{em}} \tag{31}$$

where ρ denotes the material density, C and k respectively the specific heat and thermal conductivity, all temperature dependent.

$\dot{Q}_{\mathrm{em}}$ denotes the local heat rate, generated by the eddy currents, and integrated over one period:

$$\bar{\dot{Q}}_{\mathrm{em}} = \frac{1}{T}\int_0^T \sigma |\mathbf{E}|^2 dt \tag{32}$$

The boundary conditions can be of various kinds: prescribed heat flux or temperature, convection or radiation.

4.3 Finite Element Numerical Approximation

We define Ω as being a two-dimensional axisymmetrical domain which covers the part to be heated Ω_{part}, the inductor Ω_{inductor} and a finite volume of air air surrounding the inductor and the part. We have chosen here to carry out coupling between the part and the inductor for the electromagnetic computations using finite elements rather than boundary elements; the air domain thus needs to be wide enough in order to model accurately electromagnetic wave propagation.

The domain is discretized using second order triangular finite elements (6-nodes triangles). The unknown fields - electric field E_θ for the electromagnetic computations, temperature field T for the thermal computations and velocity field $\mathbf{V}$ for the mechanical computations - can thus be approximated over the whole domain by the classical finite element approximation:

$$E(t,r,z) = \sum_n E_n(t) N_n(\xi) \tag{33}$$

where $E : n(t)$ denotes the approximated value of the θ-component of the electric field at the node n and at time t. The discretized expression of the variational formulation is:

$$[C^{\text{em}}] \left\{ \frac{\partial E}{\partial t} \right\} + [K^{\text{em}}] \{E(t)\} = \{B^{\text{em}}\} \tag{34}$$

The temperature distribution is discretized according in the same way as in Eq.(18).

Numerical models in induction heating often solve a harmonic model. This assumption is quite restrictive when one deals with non-linear magnetic materials. We have thus chosen to solve the time-dependent model. We need therefore to integrate numerically in time the electromagnetic and thermal equations.

We detail here the selected time integration scheme for the electromagnetic equation. The procedure is the same for the thermal equation. We use a second-order two time step finite difference scheme:

Step 1: the system is solved at time t^* such that $t < t^* < t + \delta t_2$ with:

$$t^* = \alpha_1 (t - \delta t_1) + \alpha_2 t + \alpha_3 (t + \delta t_2) \tag{35}$$

with $\alpha_1 + \alpha_2 + \alpha_3 = 0$.

The electric field E^* at time t^* and its time derivative write:

$$E^* = \alpha_1 E^{t-\delta t_1} + \alpha_2 E^t + \alpha_3 E^{t+\delta t_2} \tag{36}$$

$$\frac{\partial E^*}{\partial t} = \gamma \frac{E^{t+\delta t_2}}{\delta t_2} + (\gamma - 1) \frac{E^{t-\delta t_2} - E^t}{\delta t_1} \tag{37}$$

The following system Eq.(38) is written at time t^*. E^* and its derivative are replaced by Eqs.(36) and (37) and the system are solved for the unknown variable E^*:

$$\left(\frac{\gamma}{\alpha_3 \delta t_2}[C^{\mathrm{em}}]^* + [K^{\mathrm{em}}]^*\right) E^* = \{B^{\mathrm{em}}\}^* + c_1 [C^{\mathrm{em}}]^* E^t + c_0 [C^{\mathrm{em}}]^* E^{t-\delta t_1} \quad (38)$$

where $c_1 = \dfrac{\gamma}{\delta t_2} + \dfrac{\gamma - 1}{\delta t_1} + \dfrac{\gamma \alpha_2}{\alpha_3 \delta t_2}$, $c_2 = \dfrac{\gamma \alpha_1}{\alpha_3 \delta t_2} - \dfrac{\gamma - 1}{\delta t_1}$.

The two time steps scheme we have requires the solving of a non-linear equation, as the matrix $[C]$ is dependant on the magnetic field. In order to avoid an additional non-linearity, the matrix is linearized and is approximated using its values at time t and $t - \delta t_1$:

$$[C]^* = \left(\alpha_1 - \alpha_3 \frac{\delta t_2}{\delta t_1}\right)[C]_{t-\delta t_1} + \left(\alpha_2 + \alpha_3 \left(1 + \frac{\delta t_2}{\delta t_1}\right)\right)[C]_{t-\delta t_1} \quad (39)$$

Step 2: computation of:

$$\{E\}^{t+\delta t_2} = \frac{1}{\alpha_3}\left(\{E\}^* - \alpha_1 \{E\}^{t-\delta t_1} - \alpha_2 \{E\}^t\right) \quad (40)$$

4.4 The Electromagnetic/Thermal Coupling Procedure

Physical problems arising from heat transfer and electromagnetism have in common the fact that they are both time-dependent. Their specific time-scales are however very different. The specific time scale of an electromagnetic problem is related to the wave-associated period –typically 10^2s for a 100 Hz frequency down to 10^8s for a 100 MHz frequency– whereas the specific time scale for heat transfer averages normally one second.

A direct model based on finite elements has been developed in our laboratory to cope with these specificities. The model includes a specific coupling procedure for solving:

- the Maxwell equations - in order to access the electromagnetic fields giving the eddy currents dissipated in the material (main source term for the heat transfer equation),
- the heat transfer equation - leading to temperature evolution in the material.

The coupling between the electromagnetic and thermal computations relies on a convergence test over the mean heating power and on tests over the variations of the magnetic parameters that determine respectively the passage from an electrical to thermal resolution or inversely from a thermal to an electric one.

Once the electromagnetic field has been calculated, the rate of heat generation $\dot{Q}_{em}$ for the heat equation needs to be evaluated at every integration points. As the electromagnetic time step is far smaller than the thermal one, we do not consider the instantaneous Joule power calculated at a given time at every integration points. We rather consider a mean Joule power averaged over one period of the electromagnetic field:

$$Q_{em}(nT, \text{int}) = \frac{1}{T} \int_{(n-1)T}^{nT} \sigma(\text{int}, t)|E(\text{int}, t)|^2 dt \tag{41}$$

where *int* is the considered integration point, T is the period of the power supply currents, n is number of periods considered and $E_\theta(\text{int}, t)$ is the value at time t of the electric field interpolated at the integration point *int*.

At the end of each electromagnetic period, the newly calculated mean power is compared to the one calculated at the previous period until it stabilizes. Thermal computations are started with the stabilized thermal source power calculated at $(n+1)$ T if the following convergence test (33) is conducted at every integration points:

$$\frac{\bar{Q}_{em}((n+1)T) - \bar{Q}_{em}(nT)}{\bar{Q}_{em}(nT)} < \varepsilon \tag{42}$$

where ε is the user-supplied convergence parameter.

These thermal computations are valid as long as the variation of the physical magnetic parameters such as the magnetic permeability and the electric conductivity do not exceed 5%. Their variations with temperature are tested after each new thermal computation. The following criteria are tested for every mesh element:

$$\begin{aligned} \frac{\sigma(T_{max}^{n+1}) - \sigma(T_{max}^{n})}{\sigma(T_{max}^{n})} &< 5\% \\ \frac{\mu(T_{max}^{n+1}) - \mu(T_{max}^{n})}{\mu(T_{max}^{n})} &< 5\% \end{aligned} \tag{43}$$

where T_{max}^{n+1} is the maximum value of the temperature field in a given element at time $t + dt_{ther}$ and T_{max}^{n} is the maximum value of the temperature field in the same given element at current time t. When the maximum relative variations reach the threshold of 5%, the previously calculated mean heat power is assumed to be irrelevant, and a new electromagnetic calculation is carried out.

For their part, mechanical computations are carried out at the same time steps than thermal computations.

4.5 Results

The first case deals with a static long inductor in ferromagnetic EN3 mild steel annealed at 930 $°C$. In this case, the work piece is heated by an inductor of the same length as the work piece. The geometry consists in a cylindrical

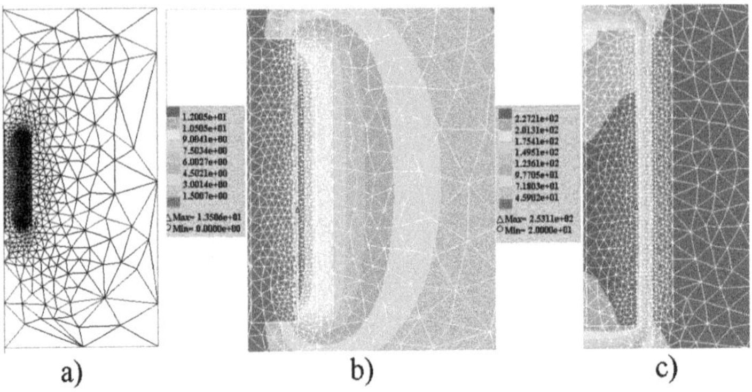

Fig. 2. a) Finite element mesh; b) Electric field; c) Temperature field

EN3 part surrounded with Kaowool insulation, a ceramic tube and finally a three-layer coil. The mesh used for this case is displayed in Fig. 2a. The multi-turn coil is modeled as a continuous single coil with a uniform initial current density. Two thermocouples have been placed: the first one on the surface on the middle of the part and the second one in a hole bored in the center of the part so as to measure temperature inside the part.

Process parameters are:

- frequency: 500Hz
- current density: 8.10^8 Amps/m2
- electromagnetic time step: T/64=3.125 10^{-5}s
- thermal time step: 0.5s

In order to enable a visualization of the dissipated joule power, an effective electrical field has been defined at a given node as the integration over an electromagnetic period of the square of the real instantaneous electrical field:

$$E_{eff} = \sqrt{\frac{1}{T} \int_{nT}^{(n+1)T} E^2(t)dt} \tag{44}$$

This effective electrical field is also proportional to the amplitude of the electric field in a harmonic complex formulation. Figure 2b shows the effective electric fields isovalues. Due to the high value of the permeability, we can notice the concentration of the magnetic field along the work piece surface; this is a well-known physical effect since magnetic field lines tend to concentrate on material surface when the magnetic permeability gets high.

We can see on Fig. 2c how the heat propagates here rather uniformly from the surface towards the centre. This is typical of a global homogeneous heating application.

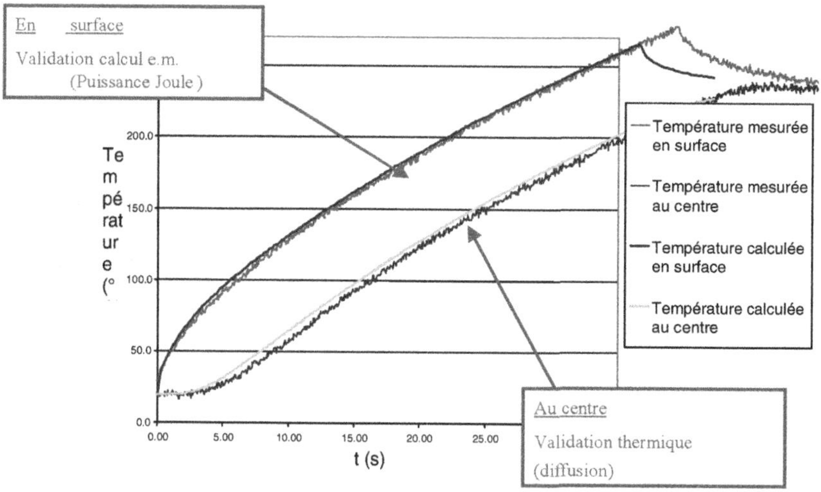

Fig. 3. Experimental validation

Figure 3 shows the excellent agreement between experimental and computed temperatures on the surface and inside the work piece. On one hand, the computed temperatures on the surface provide a good way to validate the dissipated heat due to eddy currents inside the work piece - and thus the electric field computation and the whole electromagnetic procedure. On the other hand, the good agreement observed between the experimental and numerically computed delay on heat propagation to the internal part of the work piece validates the heat transfer diffusion model and the thermal coupling procedure.

This case deals with a moving short inductor - a case representative of what takes place in an induction heat treatment process. The inductor displacement is modeled through a continuous variation of physical properties of the mesh. Figure 4 shows the geometry and the initial location of the inductor in the mesh. Figure 5 shows an example of electric field distribution at a given time step; in this case, the electric field is non-symmetric since the inductor location changes continuously.

The numerical model presented here is described more in detail in [7]. Extension to parallel computations is detailed in [8].

This model has also been used for magnetic parameters identification as detailed in [9]. A global optimization procedure has also been developed to achieve better control of induction heating processes. This strategy can be extended to various multiphysics coupled problems; it is detailed in [10].

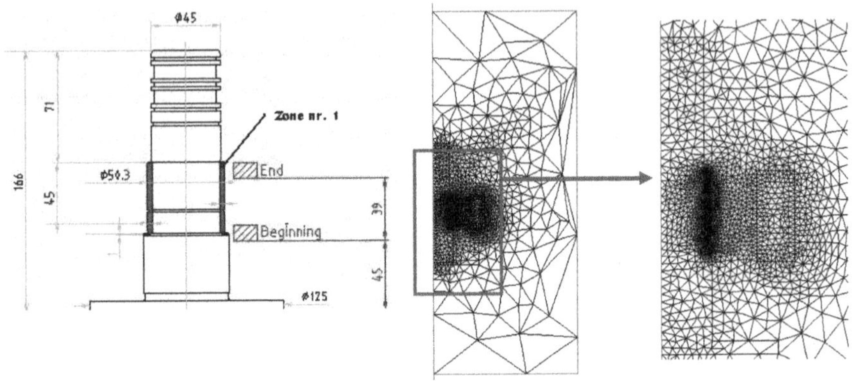

Fig. 4. Moving inductor case: geometry and initial mesh

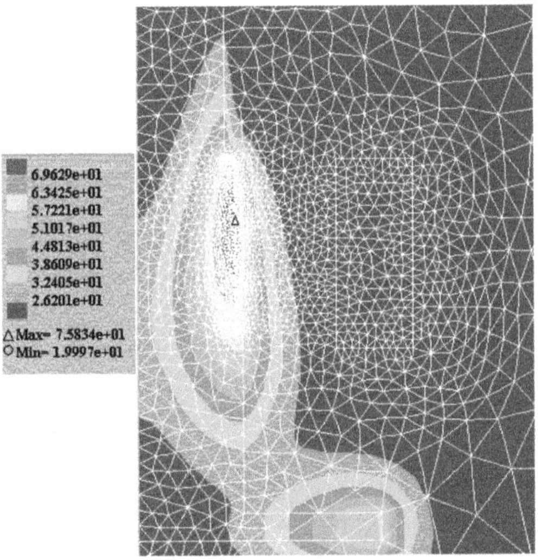

Fig. 5. Moving inductor case: electric field isovalues

5 Coupling with Microstructure Evolution

5.1 Introduction

The microstructure of the materials changes significantly during forming operations. For instance in metals, dynamic or static recovery and recrystallization can take place. These evolutions need to be modeled when the final microstructure is to be optimized, or when the behavior of the metal must be described accurately during forming. Several strategies can be developed to compute microstructure evolutions in forming processes with large spatial heterogeneity of strain rate, strain and temperature fields. The finite element method allows to model microstructure evolutions in complementary ways.

5.2 Macroscopic Approach

In a first classical way, one can assume that each node covers a material volume sufficiently large to be summarized by a Homogeneous Equivalent Medium (HEM). Laws for microstructure evolution of homogeneous media are then applied. This strategy is adequate for modeling a large volume of material, for instance at the scale of a component. A well-known example of a microstructure model is the Avrami equation [11]. It is typically used in an integrated form to describe the recrystallization phenomena for metals as a function of time t:

$$X = 1 - \exp\left(-\ln\left(\frac{t}{t_{0.5}}\right)^k\right) \tag{45}$$

where X is the state variable describing the recrystallized volume fraction, $t_{0.5}$ is the half recrystallization time, and k the Avrami exponent. Such microstructure evolution laws are typically obtained from isothermal experiments. However, a typical forming process induces significant temperature evolutions. It is assumed that non isothermal path can be divided into a number of isothermal steps. Using the results of isothermal tests to calculate the microstructure evolution for any thermal path implies that an integration rule can be applied. Additivity rules are often used in recrystallization studies, the procedure uses a fictitious time method [12], [22] to discretize the microstructure evolution equation. Using for instance the Avrami equation, the algorithm is as follows: at step i of the computation, the recrystallized volume fraction Xi at time ti has been computed; the fictitious time t^*_{i+1} is calculated as the total time needed to obtain a recrystallized volume fraction X_i at constant temperature T_{i+1}. In order to calculate the next recrystallized volume fraction X_{i+1}, Avrami equation is used for temperature T_{i+1} and time $(t^*_{i+1} + t_{i+1} - t_i)$.

This strategy has been used with the 3D Finite Element code Forge3® for forging simulations. Figure 6 illustrates the case of a 316L austenitic stainless steel part deformed at $1273K$ and annealed at $1373K$ during 50 seconds. The simulation allows us to take into account both the strain and temperature variations throughout the part, leading to a complex map of statically recrystallized volume fractions, ranging from 10 to 100%, as shown in Fig. 6. The challenge is now to model microstructure evolutions over multiple interpass and forming steps; this requires to go back and refine the description of the physical phenomena integrated in the macroscopic evolution laws.

5.3 Multi-Scale Coupling and the Digital Material

A second approach may be taken at a smaller scale, within a HEM. It consists in meshing a heterogeneous microstructure, possibly of complex topology and

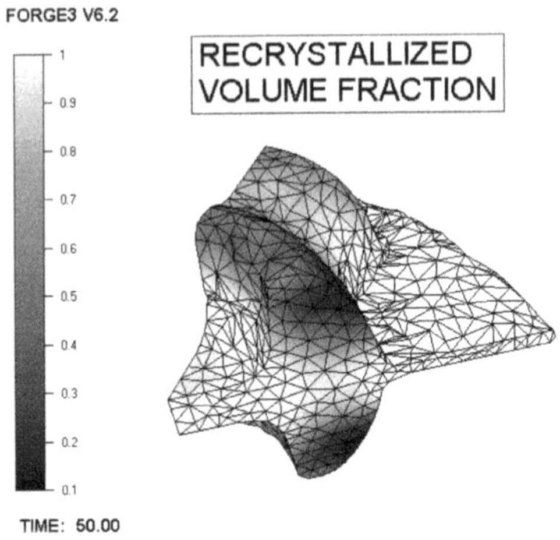

Fig. 6. Recrystallized volume fraction maps of a forged 316l part

geometry, for studying the detailed relationships between the microstructure and its evolution, or even the associated constitutive law.

This is the case for instance when dealing with the plastic deformation of a steel during phase transformation (e.g. [13]). An austenite and ferrite composite is in fact a two-phase HEM which needs to be analyzed in order to determine the mechanical behavior of the steel and the strain distribution within each phase. Analytical bounds and estimates for the mechanical behavior of a two-phase aggregate imply strong topological assumptions. The calculation of the mechanical response of these composite materials can be calculated using the F. E. M. with a representative description of the microstructure, and well posed boundary conditions. The HEM must contain enough grains so that the global behavior is statistically representative of material response. Predictions of such FEM simulations can then be compared with analytical models.

An example of a two-phase topology is shown in Fig. 7 with the corresponding 3D finite element discretization. A calculation of strain distribution for this two-phase ensemble is presented in Fig. 8.

The F.E. discretization allows us to capture strong strain gradients at the interface between the two phases. Such digital multiphase volume calculations can first be used to validate appropriate analytical models. These simpler models can be used as constitutive laws in large scale finite element simulations and coupled with the phase transformation diagram to calculate local percentage of each phase. This complete numerical model allows us to calculate thermal, mechanical and metallurgical evolutions in forming processes inducing large strain such as forging or rolling.

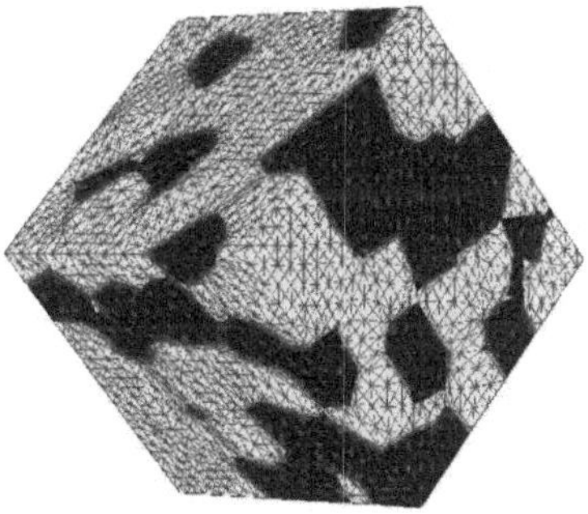

Fig. 7. Austenite-ferrite topology and corresponding 3d finite element mesh

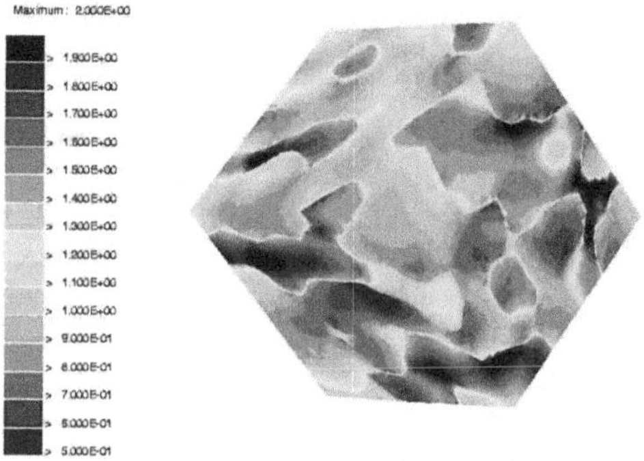

Fig. 8. Strain field calculated for the two-phase material volume of Fig. 7

This digital material approach can be further refined to integrate the crystallographic textures. A statistically representative finite element polycrystal can be created by considering each finite element as a single crystal. Such a strategy proves to be very efficient when the mechanical response is dictated by the heterogeneity of the elastic-plastic behavior throughout a polycrystal, as in fatigue [14].

6 Conclusions

We have reviewed several coupling problems where the classical mechanical treatment must be completed in order to take also into account other physical phenomenons: thermal behavior with possible localization of the strain rate, fluid-solid coupling, electro magnetic heating and microstructure evolution. The simple examples which are presented show that the domain is still under development, and that much effort must be devoted to these fields before it can be used routinely in industry for any metallic material.

References

1. Cornfield GC, Johnson RH (1973) Theoretical prediction of plastic flow in hot rolling including the effect of various temperature distribution. J Iron Steel Inst 211:567
2. Lee CH, Kobayashi S (1973) New solutions to rigid plastic deformation problems using a matrix method. Trans ASME, J Eng Ind 95:865
3. Zienkiewicz OC, Valliapan S, King IP (1969) Elasto-solution of engineering problems: initial stress, finite element approach. Int J Num Meth Engng 1:75–100
4. Surdon G., Chenot J-L (1986) Finite element calculations of three-dimensional hot forging. International Conference on Numerical Methods in Industrial Forming Processes Numiform'86, ed. by Mattiasson K *et al.*, Balkema AA, Rotterman, 287–292
5. Wagoner, RH, Chenot J-L (2001) Metal forming analysis. Cambridge University Press, Cambridge
6. Davies EJ (1990) Conduction and Induction Heating, P. Peregrinus Ltd., London
7. Bay F, Labbe V, Favennec Y, Chenot J-L (2000) A numerical model for induction heating processes coupling electromagnetism and thermomechanics. Int J Num Meth Engng 58:839–867
8. Labbe V, Favennec Y, Bay F (2002) Numerical modeling of heat treatment with induction heating processes using a parallel finite element software. 5th Conf. World Congress on Computational Mechanics, Vienna, Austria
9. Favennec Y, Labbe V, Tillier Y, Bay F (2000) Identification of magnetic parameters through inverse analysis coupled with finite element modeling. IEEE Transactions on Magnetics 38(6):3607–3619
10. Favennec Y, Labbe V, Bay F (2004) The ultraweak time coupling in nonlinear modeling and related optimization problems. Int J Num Meth Engng 60:3
11. Avrami M (1939) Kinetics of phase change. I. General theory, Journal of Chemical Physics 7:1103–1112; (1940) II. Transformation-time relations for random distribution of nuclei. Journal of Chemical Physics 8:212–224; (1941) III, Granulation, phase change and microstructure. Journal of Chemical Physics 9:177–18
12. Pumphrey WI, Jones FW (1948) Inter-relation of hardenability and isothermal transformation data. ISIJ 159:137–144
13. Thibaux P (2001) Comportement mécanique d'un acier C-Mn microallié lors du laminage intercritique. PhD Thesis, Ecole des Mines de Paris, CEMEF, Sophia Antipolis, France
14. Turkmen HS, Logé RE, Dawson PR, Miller MP (2003) On the mechanical behavior of AA 7075-T6 during cyclic loading. International Journal of Fatigue 25(4):267–281

Enhanced Rotation-Free Basic Shell Triangle. Applications to Sheet Metal Forming

Eugenio Oñate[1], Fernando G. Flores[2] and Laurentiu Neamtu[3]

[1] International Center for Numerical Methods in Engineering (CIMNE)
Technical University of Catalonia (UPC)
Edificio C1 Gran Capitán s/n, 08034 Barcelona, Spain
onate@cimne.upc.edu

[2] National University of Cordoba
Casilla de Correo 916
5000 Córdoba, Argentina
fflores@gtwing.efn.uncor.edu

[3] Quantech ATZ SA Gran Capitán 2–4, 08034 Barcelona, Spain
laur@quantech.es

Summary. An enhanced rotation-free three node triangular shell element (termed EBST) is presented. The element formulation is based on a quadratic interpolation of the geometry in terms of the six nodes of a patch of four triangles associated to each triangular element. This allows to compute an assumed constant curvature field and an assumed linear membrane strain field which improves the in-plane behaviour of the element. A simple and economic version of the element using a single integration point is presented. The implementation of the element into an explicit dynamic scheme is described. The efficiency and accuracy of the EBST element and the explicit dynamic scheme are demonstrated in many examples of application including the analysis of a cylindrical panel under impulse loading and sheet metal stamping problems.

1 Introduction

Triangular shell elements are very useful for the solution of large scale shell problems occurring in many practical engineering situations. Typical examples are the analysis of shell roofs under static and dynamic loads, sheet stamping processes, vehicle dynamics and crash-worthiness situations. Many of these problems involve high geometrical and material non linearities and time changing frictional contact conditions. These difficulties are usually increased by the need of discretizing complex geometrical shapes. Here the use of shell triangles and non-structured meshes becomes a critical necessity. Despite recent advances in the field [1]–[6] there are not so many simple shell

Eugenio Oñate and Roger Owen (eds.), Computational Plasticity, 239–265.

triangles which are capable of accurately modelling the deformation of a shell structure under arbitrary loading conditions.

A promising line to derive simple shell triangles is to use the nodal displacements as the only unknowns for describing the shell kinematics. This idea goes back to the original attempts to solve thin plate bending problems using finite difference schemes with the deflection as the only nodal variable [7]–[9].

In past years some authors have derived a number of thin plate and shell triangular elements free of rotational degrees of freedom (d.o.f.) based on Kirchhoff's theory [10]–[26]. In essence all methods attempt to express the curvatures field over an element in terms of the displacements of a collection of nodes belonging to a patch of adjacent elements. Oñate and Cervera [14] proposed a general procedure of this kind combining finite element and finite volume concepts for deriving thin plate triangles and quadrilaterals with the deflection as the only nodal variable and presented a simple and competitive rotation-free three d.o.f. triangular element termed BPT (for Basic Plate Triangle). These ideas were extended in [20] to derive a number of rotation-free thin plate and shell triangles. The basic ingredients of the method are a mixed Hu-Washizu formulation, a standard discretization into three node triangles, a linear finite element interpolation of the displacement field within each triangle and a finite volume type approach for computing constant curvature and bending moment fields within appropriate non-overlapping control domains. The so called "cell-centered" and "cell-vertex" triangular domains yield different families of rotation-free plate and shell triangles. Both the BPT plate element and its extension to shell analysis (termed BST for Basic Shell Triangle) can be derived from the cell-centered formulation. Here the "control domain" is an individual triangle. The constant curvatures field within a triangle is computed in terms of the displacements of the six nodes belonging to the four elements patch formed by the chosen triangle and the three adjacent triangles. The cell-vertex approach yields a different family of rotation-free plate and shell triangles. Details of the derivation of both rotation-free triangular shell element families can be found in [20].

An extension of the BST element to the non linear analysis of shells was implemented in an explicit dynamic code by Oñate *et al.* [25] using an updated Lagrangian formulation and a hypo-elastic constitutive model. Excellent numerical results were obtained for non linear dynamics of shells involving frictional contact situations and sheet stamping problems [17,18,19,25].

A large strain formulation for the BST element using a total Lagrangian description was presented by Flores and Oñate [23]. A recent extension of this formulation is based on a quadratic interpolation of the geometry of the patch formed by the BST element and the three adjacent triangles [26]. This yields a linear displacement gradient field over the element from which linear membrane strains and constant curvatures can be computed within the BST element.

In this chapter an enhanced version of the BST element (termed EBST element) is derived using an assumed linear field for the membrane strains and an assumed constant curvature field. Both assumed fields are obtained from the quadratic interpolation of the patch geometry following the ideas presented in [26]. Details of the element formulation are given. An efficient version of the EBST element using one single quadrature point for integration of the tangent matrix is presented. An explicit scheme adequate for dynamic analysis is briefly described.

The efficiency and accuracy of the EBST element is validated in a number of examples of application including the non linear analysis of a cylindrical shell under an impulse loading and practical sheet stamping problems.

2 Basic Thin Shell Equations Using a Total Lagrangian Formulation

2.1 Shell Kinematics

A summary of the most relevant hypothesis related to the kinematic behaviour of a thin shell are presented. Further details may be found in the wide literature dedicated to this field [8,9].

Consider a shell with undeformed middle surface occupying the domain Ω^0 in R^3 with a boundary Γ^0. At each point of the middle surface a thickness h^0 is defined. The positions $\mathbf{x}^0$ and $\mathbf{x}$ of a point in the undeformed and the deformed configurations can be respectively written as a function of the coordinates of the middle surface $\boldsymbol{\varphi}$ and the normal $\mathbf{t}_3$ at the point as

$$\mathbf{x}^0(\xi_1,\xi_2,\zeta) = \boldsymbol{\varphi}^0(\xi_1,\xi_2) + \lambda \mathbf{t}_3^0 \tag{1}$$

$$\mathbf{x}(\xi_1,\xi_2,\zeta) = \boldsymbol{\varphi}(\xi_1,\xi_2) + \zeta\lambda \mathbf{t}_3 \tag{2}$$

where ξ_1, ξ_2 are arc-length curvilinear principal coordinates defined over the middle surface of the shell and ζ is the distance from the point to the middle surface in the undeformed configuration. The product $\zeta\lambda$ is the distance from the point to the middle surface measured on the deformed configuration. The parameter λ relates the thickness at the present and initial configurations as:

$$\lambda = \frac{h}{h^0} \tag{3}$$

This approach implies a constant strain in the normal direction. Parameter λ will not be considered as an independent variable and will be computed from purely geometrical considerations (*isochoric* behaviour) via a staggered iterative update. Besides this, the usual plane stress condition of thin shell theory will be adopted.

A convective system is computed at each point as

$$\mathbf{g}_i(\xi) = \frac{\partial \mathbf{x}}{\partial \xi_i} \qquad i = 1,2,3 \tag{4}$$

$$\mathbf{g}_{\alpha}(\xi) = \frac{\partial\left(\boldsymbol{\varphi}(\xi_1,\xi_2) + \zeta\lambda\mathbf{t}_3\right)}{\partial\xi_{\alpha}} = \boldsymbol{\varphi}_{\prime\alpha} + \zeta\left(\lambda\mathbf{t}_3\right)_{\prime\alpha} \quad \alpha = 1,2 \tag{5}$$

$$\mathbf{g}_{3}(\xi) = \frac{\partial\left(\boldsymbol{\varphi}(\xi_1,\xi_2) + \zeta\lambda\mathbf{t}_3\right)}{\partial\zeta} = \lambda\mathbf{t}_3 \tag{6}$$

This can be particularized for the points on the middle surface as

$$\mathbf{a}_{\alpha} = \mathbf{g}_{\alpha}(\zeta = 0) = \boldsymbol{\varphi}_{\prime\alpha} \tag{7}$$

$$\mathbf{a}_{3} = \mathbf{g}_{3}(\zeta = 0) = \lambda\mathbf{t}_3 \tag{8}$$

The covariant (first fundamental form) metric tensor of the middle surface is

$$a_{\alpha\beta} = \mathbf{a}_{\alpha}\cdot\mathbf{a}_{\beta} = \boldsymbol{\varphi}_{\prime\alpha}\cdot\boldsymbol{\varphi}_{\prime\beta} \tag{9}$$

The Green-Lagrange strain vector of the middle surface points (membrane strains) is defined as

$$\boldsymbol{\varepsilon}_m = \left[\varepsilon_{m_{11}}, \varepsilon_{m_{12}}, \varepsilon_{m_{12}}\right]^T \tag{10}$$

with

$$\varepsilon_{m_{ij}} = \frac{1}{2}(a_{ij} - a_{ij}^0) \tag{11}$$

The curvatures (second fundamental form) of the middle surface are obtained by

$$\kappa_{\alpha\beta} = \frac{1}{2}\left(\boldsymbol{\varphi}_{\prime\alpha}\cdot\mathbf{t}_{3^{\prime}\beta} + \boldsymbol{\varphi}_{\prime\beta}\cdot\mathbf{t}_{3^{\prime}\alpha}\right) = -\mathbf{t}_3\cdot\boldsymbol{\varphi}_{\prime\alpha\beta} \quad , \quad \alpha,\beta = 1,2 \tag{12}$$

The deformation gradient tensor is

$$\mathbf{F} = [\mathbf{x}_{\prime 1}, \mathbf{x}_{\prime 2}, \mathbf{x}_{\prime 3}] = \left[\boldsymbol{\varphi}_{\prime 1} + \zeta\left(\lambda\mathbf{t}_3\right)_{\prime 1}\ \boldsymbol{\varphi}_{\prime 2} + \zeta\left(\lambda\mathbf{t}_3\right)_{\prime 2}\ \lambda\mathbf{t}_3\right] \tag{13}$$

The product $\mathbf{F}^T\mathbf{F} = \mathbf{U}^2 = \mathbf{C}$ (where $\mathbf{U}$ is the right stretch tensor, and $\mathbf{C}$ the right Cauchy-Green deformation tensor) can be written as

$$\mathbf{U}^2 = \begin{bmatrix} a_{11} + 2\kappa_{11}\zeta\lambda & a_{12} + 2\kappa_{12}\zeta\lambda & 0 \\ a_{12} + 2\kappa_{12}\zeta\lambda & a_{22} + 2\kappa_{22}\zeta\lambda & 0 \\ 0 & 0 & \lambda^2 \end{bmatrix} \tag{14}$$

In the derivation of expression (14) the derivatives of the thickness ratio $\lambda_{\prime a}$ and the terms associated to ζ^2 have been neglected.

Equation (14) shows that $\mathbf{U}^2$ is not a unit tensor at the original configuration for curved surfaces ($\kappa_{ij}^0 \neq 0$). The changes of curvature of the middle surface are computed by

$$\chi_{ij} = \kappa_{ij} - \kappa_{ij}^0 \tag{15}$$

Note that $\delta\chi_{ij} = \delta\kappa_{ij}$.

For computational convenience the following approximate expression (which is exact for initially flat surfaces) will be adopted

$$\mathbf{U}^2 = \begin{bmatrix} a_{11} + 2\chi_{11}\zeta\lambda & a_{12} + 2\chi_{12}\zeta\lambda & 0 \\ a_{12} + 2\chi_{12}\zeta\lambda & a_{22} + 2\chi_{22}\zeta\lambda & 0 \\ 0 & 0 & \lambda^2 \end{bmatrix} \tag{16}$$

This expression is useful to compute different Lagrangian strain measures. An advantage of these measures is that they are associated to material fibres, what makes it easy to take into account material anisotropy. It is also useful to compute the eigen decomposition of $\mathbf{U}$ as

$$\mathbf{U} = \sum_{\alpha=1}^{3} \lambda_\alpha \ \mathbf{r}_\alpha \otimes \mathbf{r}_\alpha \tag{17}$$

where λ_α and $\mathbf{r}_\alpha$ are the eigenvalues and eigenvectors of $\mathbf{U}$.

The resultant stresses (axial forces and moments) are obtained by integrating across the original thickness the second Piola-Kirchhoff stress vector $\boldsymbol{\sigma}$ using the actual distance to the middle surface for evaluating the bending moments. This gives

$$\boldsymbol{\sigma}_m \equiv [N_{11}, N_{22}, N_{12}]^T = \int_{h^0} \boldsymbol{\sigma} d\zeta \tag{18}$$

$$\boldsymbol{\sigma}_b \equiv [M_{11}, M_{22}, M_{12}]^T = \int_{h^0} \boldsymbol{\sigma}\lambda\zeta \ d\zeta \tag{19}$$

With these values the virtual work can be written as

$$\int\int_{A^0} \left[\delta\boldsymbol{\varepsilon}_m^T \boldsymbol{\sigma}_m + \delta\boldsymbol{\kappa}^T \boldsymbol{\sigma}_b\right] dA = \int\int_{A^0} \delta\mathbf{u}^T \mathbf{t} dA \tag{20}$$

where $\delta\mathbf{u}$ are virtual displacements, $\delta\boldsymbol{\varepsilon}_m$ is the virtual Green-Lagrange membrane strain vector, $\delta\boldsymbol{\kappa}$ are the virtual curvatures and $\mathbf{t}$ are the surface loads. Other load types can be easily included into (20).

2.2 Constitutive Models

In order to treat plasticity at finite strains an adequate stress-strain pair must be used. The Hencky measures will be adopted here. The (logarithmic) strains are defined as

$$\mathbf{E}_{\ln} = \begin{bmatrix} \varepsilon_{11} & \varepsilon_{21} & 0 \\ \varepsilon_{12} & \varepsilon_{22} & 0 \\ 0 & 0 & \varepsilon_{33} \end{bmatrix} = \sum_{\alpha=1}^{3} \ln(\lambda_\alpha)\, \mathbf{r}_\alpha \otimes \mathbf{r}_\alpha \tag{21}$$

For the type of problems dealt within the paper we use an elastic-plastic material associated to thin rolled metal sheets.

In the case of metals, where the elastic strains are small, the use of a logarithmic strain measure reasonably allows to adopt an additive decomposition of elastic and plastic components as

$$\mathbf{E}_{\ln} = \mathbf{E}^e_{\ln} + \mathbf{E}^p_{\ln} \tag{22}$$

A linear relationship between the (plane) Hencky stresses and the logarithmic elastic strains is chosen giving

$$\mathbf{T} = \mathbf{H}\,\mathbf{E}^e_{\ln} \tag{23}$$

where $\mathbf{H}$ is the constitutive matrix. The constitutive equations are integrated using a standard return algorithm. The following Mises-Hill [30] yield function with non-linear isotropic hardening is chosen

$$(G+H)\;T_{11}^2 + (F+H)\;T_{22}^2 - 2H\;T_{11}T_{22} + 2N\;T_{12}^2 = \sigma_0\,(e_0 + e^p)^n \tag{24}$$

where F, G, H and N define the non-isotropic shape of the yield surface and the parameters σ_0, e_0 and n define its size as a function of the effective plastic strain e^p.

The simple Mises-Hill yield function allows, as a first approximation, to treat rolled thin metal sheets with planar and transversal anisotropy.

The Hencky stress tensor $\mathbf{T}$ can be easily particularized for the plane stress case.

We define the rotated Hencky and second Piola-Kirchhoff stress tensors as

$$\mathbf{T}_L = \mathbf{R}_L^T\;\mathbf{T}\;\mathbf{R}_L \tag{25}$$

$$\mathbf{S}_L = \mathbf{R}_L^T\;\mathbf{S}\;\mathbf{R}_L \tag{26}$$

where $\mathbf{R}_L$ is the rotation tensor obtained from the eigenvectors of $\mathbf{U}$ given by

$$\mathbf{R}_L = \left[\mathbf{r}_1 \quad , \mathbf{r}_2 \quad , \mathbf{r}_3\right] \tag{27}$$

The relationship between the rotated Hencky and Piola-Kirchhoff stresses is $(\alpha, \beta = 1, 2)$

$$\begin{aligned}
[S_L]_{\alpha\alpha} &= \frac{1}{\lambda_\alpha^2}\,[T_L]_{\alpha\alpha} \\
[S_L]_{\alpha\beta} &= \frac{\ln\left(\lambda_\alpha/\lambda_\beta\right)}{\frac{1}{2}\left(\lambda_\alpha^2 - \lambda_\beta^2\right)}\,[T_L]_{\alpha\beta}
\end{aligned} \tag{28}$$

The second Piola-Kirchhoff stress tensor can be computed by

$$\mathbf{S} = \sum_{\alpha=1}^{2}\sum_{\beta=1}^{2}[S_L]_{\alpha\beta}\;\mathbf{r}_\alpha \otimes \mathbf{r}_\beta \tag{29}$$

The second Piola-Kirchhoff stress vector $\boldsymbol{\sigma}$ used in Eqs.(18–19) can be readily extracted from the $\mathbf{S}$ tensor.

3 Enhanced Basic Shell Triangle

The main features of the element formulation (termed EBST for Enhanced Basic Shell Triangle) are the following:

1. The geometry of the patch formed by an element and the three adjacent elements is *quadratically interpolated* from the position of the six nodes in the patch (Fig. 1).
2. The membrane strains are assumed to vary *linearly* within the central triangle and are expressed in terms of the (continuous) values of the deformation gradient at the mid side points of the triangle.
3. An assumed *constant curvature* field within the central triangle is chosen. This is computed in terms of the values of the (continuous) deformation gradient at the mid side points.

Details of the derivation of the EBST element are given below.

3.1 Definition of the Element Geometry and Computation of Membrane Strains

A quadratic approximation of the geometry of the four elements patch is chosen using the position of the six nodes in the patch. It is useful to define the patch in the isoparametric space using the nodal positions given in the Table 1 (see also Fig. 1).

Table 1. Isoparametric coordinates of the six nodes in the patch of Fig. 1

	1	2	3	4	5	6
ξ	0	1	0	1	-1	1
η	0	0	1	1	1	-1

The quadratic interpolation is defined by

$$\boldsymbol{\varphi} = \sum_{i=1}^{6} N_i \boldsymbol{\varphi}_i \tag{30}$$

with ($\zeta = 1 - \xi - \eta$)

$$\begin{array}{ll} N_1 = \zeta + \xi\eta & N_4 = \frac{\zeta}{2}(\zeta - 1) \\ N_2 = \xi + \eta\zeta & N_5 = \frac{\xi}{2}(\xi - 1) \\ N_3 = \eta + \zeta\xi & N_6 = \frac{\eta}{2}(\eta - 1) \end{array} \tag{31}$$

This interpolation allows to computing the displacement gradients at selected points in order to use an assumed strain approach. The computation of the gradients is performed at the mid side points of the central element of the patch denoted by G_1, G_2 and G_3 in Fig. 1. This choice has the following advantages.

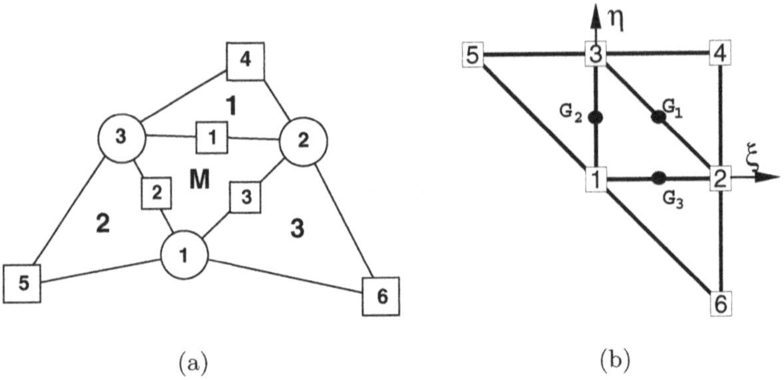

Fig. 1. (a) Patch of three node triangular elements including the central triangle (M) and three adjacent triangles (1, 2 and 3); (b) Patch of elements in the isoparametric space

- Gradients at the three mid side points depend only on the nodes belonging to the two elements adjacent to each side. This can be easily verified by sampling the derivatives of the shape functions at each mid-side point.
- When gradients are computed at the common mid-side point of two adjacent elements, the same values are obtained, as the coordinates of the same four points are used. This in practice means that the gradients at the mid-side points are independent of the element where they are computed. A side-oriented implementation of the finite element will therefore lead to a unique evaluation of the gradients per side.

The Cartesian derivatives of the shape functions are computed at the original configuration by the standard expression

$$\begin{bmatrix} N_{i,1} \\ N_{i,2} \end{bmatrix} = \mathbf{J}^{-1} \begin{bmatrix} N_{i,\xi} \\ N_{i,\eta} \end{bmatrix} \tag{32}$$

where the Jacobian matrix at the original configuration is

$$\mathbf{J} = \begin{bmatrix} \boldsymbol{\varphi}^0_{'\xi} \cdot \mathbf{t}_1 & \boldsymbol{\varphi}^0_{'\eta} \cdot \mathbf{t}_1 \\ \boldsymbol{\varphi}^0_{'\xi} \cdot \mathbf{t}_2 & \boldsymbol{\varphi}^0_{'\eta} \cdot \mathbf{t}_2 \end{bmatrix} \tag{33}$$

The deformation gradients on the middle surface, associated to an arbitrary spatial Cartesian system and to the material cartesian system defined on the middle surface are related by

$$[\boldsymbol{\varphi}_{'1}, \boldsymbol{\varphi}_{'2}] = [\boldsymbol{\varphi}_{'\xi}, \boldsymbol{\varphi}_{'\eta}] \; \mathbf{J}^{-1} \tag{34}$$

The membrane strains within the central triangle are obtained using a linear assumed strain field $\hat{\boldsymbol{\varepsilon}}_m$, i.e.

$$\boldsymbol{\varepsilon}_m = \hat{\boldsymbol{\varepsilon}}_m \tag{35}$$

with

$$\hat{\boldsymbol{\varepsilon}}_m = (1-2\zeta)\boldsymbol{\varepsilon}_m^1 + (1-2\xi)\boldsymbol{\varepsilon}_m^2 + (1-2\eta)\boldsymbol{\varepsilon}_m^3 = \sum_{i=1}^{3} \bar{N}_i \boldsymbol{\varepsilon}_m^i \tag{36}$$

where $\boldsymbol{\varepsilon}_m^i$ are the membrane strains computed at the three mid side points G_i ($i = 1, 2, 3$ see Fig. 1). In Eq.(36) $\bar{N}_1 = (1 - 2\zeta)$, etc.

The gradient at each mid side point is computed from the quadratic interpolation (30):

$$(\boldsymbol{\varphi}_{'\alpha})_{G_i} = \boldsymbol{\varphi}^i_{'\alpha} = \left[\sum_{j=1}^{3} N^i_{j,\alpha}\boldsymbol{\varphi}_j\right] + N^i_{i+3,\alpha}\boldsymbol{\varphi}_{i+3} \quad , \quad \alpha = 1,2 \quad , \quad i = 1,2,3 \tag{37}$$

Substituting Eq.(11) into (36) and using Eq.(9) gives the membrane strain vector as

$$\boldsymbol{\varepsilon}_m = \sum_{i=1}^{3} \frac{1}{2}\bar{N}_i \begin{Bmatrix} \boldsymbol{\varphi}^i_{'1} \cdot \boldsymbol{\varphi}^i_{'1} - 1 \\ \boldsymbol{\varphi}^i_{'2} \cdot \boldsymbol{\varphi}^i_{'2} - 1 \\ 2\boldsymbol{\varphi}^i_{'1} \cdot \boldsymbol{\varphi}^i_{'2} \end{Bmatrix} \tag{38}$$

and the virtual membrane strains as

$$\delta\boldsymbol{\varepsilon}_m = \sum_{i=1}^{3} \bar{N}_i \begin{Bmatrix} \boldsymbol{\varphi}^i_{'1} \cdot \delta\boldsymbol{\varphi}^i_{'1} \\ \boldsymbol{\varphi}^i_{2} \cdot \delta\boldsymbol{\varphi}^i_{'2} \\ \delta\boldsymbol{\varphi}^i_{'1} \cdot \boldsymbol{\varphi}^i_{'2} + \boldsymbol{\varphi}^i_{'1} \cdot \delta\boldsymbol{\varphi}^i_{2} \end{Bmatrix} \tag{39}$$

We note that the gradient at each mid side point G_i depends only on the coordinates of the three nodes of the central triangle and on those of an additional node in the patch, associated to the side i where the gradient is computed.

Combining Eqs.(39), (37) and (30) gives

$$\delta\boldsymbol{\varepsilon}_m = \mathbf{B}_m \delta\mathbf{a}^p \tag{40a}$$

with

$$\underset{18\times 1}{\delta\mathbf{a}^p} = [\delta\mathbf{u}_1^T, \delta\mathbf{u}_2^T, \delta\mathbf{u}_3^T, \delta\mathbf{u}_4^T, \delta\mathbf{u}_5^T, \delta\mathbf{u}_6^T]^T \tag{40b}$$

where $\delta\mathbf{a}^p$ is the patch displacement vector and $\mathbf{B}_m$ is the membrane strain matrix. An explicit form of this matrix is given in [26].

Note that the membrane strains within the EBST element are a function of the displacements of the six patch nodes.

3.2 Computation of Curvatures

We will assume the following constant curvature field within each element

$$\kappa_{\alpha\beta} = \hat{\kappa}_{\alpha\beta} \tag{41}$$

where $\hat{\kappa}_{\alpha\beta}$ is the assumed constant curvature field defined by

$$\hat{\kappa}_{\alpha\beta} = -\frac{1}{A_M^0}\int_{A_M^0} \mathbf{t}_3 \cdot \boldsymbol{\varphi}'_{\beta\alpha}\, dA^0 \tag{42}$$

where A_M^0 is the area (in the original configuration) of the central element in the patch.

Substituting Eq.(42) into (41) and integrating by parts the area integral gives the curvature vector within the element in terms of the following line integral

$$\boldsymbol{\kappa} = \begin{Bmatrix} \kappa_{11} \\ \kappa_{22} \\ 2\kappa_{12} \end{Bmatrix} = \frac{1}{A_M^0}\oint_{\Gamma_M^0} \begin{bmatrix} -n_1 & 0 \\ 0 & -n_2 \\ -n_2 & -n_1 \end{bmatrix} \begin{bmatrix} \mathbf{t}_3 \cdot \boldsymbol{\varphi}'_1 \\ \mathbf{t}_3 \cdot \boldsymbol{\varphi}'_2 \end{bmatrix} d\Gamma \tag{43}$$

where n_i are the components (in the local system) of the normals to the element sides in the initial configuration Γ_M^0. The integration by parts of Eq.(42) is typical in finite volume methods for computing second derivatives over volumes by line integrals of gradient terms [28,29].

For the definition of the normal vector $\mathbf{t}_3$, the linear interpolation over the central element is used. In this case the tangent plane components are

$$\boldsymbol{\varphi}'_\alpha = \sum_{i=1}^{3} L_{i,\alpha}^M \boldsymbol{\varphi}_i \quad , \quad \alpha = 1,2 \tag{44a}$$

$$\mathbf{t}_3 = \frac{\boldsymbol{\varphi}'_1 \times \boldsymbol{\varphi}'_2}{|\boldsymbol{\varphi}'_1 \times \boldsymbol{\varphi}'_2|} = \lambda\, \boldsymbol{\varphi}_1 \times \boldsymbol{\varphi}_2 \tag{44b}$$

From these expressions it is also possible to compute in the original configuration the element area A_M^0, the outer normals $(n_1, n_2)^i$ at each side and the side lengths l_i^M. Equation (44b) also allows to evaluate the thickness ratio λ in the deformed configuration and the actual normal $\mathbf{t}_3$.

The numerical evaluation of the line integral in Eq.(43) results in a sum over the integration points at the element boundary which are, in fact, the same points used for evaluating the gradients when computing the membrane strains. As one integration point is used over each side, it is not necessary to distinguish between sides (i) and integration points (G_i). In this way the curvatures can be computed by

$$\boldsymbol{\kappa} = \frac{1}{A_M^0}\sum_{i=1}^{3} l_i^M \begin{bmatrix} -n_1 & 0 \\ 0 & -n_2 \\ -n_2 & -n_1 \end{bmatrix} \begin{bmatrix} \mathbf{t}_3 \cdot \boldsymbol{\varphi}'_1 \\ \mathbf{t}_3 \cdot \boldsymbol{\varphi}'_2 \end{bmatrix} d\Gamma \tag{45}$$

Eq.(45) is now expressed in terms of the shape functions of the 3-noded triangle L_i^M (which coincide with the area coordinates [4]). Noting the property of the area coordinates

$$\nabla L_i^M = \begin{bmatrix} L_{i,x}^M \\ L_{i,y}^M \end{bmatrix} = -\frac{l_i^M}{2A_M}\begin{bmatrix} n_x^i \\ n_y^i \end{bmatrix} \tag{46}$$

the expression for the curvature can be expressed as

$$\boldsymbol{\kappa} = 2\sum_{i=1}^{3} \begin{bmatrix} L_{i,1}^M & 0 \\ 0 & L_{i,2}^M \\ L_{i,2}^M & L_{i,1}^M \end{bmatrix} \begin{bmatrix} \mathbf{t}_3 \cdot \boldsymbol{\varphi}_{\prime 1}^i \\ \mathbf{t}_3 \cdot \boldsymbol{\varphi}_{\prime 2}^i \end{bmatrix} \tag{47}$$

The gradient $\boldsymbol{\varphi}_{\prime\alpha}^i$ is evaluated at each side G_i from the quadratic interpolation

$$\begin{bmatrix} \boldsymbol{\varphi}_{\prime 1}^i \\ \boldsymbol{\varphi}_{\prime 2}^i \end{bmatrix} = \begin{bmatrix} N_{1,1}^i & N_{2,1}^i & N_{3,1}^i & N_{i+3,1}^i \\ N_{1,2}^i & N_{2,2}^i & N_{3,2}^i & N_{i+3,2}^i \end{bmatrix} \begin{bmatrix} \boldsymbol{\varphi}_1 \\ \boldsymbol{\varphi}_2 \\ \boldsymbol{\varphi}_3 \\ \boldsymbol{\varphi}_{i+3} \end{bmatrix} \tag{48}$$

This is a basic difference with respect of the computation of the curvature field in the original Basic Shell Triangle (BST) where the gradient at the side mid-point is computed as the average value between the values at two adjacent elements [20, 23, 26, 27].

Note again than at each side the gradients depend only on the positions of the three nodes of the central triangle and of an extra node $(i+3)$, associated precisely to the side (G_i) where the gradient is computed.

Direction $\mathbf{t}_3$ in Eq.(47) can be seen as a reference direction. If a different direction than that given by Eq.(44b) is chosen at an angle θ with the former, this has an influence of order θ^2 in the projection. This justifies Eq.(44b) for the definition of $\mathbf{t}_3$ as a function exclusively of the three nodes of the central triangle, instead of using the 6-node isoparametric interpolation.

The variation of the curvatures can be obtained as

$$\begin{aligned} \delta\boldsymbol{\kappa} = 2\sum_{i=1}^{3} \begin{bmatrix} L_{i,1}^M & 0 \\ 0 & L_{i,2}^M \\ L_{i,2}^M & L_{i,1}^M \end{bmatrix} \left\{ \sum_{i=1}^{3} \begin{bmatrix} N_{j,1}^i(\mathbf{t}_3 \cdot \delta\mathbf{u}_j) \\ N_{j,2}^i(\mathbf{t}_3 \cdot \delta\mathbf{u}_j) \end{bmatrix} + \begin{bmatrix} N_{i+3,1}^i(\mathbf{t}_3 \cdot \delta\mathbf{u}^{i+3}) \\ N_{i+3,2}^i(\mathbf{t}_3 \cdot \delta\mathbf{u}^{i+3}) \end{bmatrix} \right\} - \\ - \sum_{i=1}^{3} \begin{bmatrix} (L_{i,1}^M \rho_{11}^1 + L_{i,2}^M \rho_{11}^2) \\ (L_{i,1}^M \rho_{22}^1 + L_{i,2}^M \rho_{22}^2) \\ (L_{i,1}^M \rho_{12}^1 + L_{i,2}^M \rho_{12}^2) \end{bmatrix} (\mathbf{t}_3 \cdot \delta\mathbf{u}_i) = \mathbf{B}_b \delta\mathbf{a}^p \end{aligned} \tag{49}$$

In Eq.(49)

$$\mathbf{B}_b = [\mathbf{B}_{b_1}, \mathbf{B}_{b_2}, \cdots, \mathbf{B}_{b_6}] \tag{50}$$

Details of the derivation of the curvature matrix $\mathbf{B}_b$ are given in [26, 27].

3.3 The EBST1 Element

A simplified and yet very effective version of the EBST element can be obtained by using *one point quadrature* for the computation of all the element integrals. This element is termed EBST1. Note that this only affects the membrane stiffness matrices and it is equivalent to using a assumed constant membrane strain field defined by an average of the metric tensors computed at each side.

Numerical experiments have shown that both the EBST and the EBST1 elements are free of spurious energy modes.

4 Boundary Conditions

Elements at the domain boundary, where an adjacent element does not exist, deserve a special attention. The treatment of essential boundary conditions associated to translational constraints is straightforward, as they are the natural degrees of freedom of the element. The conditions associated to the normal vector are crucial in the bending formulation. For clamped sides or symmetry planes, the normal vector $\mathbf{t}_3$ must be kept fixed (clamped case), or constrained to move in the plane of symmetry (symmetry case). The former case can be seen as a special case of the latter, so we will consider symmetry planes only. This restriction can be imposed through the definition of the tangent plane at the boundary, including the normal to the plane of symmetry $\boldsymbol{\varphi}^0_{'n}$ that does not change during the process.

The tangent plane at the boundary (mid-side point) is expressed in terms of two orthogonal unit vectors referred to a local-to-the-boundary Cartesian system (see Fig. 2) defined as

$$\left[\boldsymbol{\varphi}^0_{'n},\ \bar{\boldsymbol{\varphi}}_{'s}\right] \tag{51}$$

where vector $\boldsymbol{\varphi}^0_{'n}$ is fixed during the process while direction $\bar{\boldsymbol{\varphi}}_{'s}$ emerges from the intersection of the symmetry plane with the plane defined by the central element (M). The plane (gradient) defined by the central element in the selected original convective Cartesian system $(\mathbf{t}_1, \mathbf{t}_2)$ is

$$\left[\boldsymbol{\varphi}^M_{'1},\ \boldsymbol{\varphi}^M_{'2}\right] \tag{52}$$

the intersection line (side i) of this plane with the plane of symmetry can be written in terms of the position of the nodes that define the side (j and k) and the original length of the side l_i^M, i.e.

$$\boldsymbol{\varphi}^i_{'s} = \frac{1}{l_i^M}\left(\boldsymbol{\varphi}_k - \boldsymbol{\varphi}_j\right) \tag{53}$$

That together with the outer normal to the side $\mathbf{n}^i = [n_1, n_2]^T = [\mathbf{n}\cdot\mathbf{t}_1, \mathbf{n}\cdot\mathbf{t}_2]^T$ (resolved in the selected original convective Cartesian system) leads to

$$\begin{bmatrix} \boldsymbol{\varphi}^{iT}_{'1} \\ \boldsymbol{\varphi}^{iT}_{'2} \end{bmatrix} = \begin{bmatrix} n_1 & -n_2 \\ n_2 & n_1 \end{bmatrix} \begin{bmatrix} \boldsymbol{\varphi}^{iT}_{'n} \\ \boldsymbol{\varphi}^{iT}_{'s} \end{bmatrix} \tag{54}$$

where, noting that λ is the determinant of the gradient, the normal component of the gradient $\boldsymbol{\varphi}^i_{'n}$ can be approximated by

$$\boldsymbol{\varphi}^i_{'n} = \frac{\boldsymbol{\varphi}^0_{'n}}{\lambda|\boldsymbol{\varphi}^i_{'s}|} \tag{55}$$

For a simple supported (hinged) side, the problem is not completely defined. The simplest choice is to neglect the contribution to the side rotations

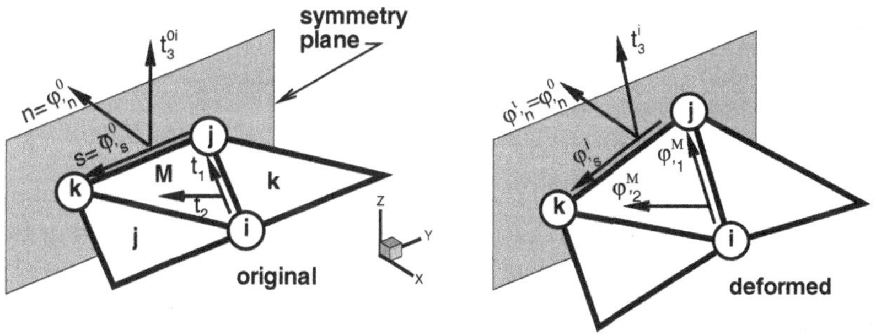

Fig. 2. Local Cartesian system for the treatment of symmetry boundary conditions

from the adjacent element missing in the patch in the evaluation of the curvatures via Eq.(43) [20, 23, 26]. This is equivalent to assume that the gradient at the side is equal to the gradient in the central element, i.e.

$$\left[\boldsymbol{\varphi}^{i}_{\prime 1},\ \boldsymbol{\varphi}^{i}_{\prime 2}\right] = \left[\boldsymbol{\varphi}^{M}_{\prime 1},\ \boldsymbol{\varphi}^{M}_{\prime 2}\right] \tag{56}$$

More precise changes can be however introduced to account for the different natural boundary conditions. One may assume that the curvature normal to the side is zero, and consider a contribution of the missing side to introduce this constraint. As the change of curvature parallel to the side is also zero along the hinged side, this obviously leads to zero curvatures in both directions.

We note finally that for the membrane formulation of element EBST, the gradient at the mid-side point of the boundary is assumed equal to the gradient of the main triangle.

More details on the specification of the boundary conditions on the EBST element can be found in [26, 27].

5 Explicit Solution Scheme

For simulations including large non-linearities, such as those occuring in sheet metal forming processes involving frictional contact conditions on complex geometries or large instabilities, convergence is difficult to achieve with implicit schemes. In those cases an explicit solution algorithm is typically most advantageous. This scheme provides the solution for dynamic problems and also for quasi-static problems if an adequate damping is chosen.

The dynamic equations of motion to solve are of the form

$$\mathbf{r}(\mathbf{u}) + \mathbf{D}\dot{\mathbf{u}} + \mathbf{M}\ddot{\mathbf{u}} = 0 \tag{57}$$

where $\mathbf{M}$ is the mass matrix, $\mathbf{D}$ is the damping matrix and the dot means the time derivative. The solution is performed using the *central difference method.*

To make the method competitive a diagonal (lumped) $\mathbf{M}$ matrix is typically used and $\mathbf{D}$ is taken proportional to $\mathbf{M}$. As usual, mass lumping is performed by assigning one third of the triangular element mass to each node in the central element.

The explicit solution scheme can be summarized as follows. At each time step n where displacements have been computed:

1. Compute the internal forces $\mathbf{r}^n$. This follows the steps described in Box 1.
2. Compute the accelerations at time t_n

$$\ddot{\mathbf{u}}^n = \mathbf{M}_d^{-1}[\mathbf{r}^n - \mathbf{D}\dot{\mathbf{u}}^{n-1/2}] \tag{58}$$

   where $\mathbf{M}_d$ is the diagonal (lumped) mass matrix.
3. Compute the velocities at time $t_{n+1/2}$

$$\dot{\mathbf{u}}^{n+1/2} = \dot{\mathbf{u}}^{n-1/2} + \ddot{\mathbf{u}}^n \delta t \tag{59}$$

4. Compute the displacements at time t_{n+1}

$$\mathbf{u}^{n+1} = \mathbf{u}^n + \dot{\mathbf{u}}^{n+1/2}\delta t \tag{60}$$

5. Update the shell geometry
6. Check frictional contact conditions

6 Example 1. Cylindrical Panel under Impulse Loading

The geometry of the cylinder and the material properties are shown in Fig. 3. A prescribed initial normal velocity of $v_o = -5650$ in/sec is applied to the points in the region shown modelling the effect of the detonation of an explosive layer. The panel is assumed to be clamped along all the boundary. One half of the cylinder is discretized only due to symmetry conditions. Three different meshes of 6×12, 12×32 and 18×48 triangles were used for the analysis. The deformed configurations for $time = 1msec$ are shown for the three meshes in Fig. 3.

The analysis was performed assuming an elastic-perfect plastic material behaviour ($\sigma_y = k_y \quad k'_y = 0$). A study of the convergence of the solution with the number of thickness layers showed again that four layers suffice to capture accurately the non linear material response [25].

A comparison of the results obtained with the BST and EBST1 elements using the coarse mesh and the finer mesh is shown in Fig. 4 where experimental results reported in [32] have also been plotted for comparison purposes. Good agreement between the numerical and experimental results is obtained. Figs. 4 show the time evolution of the vertical displacement of two reference points along the center line located at $y = 6.28$in and $y = 9.42$in, respectively. For the finer mesh results between both elements are almost identical. For the coarse mesh it can been seen that the BST element is more flexible than the EBST1.

1. Generate the actual configuration $\boldsymbol{\varphi}^{n+1} = \boldsymbol{\varphi}^{n} + \Delta\mathbf{u}^{n}$
2. Compute the metric tensor $a_{\alpha\beta}^{n+1}$ and the curvatures $\kappa_{\alpha\beta}^{n+1}$. Then at each layer k compute the (approximate) right Cauchy-Green tensor. From Eq.(14)

$$\mathbf{C}_k^{n+1} = \mathbf{a}^{n+1} + z_k \boldsymbol{\chi}^{n+1} \tag{61}$$

3. Compute the total (21) and elastic (22) deformations at each layer k

$$\boldsymbol{\varepsilon}_k^{n+1} = \frac{1}{2} \ln \mathbf{C}_k^{n+1} \tag{62}$$

$$[\boldsymbol{\varepsilon}_e]_k^{n+1} = \boldsymbol{\varepsilon}_k^{n+1} - [\boldsymbol{\varepsilon}_p]_k^{n}$$

4. Compute the trial Hencky elastic stresses (23) at each layer k

$$\mathbf{T}_k^{n+1} = \mathbf{H}\,[\boldsymbol{\varepsilon}_e]_k^{n+1} \tag{63}$$

5. Check the plasticity condition and return to the plasticity surface. If necessary correct the plastic strains $[\boldsymbol{\varepsilon}_p]_k^{n+1}$ at each layer
6. Compute the second Piola-Kirchhoff stress vector $\boldsymbol{\sigma}_k^{n+1}$ and the generalized stresses

$$\boldsymbol{\sigma}_m^{n+1} = \frac{h^0}{N_L} \sum_{k=1}^{N_L} \boldsymbol{\sigma}_k^{n+1} w_k$$

$$\boldsymbol{\sigma}_b^{n+1} = \frac{h^0}{N_L} \sum_{k=1}^{N_L} \boldsymbol{\sigma}_k^{n+1} z_k w_k \tag{64}$$

Where w_k is the weight of the through-the-thickness integration point. Recall that z_k is the current distance of the layer to the mid-surface and not the original distance. However, for small strain plasticity this distinction is not important.
This computation of stresses is exact for an elastic problem.

7. Compute the residual force vector from

$$\mathbf{r}_i^e = \iint_A L_i \mathbf{t}\, dA - \iint_{A^\circ} (\mathbf{B}_{m_i}^T \boldsymbol{\sigma}_m + \mathbf{B}_{b_i}^T \boldsymbol{\sigma}_b) dA \tag{65}$$

Box 1. Computation of the internal forces vector

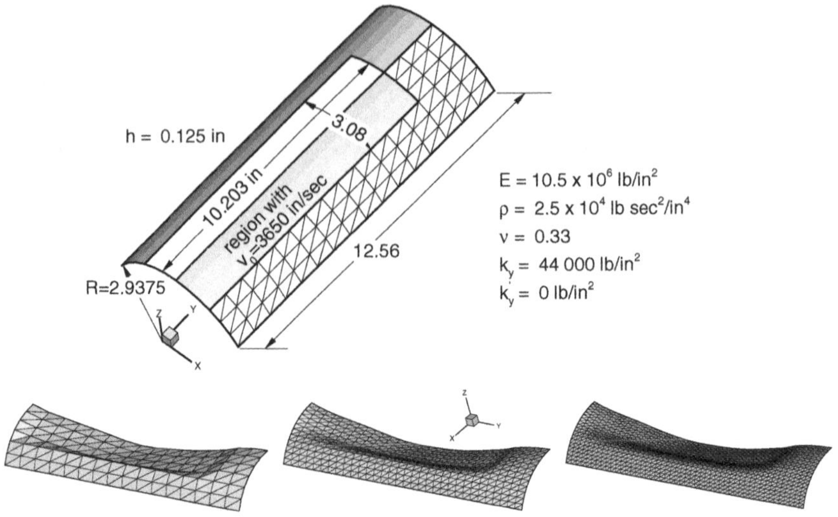

Fig. 3. Cylindrical panel under impulse loading. Geometry and material properties. Deformed meshes for $time = 1msec$

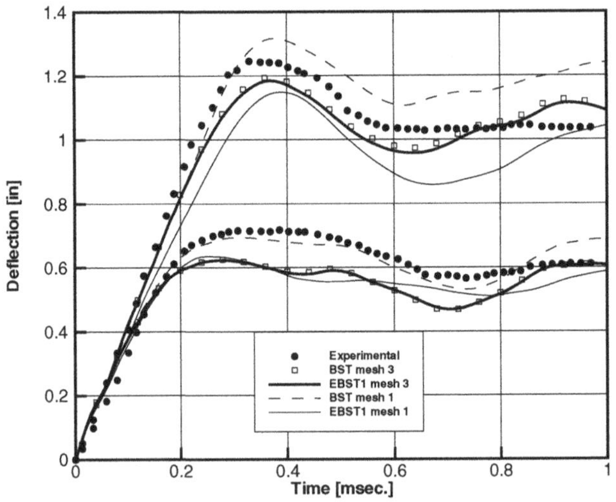

Fig. 4. Cylindrical panel under impulse loading. Time evolution of the displacement of two points along the crown line. Upper lines $y = 6.28$in. Lower lines $y = 9.42$ in. Comparison of results obtained with BST and EBST1 elements (mesh 1: 6×12 elements and mesh 3: 18×48 elements) and experimental values

The numerical values of the vertical displacement at the two reference points obtained with the BST and EBST1 elements after a time of 0.4 ms using the 16 × 32 mesh are compared in Table 2 with a numerical solution obtained by Stolarski *et al.* [31] using a curved triangular shell element and the 16 × 32 mesh. Experimental results reported in [32] are also given for comparison. It is interesting to note the reasonable agreement of the results for $y = 6.28$in. and the discrepancy of present and other published numerical solutions with the experimental value for $y = 9.42$in.

Table 2. Cylindrical panel under impulse load. Comparison of vertical displacement values of two central points for $t = 0.4$ ms

	Vertical displacement (in.)	
element/mesh	$y = 6.28$in	$y = 9.42$in
BST (6 × 12 el.)	-1.310	-0.679
BST (18 × 48 el.)	-1.181	-0.587
EBST1 (6 × 12 el.)	-1.147	-0.575
EBST1 (18 × 48 el.)	-1.171	-0.584
Stolarski *et al.* [31]	-1.183	-0.530
Experimental [32]	-1.280	-0.700

The deformed shapes of the transverse section for $y = 6.28$in. and the longitudinal section for $x = 0$ obtained with the both elements for the coarse and the fine meshes after 1ms. are compared with the experimental results in Figs. 5 and 6. Excellent agreement is observed for the fine mesh for both elements.

7 Application to Sheet Metal Forming Problems

The features of tghe EBST1 element make it ideal for analysis of sheet metal stamping processes. A number of examples of simulations of practical problems of this kind are presented. Numerical results have been obtained with the sheet stamping simulation code STAMPACK where the EBST1 element has been implemented [35].

7.1 S-rail Sheet Stamping

The next problem corresponds to one of the sheet stamping benchmark tests proposed in NUMISHEET'96 [33]. The analysis comprises two parts, namely, simulation of the stamping of a S-rail sheet component and springback computations once the stamping tools are removed. Figure 7 shows the deformed sheet after springback.

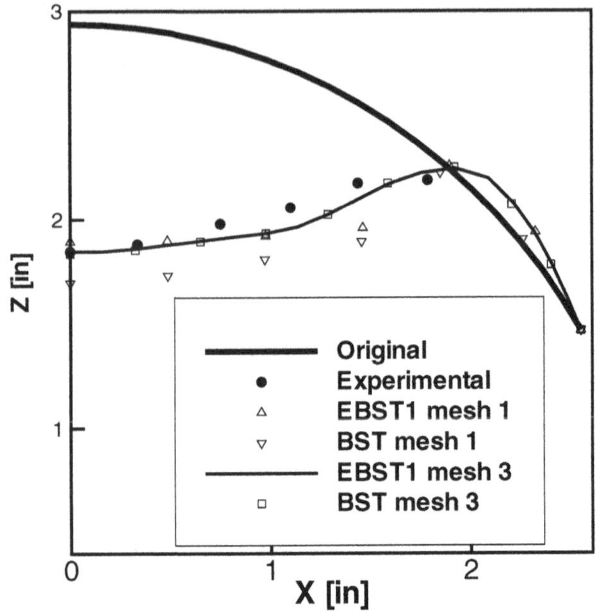

Fig. 5. Cylindrical panel under impulse loading. Final deformation ($t = 1msec$) of the panel at the cross section $y = 6.28in$. Comparison with experimental values

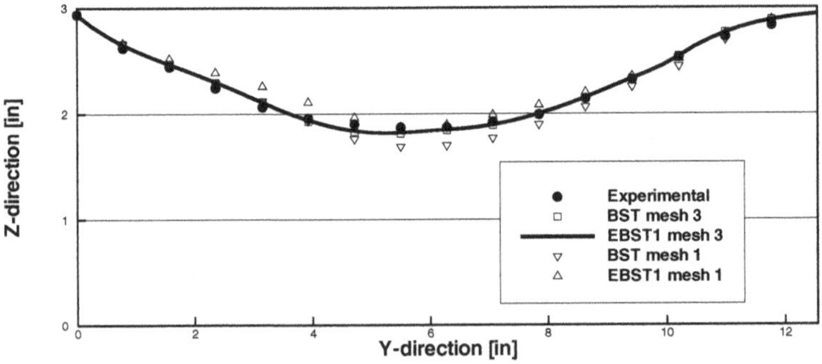

Fig. 6. Cylindrical panel under impulse loading. Final deformation ($t = 1msec$) of the panel at the crown line ($x = 0.00in$). Comparison with experimental values

The detailed geometry and material data can be found in the proceedings of the conference [33] or in the web [34]. The mesh used for the sheet has 6000 triangles and 3111 points (Fig. 7). The tools are treated as rigid bodies.

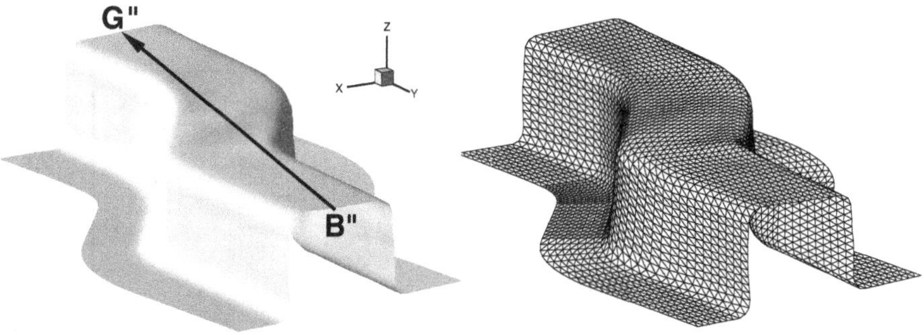

Fig. 7. Stamping of a S-rail. Final deformation of the sheet after springback obtained in the simulation. The triangular mesh of the deformed sheet is also shown

The meshes used for the sheet and the tools are those provided by the benchmark organizers. The material considered here is a mild steel (IF) with Young Modulus $E = 2.06GPa$ and Poisson ratio $\nu = 0.3$. Mises yield criterion was used for plasticity behaviour with non-linear isotropic hardening defined by $\sigma_y(e^p) = 545(0.13 + e^p)^{0.267}[MPa]$. A uniform friction of 0.15 was used for all the tools. A low (10kN) blank holder force was considered in this simulation.

Figure 8 compares the punch force during the stamping stage obtained with both BST and EBST1 elements for the simulation and experimental values. Also for reference the average values of the simulations presented in the conference are included. Explicit and implicit simulations are considered as different curves. There is a remarkable coincidence between the experimental values and the results obtained with both the BST and EBST1 elements.

Figure 9 plots the Z coordinate along line B"–G" after springback. The top surface of the sheet does not remain plane due to some instabilities due to the low blank holder force used. Results obtained with the simulations compare very well with the experimental values.

7.2 Stamping of Industrial Automotive Part

Figure 10 shows the geometry of the lateral panel of a car and the mesh of 457760 EBST1 elements used for the computation. Results of the stamping simulation are shown in Fig. 11. Note that the outpus of the simulation have been translated into graphical plots indicating the quality of the stamping process and the risk of failure in the different zones of the panel. This helps designers to taking decissions on the adequacy of the stamping process and for introducing changes in the design of the stamping tools (dies, punch, blankholders, etc.) and the process parameters if needed.

Figure 12 shows the geometry mesh and results of the stamping of a front fender part of an automotive. The initial mesh had 121960 EBST1 elements.

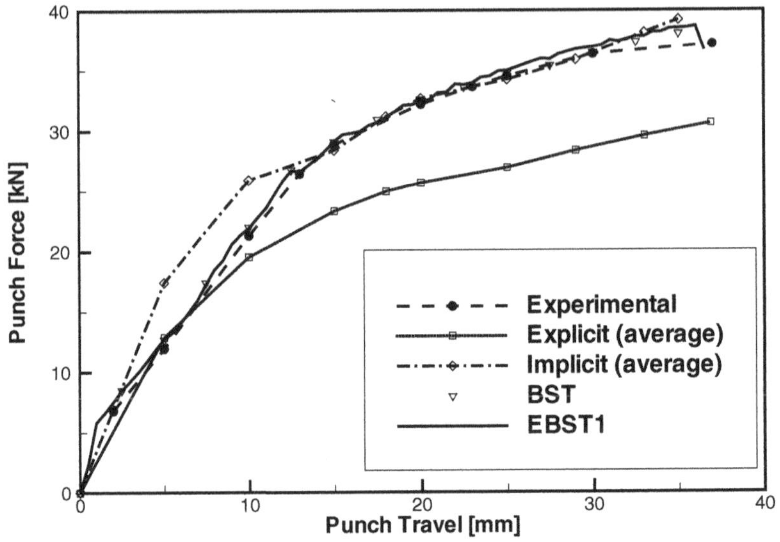

Fig. 8. Stamping of a S-rail. Punch force versus punch travel. Average of explicit and implicit results reported at the benchmark conference are also shown

Adaptive mesh refinement was used along the simulation process leading to a final mesh of 389870 elements. Finally, Figs. 13 and 14 show the same type of information for the stamping of a car tail gate. The initial and final meshes (after adaptive mesh refinement) had 186528 and 489560 EBST1 elements, respectively. The simulation results are displayed in both problems with an "engineering insight" in order to help the design and manufacturing of the stamping tools and the definition of the stamping process as previously mentioned.

8 Concluding Remarks

An enhanced rotation-free shell triangle (termed EBST) is obtained by using a quadratic interpolation of the geometry in terms of the six nodes belonging to the four elements patch associated to each triangle. This allows to computing an assumed constant curvature field and an assumed linear membrane strain field which improves the in-plane behaviour of the original element. A simple and economic version of the element using a single integration point has been presented. The efficiency of the rotation-free shell triangle has been demonstrated in examples of application including the analysis of a cylinder under impulse loading and practical sheet stamping problems.

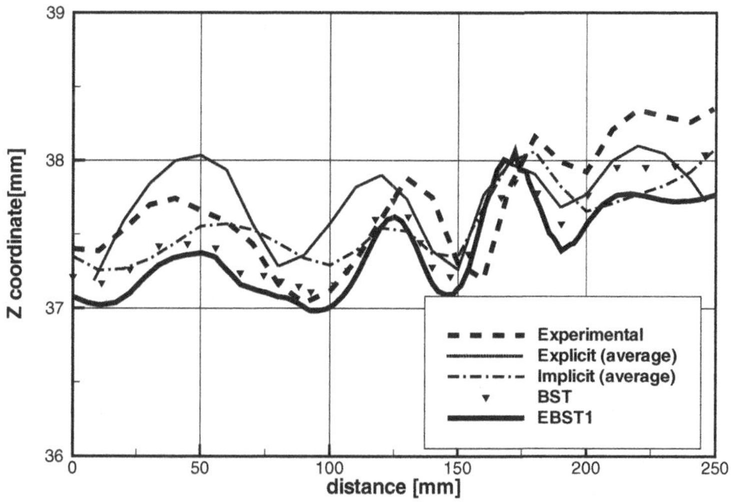

Fig. 9. Stamping of a S-rail. Z-coordinate along line B"—G" after springback. Average of explicit and implicit results reported at the benchmark conference are also shown

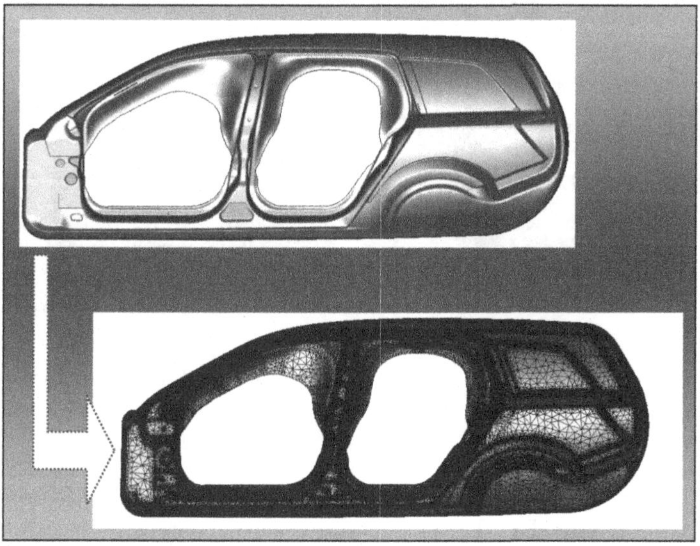

Fig. 10. Lateral panel of an automotive. Finite element mesh of 457760 triangles used for the simulation

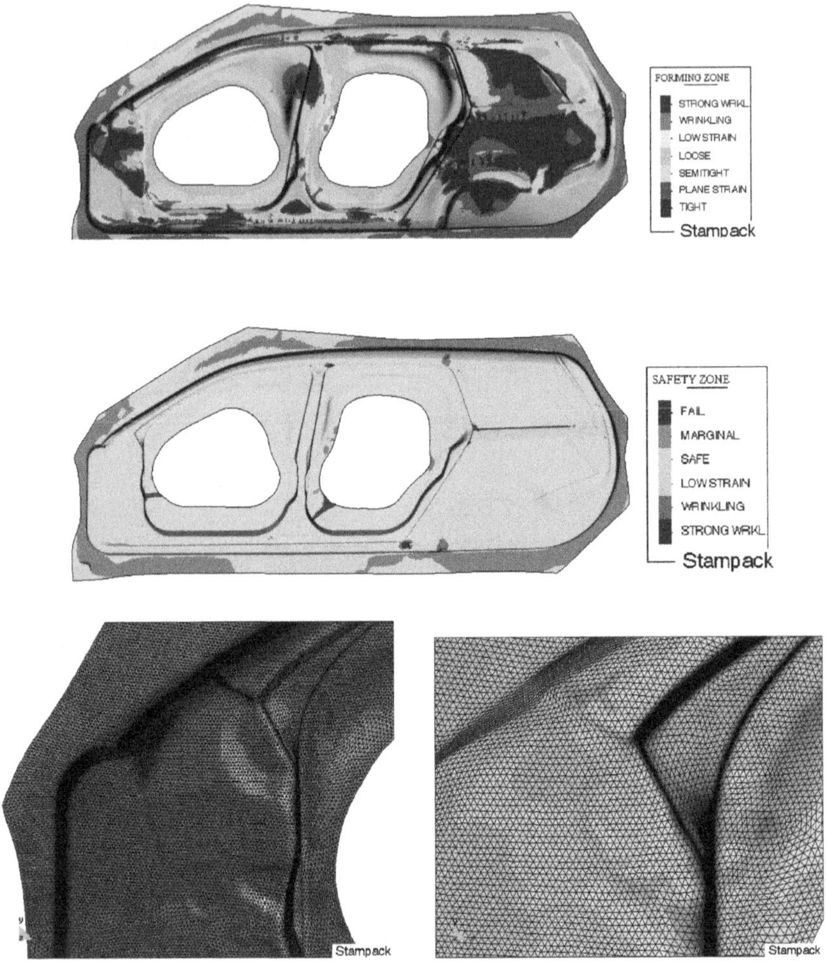

Fig. 11. Lateral panel of a car. Results of the stamping analysis

The enhanced rotation-free basic shell triangle element with a single integration point (the EBST1 element) is an excellent candidate for solving practical sheet metal stamping problems and other non linear shell problems in engineering involving complex geometry, dynamics, material and geometrical non linearities and frictional contact conditions.

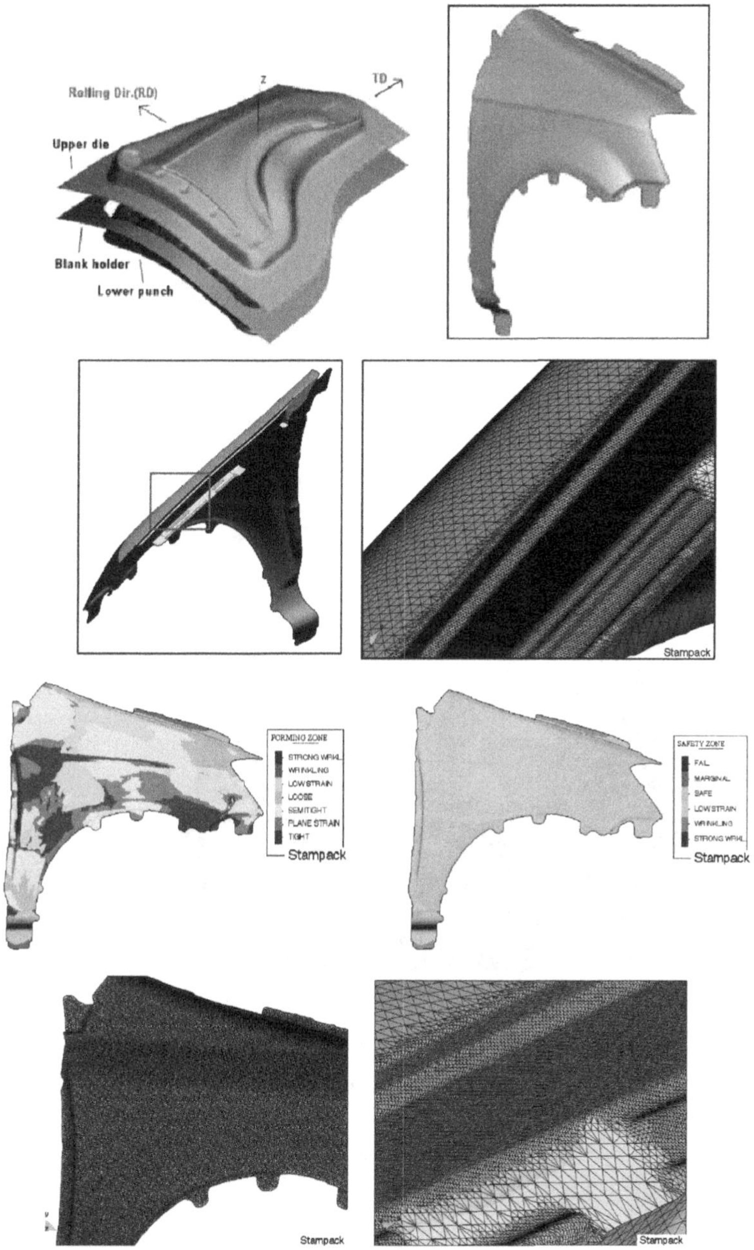

Fig. 12. Front fender. Results of the stamping analysis using an initial mesh of 121960 EBST1 elements. The final adapted mesh had 389870 elements

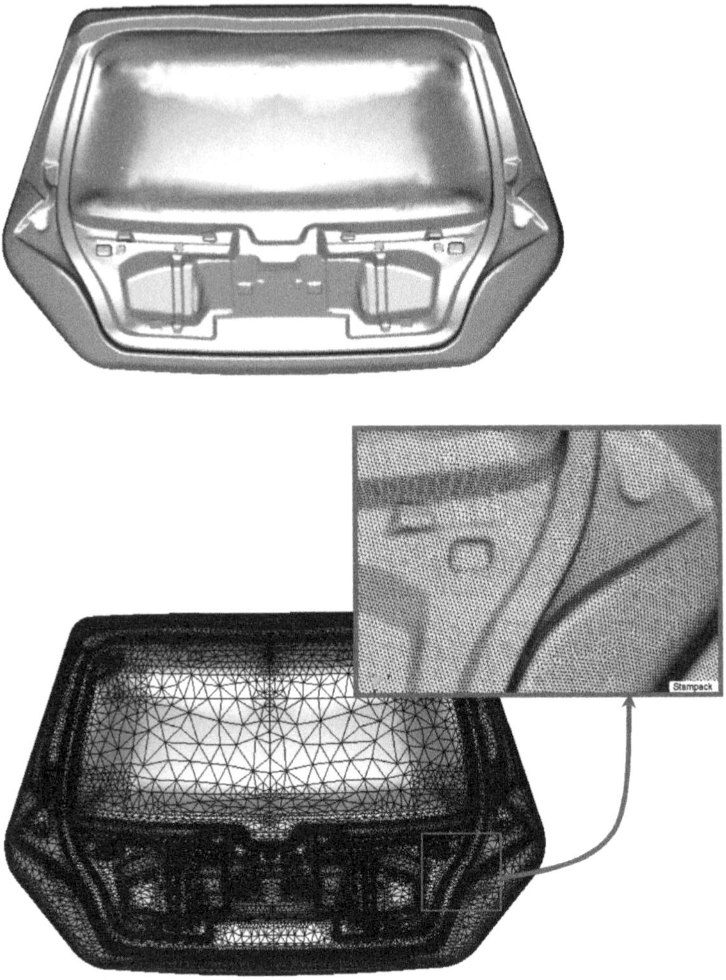

Fig. 13. Car tail gate. Geometry and final adapted mesh of 489560 EBST1 elements used for the stamping simulation

ACKNOWLEDGEMENTS

The support of the company QUANTECH (www.quantech.es) providing the code STAMPACK [35] is gratefully acknowledged.

This research was partially supported by project SEDUREC of the Consolider Programme of the Ministerio de Educación y Ciencia of Spain.

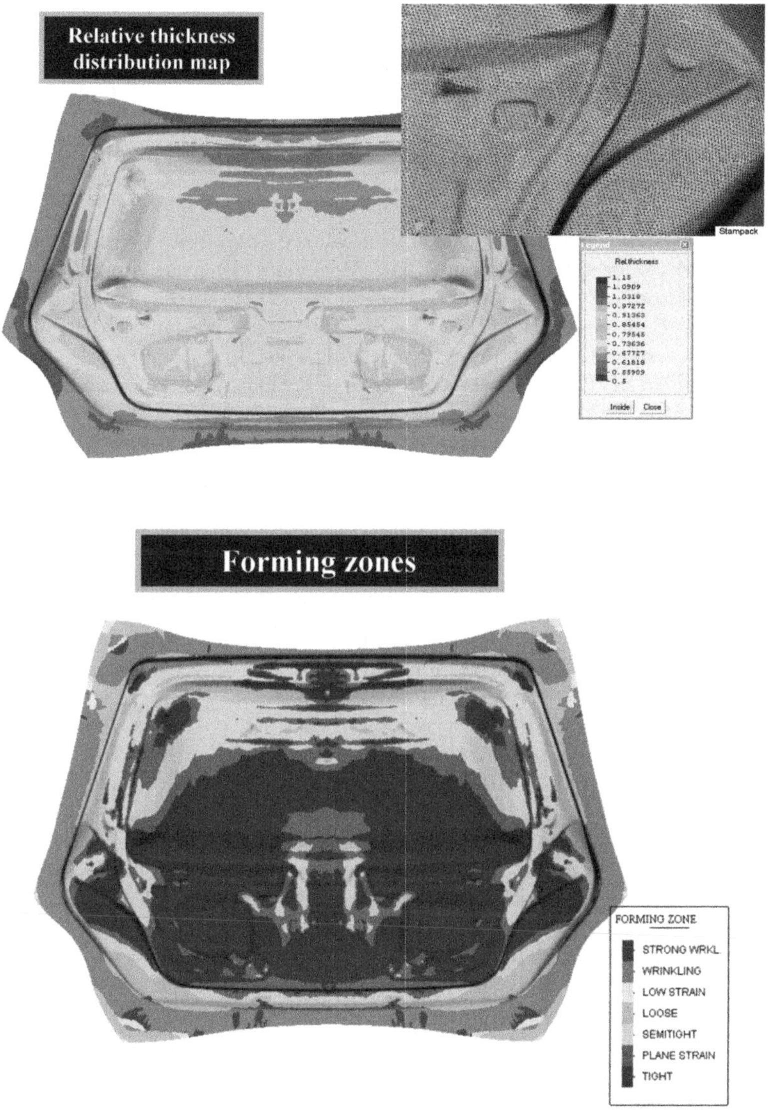

Fig. 14. Car tail gate. Map of relative thickness distribution and forming zones on the stamped part

References

1. Oñate E (1994) A review of some finite element families for thick and thin plate and shell analysis. Publication CIMNE N. 53, May
2. Flores FG, Oñate E, Zárate F (1995) New assumed strain triangles for non-linear shell analysis. Computational Mechanics 17:107–114

3. Argyris JH, Papadrakakis M, Apostolopoulou C, Koutsourelakis S (2000) The TRIC element. Theoretical and numerical investigation. Comput Meth Appl Mech Engrg 182:217–245
4. Zienkiewicz OC, Taylor RL (2005) The finite element method. Solid Mechanics. Vol II, Elsevier
5. Stolarski H, Belytschko T, Lee S-H (1995) A review of shell finite elements and corotational theories. Computational Mechanics Advances vol. 2 (2), North-Holland
6. Ramm E, Wall WA (2002) Shells in advanced computational environment. In V World Congress on Computational Mechanics, Eberhardsteiner J, Mang H, Rammerstorfer F (eds), Vienna, Austria, July 7–12. http://wccm.tuwien.ac.at.
7. Bushnell D, Almroth BO (1971) Finite difference energy method for non linear shell analysis. J Computers and Structures 1:361
8. Timoshenko SP (1971) Theory of Plates and Shells, McGraw Hill, New York
9. Ugural AC (1981) Stresses in Plates and Shells, McGraw Hill, New York
10. Nay RA, Utku S (1972) An alternative to the finite element method. Variational Methods Eng vol. 1
11. Hampshire JK, Topping BHV, Chan HC (1992) Three node triangular elements with one degree of freedom per node. Engng Comput 9:49–62
12. Phaal R, Calladine CR (1992) A simple class of finite elements for plate and shell problems. I: Elements for beams and thin plates. Int J Num Meth Engng 35:955–977
13. Phaal R, Calladine CR (1992) A simple class of finite elements for plate and shell problems. II: An element for thin shells with only translational degrees of freedom. Int J Num Meth Engng 35:979–996
14. Oñate E, Cervera M (1993) Derivation of thin plate bending elements with one degree of freedom per node. Engineering Computations 10:553–561
15. Brunet M, Sabourin F (1994) Prediction of necking and wrinkles with a simplified shell element in sheet forming. Int Conf of Metal Forming Simulation in Industry II:27–48, Kröplin B (ed)
16. Rio G, Tathi B, Laurent H (1994) A new efficient finite element model of shell with only three degrees of freedom per node. Applications to industrial deep drawing test. In Recent Developments in Sheet Metal Forming Technology, Barata Marques MJM (ed), 18th IDDRG Biennial Congress, Lisbon
17. Rojek J, Oñate E (1998) Sheet springback analysis using a simple shell triangle with translational degrees of freedom only. Int J of Forming Processes 1(3):275–296
18. Rojek J, Oñate E, Postek E (1998) Application of explicit FE codes to simulation of sheet and bulk forming processes. J of Materials Processing Technology 80-81:620–627
19. Jovicevic J, Oñate E (1999) Analysis of beams and shells using a rotation-free finite element-finite volume formulation, Monograph 43, CIMNE, Barcelona
20. Oñate E, Zárate F (2000) Rotation-free plate and shell triangles. Int J Num Meth Engng 47:557–603
21. Cirak F, Ortiz M (2000) Subdivision surfaces: A new paradigm for thin-shell finite element analysis. Int J Num Meths in Engng 47:2039-2072
22. Cirak F, Ortiz M (2001) Fully C^1-conforming subdivision elements for finite deformations thin-shell analysis. Int J Num Meths in Engng 51:813-833
23. Flores FG, Oñate E (2001) A basic thin shell triangle with only translational DOFs for large strain plasticity. Int J Num Meths in Engng 51:57–83

24. Engel G, Garikipati K, Hughes TJR, Larson MG, Mazzei L, Taylor RL (2002). Continuous/discontinuous finite element approximation of fourth-order elliptic problems in structural and continuum mechanics with applications to thin beams and plates, and strain gradient elasticity. Comput Methods Appl Mech Engrg 191:3669–3750
25. Oñate E, Cendoya P, Miquel J (2002) Non linear explicit dynamic analysis of shells using the BST rotation-free triangle. Engineering Computations 19(6):662–706
26. Flores FG, Oñate E (2005) Improvements in the membrane behaviour of the three node rotation-free BST shell triangle using an assumed strain approach. Comput Meth Appl Mech Engng 194(6-8):907–932
27. Oñate E, Flores FG (2005) Advances in the formulation of the rotation-free basic shell triangle. Comput Meth Appl Mech Engng 194(21–24):2406–2443
28. Zienkiewicz OC, Oñate E (1991) Finite Elements vs. Finite Volumes. Is there really a choice?. Nonlinear Computational Mechanics. State of the Art. Wriggers P, Wagner R (eds), Springer Verlag, Heidelberg
29. Oñate E, Cervera M, Zienkiewicz OC (1994) A finite volume format for structural mechanics. Int J Num Meth Engng 37:181–201
30. Hill R (1948) A Theory of the Yielding and Plastic Flow of Anisotropic Metals. Proc Royal Society London A193:281
31. Stolarski H, Belytschko T, Carpenter N (1984) A simple triangular curved shell element. Eng Comput 1:210–218
32. Balmer HA, Witmer EA (1964) Theoretical experimental correlation of large dynamic and permanent deformation of impulsively loaded simple structures. Air force flight Dynamic Lab Rep FDQ-TDR-64-108, Wright-Patterson AFB, Ohio, USA
33. NUMISHEET'96 Third International Conference and Workshop on Numerical Simulation of 3D Sheet Forming Processes, Lee EH, Kinzel GL, Wagoner RH (eds), Dearbon-Michigan, USA, 1996
34. `http://rclsgi.eng.ohio-state.edu/%7Elee-j-k/numisheet96/`
35. STAMPACK. A General Finite Element System for Sheet Stamping and Forming Problems, Quantech ATZ, Barcelona, Spain, 2007. (www.quantech.es)

Computational Methods in Applied Sciences

1. J. Holnicki-Szulc and C.A.M. Soares (eds.): *Advances in Smart Technologies in Structural Engineering*. 2004 ISBN 3-540-22331-2
2. J.A.C. Ambrósio (ed.): *Advances in Computational Multibody Systems*. 2005 ISBN 1-4020-3392-3
3. E. Oñate and B. Kröplin (eds.): *Textile Composites and Inflatable Structures*. 2005 ISBN 1-4020-3316-8
4. J.C. García Orden, J.M. Goicolea and J. Cuadrado (eds.): *Multibody Dynamics*. Computational Methods and Applications. 2007 ISBN 1-4020-5683-4
5. V.M.A. Leitão, C.J.S. Alves and C.A. Duarte (eds.): *Advances in Meshfree Techniques*. 2007 ISBN 1-4020-6094-6
6. C.A.M. Soares, J.A.C. Martins, H.C. Rodrigues and J.A.C. Ambrósio (eds.): *Computational Mechanics*. Solids, Structures and Coupled Problems. 2006 ISBN 978-1-4020-4978-1
7. E. Oñate and R. Owen (eds.): *Computational Plasticity*. 2007 ISBN 978-1-4020-6576-7

www.springer.com